FROM GENE TO PROTEIN:
TRANSLATION INTO BIOTECHNOLOGY

MIAMI WINTER SYMPOSIA—VOLUME 19

MIAMI WINTER SYMPOSIA—VOLUME 19

FROM GENE TO PROTEIN:
TRANSLATION
INTO BIOTECHNOLOGY

edited by

Fazal Ahmad　　Julius Schultz
The Papanicolaou Cancer Research Institute
Eric E. Smith　　William J. Whelan
University of Miami School of Medicine

Proceedings of the Miami Winter Symposium, January 1982
Sponsored by the Department of Biochemistry
University of Miami School of Medicine, Miami, Florida
Symposium Director: W. J. Whelan
and by
The Papanicolaou Cancer Research Institute, Miami, Florida
Symposium Director: J. Schultz

ACADEMIC PRESS　　　1982
A Subsidiary of Harcourt Brace Jovanovich, Publishers
New York　London
Paris　San Diego　San Francisco　São Paulo　Sydney　Tokyo　Toronto

ACADEMIC PRESS, INC.
111 Fifth Avenue, New York, New York 10003

United Kingdom Edition published by
ACADEMIC PRESS, INC. (LONDON) LTD.
24/28 Oval Road, London NW1 7DX

Library of Congress Cataloging in Publication Data
Main entry under title:

From gene to protein: Translation into Biotechnology.

  (Miami winter symposia ; v. 19)
  Proceedings of the symposium held in Miami Beach in
Jan. 1982
  Includes index.
  1. Genetic engineering--Congresses. 2. Protein
biosynthesis--Congresses. I. Ahmad, Fazal, Date
II. Series.
QH442.F75 1982   660'.6   82-18172
ISBN 0-12-045560-9

CONTENTS

The Thirteenth Lynen Lecture

Introduction

Techniques of Eukaryotic Cell Culture

Monoclonal Antibodies—Production and Uses

In Vitro Synthesis of DNA and the Generation of Protein Analogs

Cloning into Eukaryotic Cells

Increasing Levels of Gene Expression

Biological Activities of Cloned Gene Products

Horizons in Biotechnology

Free Communications

SPEAKERS AND DISCUSSANTS

M. Adesnik   New York University School of Medicine, New York, New York
A. A. Ansari   Northrop Services, Inc., Research Triangle Park, North Carolina
M. A. Apple   International Plant Research Institute, Inc., San Carlos, California
R. Axel   College of Physicians and Surgeons of Columbia University,
New York, New York
D. Baltimore   Massachusetts Institute of Technology, Cambridge, Massachussetts
D. Barngorver   Massachusetts Institute of Technology, Cambridge, Massachusetts
L. Baumbach   University of Florida, Gainesville, Florida
M. Bina   Purdue University, West Lafayette, Indiana
H. de Boer   Genentech, Inc., South San Francisco, California
D. Botstein   Massachusetts Institute of Technology, Cambridge Massachusetts
V. Braman   University of Pennsylvania, Philadelphia, Pennyslvania
R. L. Brinster   University of Pennsylvania, Philadelphia, Pennsylvania
D. W. Buck   University of Pennsylvania, Philadelphia, Pennsylvania
D. Burke   University of Warwick, Coventry, England
J. P. Burnett   Eli Lilly and Company, Indianapolis, Indiana
H. Busch   Baylor College of Medicine, Houston, Texas
J. Cahill   Revlon Health Care Group, Tuckahoe, New York
M. H. Caruthers   University of Colorado, Boulder, Colorado
S. I. Chavin   University of Rochester Medical School, Rochester, New York
V. Chowdhry   E. I. Du Pont de Nemours and Company, Wilmington, Delaware
M. Chretien   Montreal Clinical Research Institute, Montreal, Canada
C. Colby   Cetus Corporation, Berkeley, California
L. Comstock   Genentech, Inc., South San Francisco, California
C. M. Croce   The Wistar Institute of Anatomy and Biology, Philadelphia,
Pennsylvania
A. C. Cuello   University of Oxford, Oxford, England

Names in bold indicate speakers at the conference.

D. T. Denhardt University of Western Ontario, London, Ontario, Canada

R. Derynck Genentech, Inc., South San Francisco, California

C. Desaymard Albert Einstein College of Medicine, Bronx, New York

T. Van Dyke University of Florida, Gainesville, Florida

K. B. Eager University of Pennsylvania, Philadelphia, Pennsylvania

H. Faber Papanicolaou Cancer Research Institute, Miami, Florida

S. Fein University of Texas System Cancer Center, Houston, Texas

J. Fenno Genentech, Inc., South San Francisco, California

J. A. Galloway Eli Lilly and Company, Indianapolis, Indiana

D. Garfinkel University of Washington, Seattle, Washington

J. R. Geiger Olin Corporation, New Haven, Connecticut

D. H. Gelfand Cetus Corporation, Berkeley, California

A. Giusti Albert Einstein College of Medicine, Bronx, New York

D. V. Goeddel Genentech, Inc., South San Francisco, California

M. P. Gordon University of Washington, Seattle, Washington

P. W. Gray Genentech, Inc., South San Francisco, California

J. Gutterman University of Texas System Cancer Center, Houston, Texas

F. E. Hagie Genentech, Inc., South San Francisco, California

R. W. F. Hardy E. I. Du Pont de Nemours and Company, Wilmington, Delaware

J. Harford National Institutes of Health, Bethesda, Maryland

G. Haughton University of North Carolina, Chapel Hill, North Carolina

R. J. Hay American Type Culture Collection, Rockville, Maryland

D. V. Hendrick Bethesda Research Laboratories, Inc., Gaithersburg, Maryland

H. Heyneker Genentech, Inc., South San Francisco, California

P. A. Hieter National Institutes of Health, Bethesda, Maryland

R. Hintz Stanford University, Stanford, California

R. A. Hitzeman Genentech, Inc., South San Francisco, California

G. F. Hollis National Institutes of Health, Bethesda, Maryland

T. Horn Genentech, Inc., South San Francisco, California

D. H. Hoscheit Schuyler, Banner, Birch, McKie and Beckett, Washington, D.C.

T. Huang City of Hope Research Institute, Duarte, California

P. P. Hung Bethesda Research Laboratories, Inc., Gaithersburg, Maryland

K. Itakura City of Hope Research Institute, Duarte, California

S. L. Kaplan University of California, San Francisco, California

R. H. Kennett University of Pennsylvania, Philadelphia, Pennsylvania

R. Kierzek Polish Academy of Sciences, Poznan, Poland

S. King F. Eberstadt and Company, Inc., New York, New York

H. Klee University of Washington, Seattle, Washington

G. Knapp University of Alabama, Birmingham, Alabama

V. Knuf University of Washington, Seattle, Washington

P. C. Kung Centocor, Inc., Malvern, Pennsylvania

S. Kwan Albert Einstein College of Medicine, Bronx, New York

W. Kwok University of Washington, Seattle, Washington

E. Lamon University of Alabama, Birmingham, Alabama

J. W. Larrick   Stanford University, Palo Alto, California

R. M. Lawn   Genentech, Inc., South San Francisco, California

A. Leder   National Institutes of Health, Bethesda, Maryland

P. Leder   National Institutes of Health, Bethesda, Maryland

S. H. Lee   Genentech, Inc., South San Francisco, California

D. W. Leung   Genentech, Inc., South San Francisco, California

A. D. Levinson   Genentech, Inc., South San Francisco, California

N. L. Levy   Abbott Laboratories, North Chicago, Illinois

A. Leza   University of Florida, Gainesville, Florida

P. A. Liberti   Jefferson Medical College, Philadelphia, Pennsylvania

C. Lichtenstein   University of Washington, Seattle, Washington

A. Lichtler   University of Florida, Gainesville, Florida

B. J. Marafino   Genentech, Inc., South San Francisco, California

F. Marashi   University of Florida, Gainesville, Florida

J. McPherson   University of Washington, Seattle, Washington

R. B. Meagher   University of Georgia, Athens, Georgia

G. F. Merrill   University of Washington, Seattle, Washington

B. Meyer   University of Pennsylvania, Philadelphia, Pennsylvania

J. Mills   University of London King's College, London, England

C. Milstein   Medical Research Council Centre, Cambridge, England

A. Montoya   University of Washington, Seattle, Washington

U. R. Muller   East Carolina University, Greenville, North Carolina

R. Najarian   Genentech, Inc., South San Francisco, California

B. Nelkin   The Johns Hopkins University, Baltimore, Maryland

E. Nester   University of Washington, Seattle, Washington

S. Newberry   University of Pennsylvania, Philadelphia, Pennsylvania

M. Nilsen-Hamilton   The Salk Institute, San Diego, California

P. O'Hara   University of Washington, Seattle, Washington

I. Paek   College of Physicians and Surgeons of Columbia University, New York, New York

R. D. Palmiter   University of Washington, Seattle, Washington

M. Pater   New Jersey Medical School, Newark, New Jersey

D. Pennica   Genentech, Inc., South San Francisco, California

M. Plumb   University of Florida, Gainesville, Florida

R. R. Pollock   Albert Einstein College of Medicine, Bronx, New York

A. Powell   University of Washington, Seattle, Washington

J. Quesada   University of Texas System Cancer Center, Houston, Texas

V. R. Racaniello   Massachusetts Institute of Technology, Cambridge, Massachusetts

L. W. Ream   University of Washington, Seattle, Washington

P. Reczek   Harvard Medical School, Boston, Massachusetts

L. C. M. Reid   Albert Einstein College of Medicine, Bronx, New York

K. W. Renton   Dalhousie University, Halifax, Nova Scotia

R. Rickles   University of Florida, Gainesville, Florida

S. B. Roberts   Albert Einstein College of Medicine, Bronx, New York

D. M. Robins   College of Physicians and Surgeons of Columbia University, New York, New York

M. A. Root   Eli Lilly and Company, Indianapolis, Indiana

M. Rose   Massachusetts Institute of Technology, Cambridge, Massachusetts

J. J. Rossi   City of Hope Research Institute, Duarte, California

R. Samulski   University of Florida, Gainesville, Florida

G. H. Sato   University of California, La Jolla, California

W. I. Schaeffer   University of Vermont College of Medicine, Burlington, Vermont

M. D. Scharff   Albert Einstein College of Medicine, Bronx, New York

D. Schlessinger   Washington University School of Medicine, St. Louis, Missouri

J. Schultz   Papanicolaou Cancer Research Institute, Miami, Florida

P. H. Seeburg   Genentech, Inc., South San Francisco, California

H. M. Shepard   Genentech, Inc., South San Francisco, California

S. A. Sherwin   Biological Response, Frederick, Maryland

P. J. Sherwood   Genentech, Inc., South San Francisco, California

D. Shortle   Massachusetts Institute of Technology, Cambridge, Massachusetts

F. Sierra   University of Florida, Gainesville, Florida

C. Simonsen   Genentech, Inc., South San Francisco, California

R. Simpson   University of Washington, Seattle, Washington

K. Sirotkin   University of Tennessee, Knoxville, Tennessee

A. M. Skalka   Roche Institute of Molecular Biology, Nutley, New Jersey

R. G. Smith   University of Texas Medical Center, Houston, Texas

N. Stebbing   Genentech, Inc., South San Francisco, California

G. S. Stein   University of Florida, Gainesville, Florida

J. L. Stein   University of Florida, Gainesville, Florida

J. L. Strominger   Harvard University, Cambridge, Massachusetts

R. Swift   Genentech, Inc., South San Francisco, California

B. Taylor   University of Washington, Seattle, Washington

P. Thammana   Albert Einstein College of Medicine, Bronx, New York

W. G. Thilly   Massachusetts Institute of Technology, Cambridge, Massachusetts

J. N. Thomas   Massachusetts Institute of Technology, Cambridge, Massachusetts

S. C. Turner   Bethesda Research Laboratories, Inc., Gaithersburg, Maryland

A. Ullrich   Genentech, Inc., South San Francisco, California

M. Vasser   Genentech, Inc., South San Francisco, California

P. Walker   City of Hope Research Institute, Duarte, California

R. L. Ward   Christ Hospital Institute for Medical Research, Cincinnati, Ohio

P. K. Weck   Genentech, Inc., South San Francisco, California

W. J. Whelan   University of Miami School of Medicine, Miami, Florida

F. White   University of Washington, Seattle, Washington

A. Wieland   Genentech, Inc., South San Francisco, California

S. L. C. Woo   Baylor College of Medicine, Houston, Texas

D. E. Yelton   Albert Einstein College of Medicine, Bronx, New York

E. Yelverton   Genentech, Inc., South San Francisco, California

H. Young   Bethesda Research Laboratories, Inc., Gaithersburg, Maryland

D. J. Zack   Albert Einstein College of Medicine, Bronx, New York

J. B. Zeldis   Massachusetts General Hospital, Boston, Massachusetts

PREFACE

The Miami Winter Symposia are now well established on the national and international scene as one of the major annual expositions of the new biology of the 1970s and 1980s. This nineteenth volume is the record of the proceedings of the fourteenth symposium held in Miami Beach in January 1982. The theme was the translation of the new basic research findings into the practical application of biotechnology, with reviews of methodology and the applications of such methodology that lie behind the practical innovations.

The theme of the symposium was set by the Feodor Lynen Lecturer, Cèsar Milstein, whose development, with George Köhler, of monoclonal antibodies promises to be the dominating tool of medical technology in the next decade, both for diagnosis and therapy. The symposium began with reviews of techniques of eukaryotic cell culture, hybridoma technology and uses, and the *in vitro* synthesis of DNA and its use in the generation of protein analogs. Cloning into eukaryotic cells and methods of increasing the levels of gene expression were sessions that clearly reflected current areas of intensive research that have important commercial and clinical value. The formal presentations concluded with descriptions of the biological activities of cloned gene products, including reports on trials with human subjects of interferon, human insulin, and growth hormone, reports that indicate how far and how fast this field has moved in the space of less than a decade since the technologies were developed. A panel session on horizons in biotechnology concluded the meeting, with the speakers looking forward to the directions of future research and its applications.

The symposium drew a capacity audience of almost 800. We are gratified to have been so successful in our efforts to stage a timely and topical meeting, and we believe that this volume, which also includes the discussions of each presentation, will be a most useful source of reference to basic scientists and biotechnologists alike.

Our ability to organize these meetings and subsequently publish the proceedings depends heavily on the help we receive from our faculty colleagues. Special thanks are due to the committee chaired by Thomas R. Russell and the secretarial staff, notably Sandra Black, Olga Sanchez, and Pat Buchanan. We are also most indebted to Dr. Ralph W. F. Hardy for his convening of the panel on "Horizons in Biotechnology."

The symposium was made possible in part by the financial assistance of Abbott Laboratories; Beckman Instruments, Inc.; Eli Lilly and Company; Hoffmann-La Roche, Inc.; ICN Pharmaceuticals, Inc.; Merck Sharp & Dohme Research Laboratories; New England Nuclear; Smith Kline & French Laboratories; and the University of Miami School of Medicine, Departments of Dermatology and Pathology, and the Office of the Dean. Our special thanks are due to Bethesda Research Laboratories, Inc., who have become the sponsors of the Feodor Lynen Lecture.

Fazal Ahmad
Julius Schultz
Eric E. Smith
William J. Whelan

THE THIRTEENTH LYNEN LECTURE

MESSING ABOUT WITH ISOTOPES AND ENZYMES AND ANTIBODIES

César Milstein

Medical Research Council
Laboratory of Molecular Biology
Cambridge, England

"Looking back over my life,
everything seems to have happened by accident"
A.J.P. Taylor

"La oportunidad la pintan calva"
Old Spanish Proverb

I never imagined one could do amino acid seqences by the changes in electrophoretic and chromatographic mobility of radioactive peptides until I was actually doing it myself. This was in 1960, I was following Fred Sanger's instructions, and when we established the sequence of the pentapeptide involved in the phosphorylation site of phosphoglucomutase I began to really appreciate what it meant to have fun in research. And since the Lynen Lecture traditionally includes personal notes, I might as well start by telling you how it happened. This may seem pretty irrelevant to the subject of the symposium, and let me put you at ease, because it is totally irrelevant! Trouble is, in addition to being irrelevant, it is likely to be boring! The blame goes squarely on the organisers first for having decided on a lecture instead of something more entertaining, and second for choosing me as the lecturer. They should know better.

I took my degree in Argentina, at the University of Buenos Aires. I wasn't a particularly good university student. My major preoccupation was not academic matters, but the Students' Union, and the students' involvement in political and social matters. But somehow I stumbled through to the end of my exams. By then, I was working part-time in a laboratory

FROM GENE TO PROTEIN:
TRANSLATION INTO BIOTECHNOLOGY

3

of clinical biochemistry, where I learned what it meant to
work efficiently; organise your time, choose your methods
carefully, pre-define the level of accuracy, and work as fast
as is compatible with that level of accuracy.

MY ENZYME PERIOD

So I had to decide what I was going to do with my life and
with my degree. I reasoned that the way to make money was to
continue with clinical biochemistry. The alternative was to
try and do science, but this was a truly romantic hope. My
university training was poor, and strongly biased towards a
rather snobbish and remote view of what science was. It was
clear that if I wanted to have a go at science, I should not
start unless I could find someone I could trust to be a
scientist, and not a fake. More or less by accident, I
discovered that there was a biochemist called Leloir. Leloir
gave one of these Winter Symposia lectures, which was entitled
"I hate boring people with my recollections". I am sure we
all hate people being bored by our recollections. I will go
even further; I will say that I get furious when people get
bored with my recollections.

But, going back to my story, I went to see Leloir, who
could not take me, and suggested that I should go and see
Stoppani. Although I had never heard of him, he was the new
Professor of Biochemistry at the Medical School, and somehow I
felt that I was along the right tracks. Stoppani said that he
would take me as a research student, but that he could offer
no economic prospects at all, either in the form of a
fellowship, or of a future promise of any description. The
mention of those points was more than I actually expected and
the subject suggested, the role of -SH groups in aldehyde
dehydrogenase, met with my complete approval. I had no idea
of what it was all about, but it seemed to have the correct
blend of chemistry and biology. And so that's the way I
started messing about with enzymes.

When I look back to that early period, when Stoppani was
one of the few, and perhaps the only, full-time Professor of
the Faculty of Medicine in the University of Buenos Aires, a
full-time professor who probably had a salary of about the
same order of magnitude as a janitor, trying to do serious and
honest research in a laboratory with no funds at all, I must
confess that the idea that we were messing about with enzymes
seems today almost too pretentious.

My first failure, which almost cost me my position in the
lab, was to break, successively, three five-litre round bottom
flasks, out of a total of five. A fortune, one of the most

precious pieces of equipment in the laboratory. *The* most precious piece of equipment was a Warburg apparatus, which Stoppani didn't allow anybody to use but himself. My first success was the development of a workable and reproducible protocol for the preparation of yeast acetone powders from which the enzyme aldehyde dehydrogenase could be extracted in good yield. It was the Argentinian answer to the insurmountable problem of the extraction methods of the original paper which called for liquid nitrogen. We could afford neither the liquid nitrogen, nor indeed the appropriate thermos flask. Acetone and ether were all right, because they could be recovered by distillation. As you can see, I learned biochemistry the hard way.

The situation in Argentina changed quite considerably after 1955, and in a short time we were preparing our enzyme using a refrigerated centrifuge, and we could even assay activity, not by the old Warburg method but with a spectrophotometer, which we could borrow from a richer neighbouring laboratory. And then I really started getting involved in the kinetic and catalytic properties of active sites (1,2). By the time I wrote my thesis, I had a solid background of enzymology. I even had a proper job to come back to when I was awarded a fellowship from the British Council to go to work in Cambridge.

CAMBRIDGE – PART 1

In 1958, the Department of Biochemistry at Cambridge was a mixture of the old and the new. My research supervisor was Malcom Dixon, and only a few doors away from the enzyme unit, Dorothy Needham and Robin Hill were still active in the laboratory. But in his unassuming way, Sanger, who that year, only two weeks after my arrival at the laboratory, was awarded his first Nobel Prize, was a dominant influence. A high-voltage electrophoresis room had a warning at the entrance, where someone had altered the original D in the word Danger to read "Sanger – High Power"!

Dixon suggested that I work on the enzyme phosphoglucomutase, to clarify some odd observations made a number of years earlier at the department, that phosphoglucomutase required two metals for full activity, magnesium and a trivalent metal like chromium. Phosphoglucomutase had been the subject of two controversies at that time. The first had to do with the requirement of a co-enzyme for activity. The second involved activation by chromium. The first controversy was resolved in Argentina by Leloir and co-workers, who demonstrated that the true co-factor was glucose 1,6-diphosphate, which had been a previously unrecognised impurity. It was with some pleasure that I could sort out the second and much less important controversy to show that the activation by chromium and by metal chelators was of the same nature, in both cases involving the removal of heavy metals which were highly toxic to the enzyme (3).

I suspect that the choice of phosphoglucomutase on the part of Malcom Dixon had another element, which he didn't mention much. Considerable interest in this enzyme had been generated by a report by Koshland and Erwin, that the enzyme contained an active serine in an amino acid sequence that was similar to that of the proteolytic enzymes (4). This serine was phosphorylated by glucose 1,6-diphosphate, and since the serine phosphate derivative was stable to acid hydrolysis, radioactivity could be easily used to follow the derived peptides.

Fred Sanger at that time was actually looking for the development of methods which would allow amino acid seqences to be determined, using radioactive techniques. One day, we happened to be having tea at the same table, and he asked me what I was doing. He vaguely knew of my existence, because on arrival at Cambridge, he had been extremely kind, and arranged for my wife Celia to work with Kenneth Bailey. When I said that I was working on phosphoglucomutase, he immediately asked me whether I was going to work on the active site. I was extremely reluctant to do so. The radioactive preparation I estimated would take me at least 3 weeks, and the prospect of just confirming the previous sequence wasn't a very good incentive. But each time I went to Sanger's laboratory to use the only pH meter which worked reasonably well in the whole department, Fred kept asking me when I was going to make a radioactive preparation. One day it dawned on me that perhaps I didn't have to purify the [^{32}P]-labelled glucose 1,6-diphosphate. I simply had to make a crude preparation, which would only take a few hours, and then allow the purified enzyme inside a dialysis bag to equilibrate with the labelled substrate. Then by dialysing against saline, I would end up with a labelled enzyme preparation, without having to work at all. When I explained this project to Fred, I was rewarded with a grin of approval, and encouraging noises. It is possible that this idea was the one in my whole scientific career which had the biggest impact on my personal future.

A few days later, we were running an electrophoresis of a partial acid hydrolysate of the labelled enzyme. The pattern which emerged from the autoradiograph did not bear any resemblance to the pattern of the proteolytic enzymes. The amount of material we had was ample at the radioactive level, but utterly insufficient for purification and sequence analysis. And I started chasing radioactive spots to try to derive an amino acid sequence by new methods. Fred had some ideas on how to do this, and the active centre of phosphoglucomutase appeared to be the right challenge.

Although the work is largely forgotten, it is an example of how his mind was working around 1960, a couple of years before he started following radioactive spots derived from nucleic acids. The first step was to establish the correlation between the peptides (Fig. 1). So each band was eluted, and subjected again to partial hydrolysis. A band could give only phosphoserine, in which case it was a dipeptide, it could give a band identified as a dipeptide, plus phosphoserine and nothing else, in which case it would be a tripeptide,and so on. In this way all the radioactive spots present in the partial hydrolysate could be mapped as derived from a pentapeptide in which the phosphoserine was in the middle. Removal of the N-terminal residue by Edman degradation gave rise to a smaller peptide defining the polarity of the map.

The next step was to find out what the amino acids that made up the dipeptides and tripeptides were. Charged amino acids were easily spotted. Titration using mobility data showed that one of the amino acids next to the phosphoserine was histidine. We were excited about the presence of a histidine attached to a phosphoserine and we decided to test it by diagonal electrophoresis. Fred had been thinking about making use of diagonal electrophoresis to spot changes in

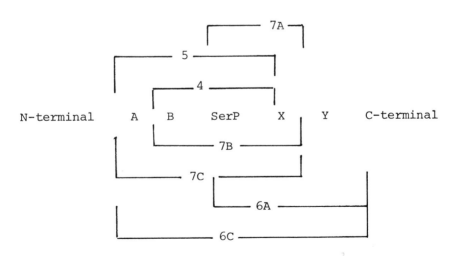

FIGURE 1. The correlation between the (^{32}P)-peptides of a partial acid hydrolysate of (^{32}P)-phosphoglucomutase. Taken from (5).

mobility of peptides, while I was reading about the
sensitivity of histidine to photo-oxidation in the presence of
methylene blue. The two things came together on a Saturday
morning when, if my memory serves me right, Fred and I were
discussing with Richard Ambler this likely histidine. The
idea was to spread the peptides on a first dimension, expose
the paper to ultraviolet light in the presence of methylene

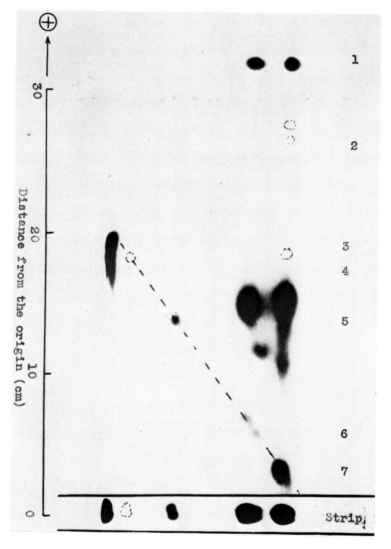

FIGURE 2. The identification of histidine next to (^{32})-
SerP by diagonal electrophoresis. Taken from (5).

blue, dry and then run again on a second dimension at the same pH to see whether the treatment had any effect on the mobility of peptides. The result was there; it was a beautiful diagonal with the peptides containing histidine moving away from the diagonal position (Fig. 2).

Using specific reactions of this type, and partition chromatography data, we came up with a pentapeptide sequence (5) which had nothing to do with the previously reported sequence, and which I had to defend in several seminars I gave in the United States on my way back to take up my position in Argentina as head of the newly-created Division de Biologia Molecular at the Instituto Nacional de Microbiologia.

THE ARGENTINIAN INTERLUDE

The Instituto Nacional de Microbiologia was a rather old institution created around the model of the Institut Pasteur. After going through a period of great neglect, it got a fresh lease of life when a new director, Ignacio Pirosky, was appointed. He obtained from the Government special concessions, which allowed him to appoint a large number of very young scientists to full-time key positions. This was a very bold and imaginative move, which immediately put him and the young people appointed in direct conflict with the sclerotic old guard. But Pirosky had strong backing from the Government, while it lasted, and in the meantime an atmosphere of great scientific excitement was developing among the young newly-appointed full-time scientists. The Division of Molecular Biology was essentially a place where fundamental research was to be done, and in the Division I could count on a group of extremely gifted people.

I started a programme on alkaline phosphatase and phosphoglyceromutase as possible candidates on which active sites around phosphoserine residues could be studied. But of course, it is a well-known fact that governments in Argentina don't last very long, especially if they are civilian governments appointed by popular ballot. So, about a year after my arrival in Argentina, we had a military coup, and a new Minister of Public Health. Predictably, this new Minister dismissed Pirosky, blaming him for all sorts of things, and young Argentinians being what they are, we became very emotionally involved in the defence of the dismissed Director. I am still surprised (a) that the officers of the Association of Scientists lasted as long as they did (1 year), (b) that the Minister himself lasted as long as he did, since in between there were two or three more sub-military coups, and

(c) that in all these dealings and political upheaval, we could do any science at all (6,7,8).

As it happened, the Chairman and the Secretary of the Association of Scientific Staff were members of my division. When they were dismissed for the most appallingly trivial and unimaginative reasons, and without consulting me, I myself resigned, and wrote a letter to Fred Sanger, asking for a job. To my delight, a reply came by return of post, and very shortly after, I was back in Cambridge.

A JUMP INTO ANTIBODIES FROM THE DISULPHIDE BRIDGES

I arrived back in Cambridge in the spring of 1963. The Laboratory of Molecular Biology had already been functioning for just over a year (Fig. 3). When Fred suggested that I might start doing some experiments with antibodies, I quickly became interested. He suggested labelling tyrosines in the

FIGURE 3. The MRC Laboratory of Molecular Biology as it was in 1963.

"active centre" of antibodies with radioactive iodine to study
their amino acid sequences, as we had been doing with enzymes.
That suited me very well, not only because I liked the idea of
iodination of tyrosines, but also becuse in Argentina, I had
been toying with the idea of reducing disulphide bonds of
antibodies, labelling them with radioactive iodoacetate, and
comparing the sequences around the labelled cysteine residues.
The idea was to find out whether two different antibodies
differed in primary structure, and, if they did, what the
difference was. The critical experiment, therefore, looked
deceptively simple.

After dozens and dozens of autoradiographs of peptide maps
of iodinated DNP antibodies and normal immunoglobulins, I
became convinced that this was a blind alley. A year ago, I
disposed of those fingerprints, in a nostalgic but ruthless
"clear-up" when I moved from one lab to another.

Fortunately, at the same time as I was doing these
iodination experiments, I was also labelling the cysteines
with ^{14}C-iodoacetate. In the laboratory just across the
corridor, the diagonal electrophoresis approach was being
applied by Brown and Hartley, to define the disulphide bonds
in proteins. Too near not to be attracted by the idea of
applying the approach to γ-globulin and antibodies.

My ignorance of immunology was absolute. I was totally
unaware of the vast literature already around at the time
concerning the relations between myeloma proteins and normal
immunoglobulins. So, when I noticed a very beautiful
difference in the disulphide bridge diagonals of the light
chains of a myeloma and a macroglobulin, I jumped with
excitement, and rushed to London to tell Sidney Cohen and
Rodney Porter that I thought I had discovered the difference
between macroglobulins and γ-globulins. Sidney dampened my
enthusiasm by saying "That macro I gave you is type II.
Before you jump to conclusions, you should try another myeloma
protein, type I".

Well, this was news to me. What were type I and type II?
Most of you probably don't know what type I and type II are
either, because the nomenclature has been totally dropped, but
you will be on familiar ground if I tell you that type I and
type II are what we now know as proteins containing κ and λ
light chains respectively. So, obligingly, he gave me a
little bit of another myeloma protein, this time 7S IgG, which
was type I, ie with κ chains as opposed to the λ chains of the
macroglobulin. Of course, that was the difference between the
two fingerprints.

By a comparison between normal light chains and those
coming from myeloma patients, I was able to build up a picture
of disulphide bridges of the light chain which, published in
1964 (9), represented the first batch of sequence data on

Bence Jones proteins. It came out just in time for me to be rather belatedly invited to the "historic" antibody workshop held in Warner Springs in the early part of 1965, where Hilschmann described for the first time the block sequence

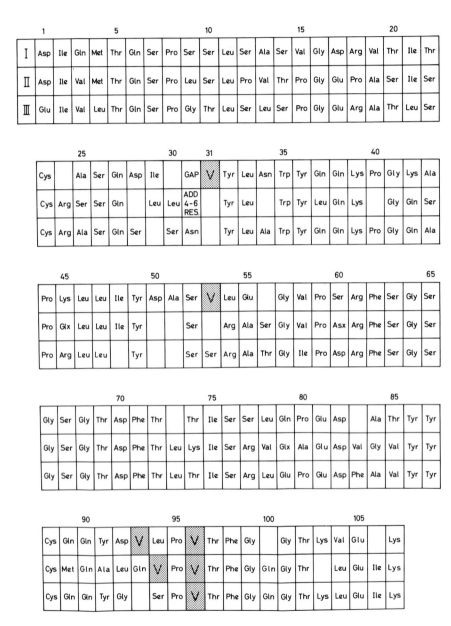

FIGURE 4.   The three basic sequences of human Kappa chains.

difference between two Bence Jones proteins demonstrating a constant and a variable part (10). My description in the 1964 paper of three disulphide bonds, one at the C-terminus being interchain, one being common to all Kappa Bence Jones proteins and light chains, and the other variable, fitted beautifully with the concept of the invariant C-terminus with one interchain and one intrachain disulphide bond and a variable N-terminus with another intrachain bond. And so, almost without realising it, I became an active member of the small bunch of the newly emerging molecular immunologists trying to understand the molecular nature of antibody diversity.

My studies of Bence Jones proteins culminated with the recognition of three subgroups of human kappa chains (11,12). Each subgroup (Fig. 4) contained a large family of chains, where the individuality of each chain was defined by amino acid sequence differences scattered along the chain, but mainly concentrated in certain regions, which are now recognised as the hypervariable regions. This also implied multiple genes for the V region. The postulate of Dreyer and Bennet (13) that V and C genes were separately encoded now seemed inescapable.

With Richard Pink, my first research student in Cambridge, and then with Frangione, we attacked the S-S bonds of heavy chains. We became fascinated by the fact that the four subclasses of human IgG appeared to have derived from a common recent ancestor. That meant, for instance, that the γ1 immunoglobulin heavy chain of the human was not the homologue of the γ1 of the mouse. In this way, we became aware of a fundamental property of evolution in multigenic families, namely that the individual components were not in constant expansion, but rather in a continuous dynamic expansion and contraction process. Thus we suggested that the divergence point for the heavy chain classes was old in evolution, but the subclass evolution, being a much more recent event, did not have a common departure point for all the species. It was immediately obvious that, if one applied the same principles to the V regions, many of the puzzling aspects we had observed could easily be explained (14). Let me quote our conclusions from our 1970 paper: "We suggest that the section of the genome involved in the coding of immunoglobulin chains undergoes an expansion-contraction evolution: that the number of individual genes coding for basic sequences is not large, and that it varies in different species and even within species at different stages of its own history. The task of providing for the endless variety of individual chains is left to somatic processes".

By then, the Division of Protein Chemistry had become the Division of Protein and Nucleic Acid Chemistry, and not surprisingly, I started to think about light chain messenger

RNA. My first attempts were made much too early, in
collaboration with Peter Fellner, a research student of Fred
Sanger's. Although the results were a clear disaster, with a
hint of hope, the interest they created in my mind was
strongly reinforced when George Brownlee felt that his success
with the 6S RNA, the longest RNA sequence at that time,
qualified him for more ambitious things. We worked

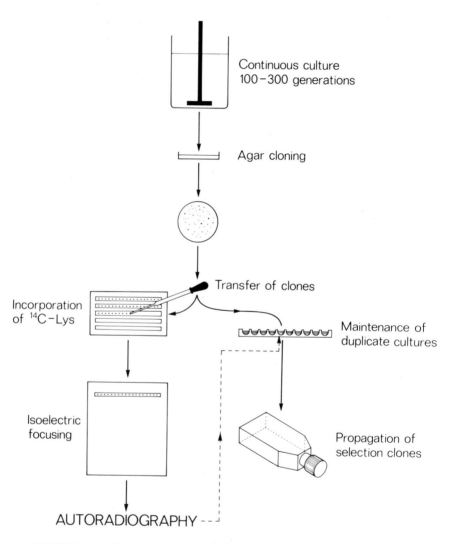

FIGURE 5. *The protocol used for the screening of 7,000
clones of P3 myeloma cells.* *A number of structural mutants
were detected, and are described in Table 1.* *Taken from (21).*

frantically on solid tumours for over a year, but the results
this time were very disappointing. We decided that we ought
to move away from solid tumours, and work with tissue culture
cells. But moving into tissue culture was not an easy
decision. It all seemed too much like witchcraft.

George Brownlee and I started to grow mouse myeloma cells
in culture which we imported from the Salk Institute. We had
two or three to choose from, and the choice was based on easy
growth on the one hand, and high production of immunoglobulin
on the other. To assess production, we added to the culture
medium a labelled amino acid, and then took the supernatant
and subjected it to cellulose acetate electrophoresis. It was
reassuring to look at a radioactive band, and the intensity of
the radioactive band gave us a very good idea of the synthetic
capacity of the individual cells. I must now resist the
temptation to expand on my collaboration with Brownlee and
later with Harrison and others, which led to the discovery of
the precursor of light chains and its significance in
secretion (15,16), and the sequence analysis of light chain
mRNA (17,18).

Somehow the culturing of myeloma cells, our continued
interest in somatic mutation and the potential of radioactive
methods clicked together when a new research student, David
Secher, was struggling with the sequence of a human myeloma
protein, and feeling a bit unhappy about it. On an exciting
Saturday morning, which was prolonged after lunch, David and I
decided to look for somatic mutants of cells in culture. Only
later were we to discover that Scharff and his collaborators
were already doing excellent work along those lines.

We would have to learn how to clone individual cells to be
able to screen for the large number of clones which would be
required for those experiments. The fact that we were total
beginners with tissue culture methods did not deter us. By
extraordinary luck, we received reinforcement in the form of
an Australian postdoc, Dick Cotton, who fortunately had no
experience in tissue culture either. But he was willing to
learn the little that we knew, and to carry on from there.
And we embarked on a search for mutants of cells in culture.
We settled for a protocol which is described in Fig. 5. The
experiments were indeed very successful, and we did manage to
obtain mutants (19). We then proceeded to sequence them, and
to understand the chemistry of the mutations (20).

The results were successful in one way, but unsuccessful
in another, because none of the mutants involved the variable
region (Table 1). But a great bonus was not to do with the
results themselves, but with the acquisition of the technology
of growing clones, and perhaps even more important, with the
establishment of continuous spinner cultures, growing
uninterrupted for over a year (21). The introduction of

TABLE I.   Spontaneous structural mutants of MOPC 21
heavy chains

Mutant	Protein Defect	Genetic defect
IF1	Last 82 residues of CH3 missing; carbohydrate difference	Ser (387)→Ter small deletion?
IF2	Whole CH1 deleted	5.5 K bases deleted including CH1 exon. Aberrant switch?
IF3	Altered sequence of residues 367–380. Deletion of rest of CH3	Frameshift (−2). Premature "ochre" termination
IF4	Asparagine 452 to Aspartic acid. ('mis-sense')	A to G transition
NSII/1	Deletion of last 67 residues	Trp (406)→Ter G to A transition ("non-sense")

tissue culture techniques into our set-up had a profound
influence on the way we were thinking about experiments. The
possibility of fusing two myeloma cells to see if V and C
regions could be scrambled (Fig. 6) was no longer out of our
reach (22). When Cotton left, Shirley Howe and I continued
with the fusion experiments, and when another total beginner
in tissue culture, Georges Kohler, arrived, he quickly derived
an azaguanine derivative of the non-secretor mutant NSI, with
which they produced a couple of very informative hybrids.

All this hybrid work was very nice, but what we really
wanted was to push forward with our mutant studies, and for

this we needed a myeloma line with antibody activity.   Georges
and I kept arguing about this every day.   He wanted a line
producing any class of immunoglobulin, providing that it had
antibody activity.   I wanted it to be an IgG - preferably an
IgG$_1$ - because I was concerned about the future difficulties
with sequencing the large number of mutants of our dreams.   We

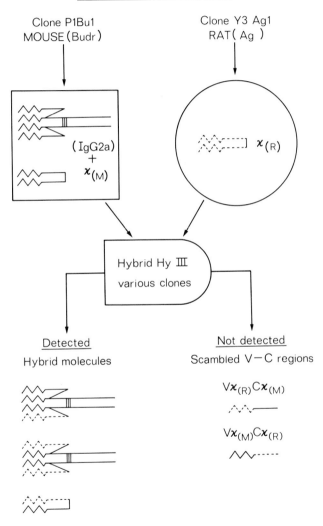

*FIGURE 6.   Codominant cis expression of antibody genes in
hybrids of myeloma cells.   The diagram, taken from (28)
presents the data described in (22).*

were lucky, because we failed. We failed to grow myelomas
with antibody activity, and also failed to find an antibody
activity associated with the IgG$_1$ myeloma MOPC 21. But we
were not prepared to give up. The frustration, the incentive
and the fact that we were working with hybrids and myeloma
cells which were ideally adapted for cell fusion purposes were
the essential ingredients for the successful derivation of the
first hybrid myeloma producing a predefined antibody (24)
(Fig. 7).

Looking back, I realise that the success of that
experiment was also, to a large degree, due to good luck. We
didn't know that a very long time in continuous culture was a
good way of making cells suitable for fusion (25). We didn't
know that myelomas were essential to the procedure, and that
clone X63 Ag8 was particularly good. We didn't know that to
perform the fusions, cells had to be growing for at least
several days well below their stationary phase (26). If the
first experiment had been experiment number 8, I wouldn't be
here today. Experiment number 8 did not work, and neither did
numbers 9, 10 and so on, until experiment number 25. It was
not until much later that we began to understand the
conditions required for reproducibility. We were lucky that
the first experiment worked. For incidental reasons, we had
all the required conditions fulfilled.

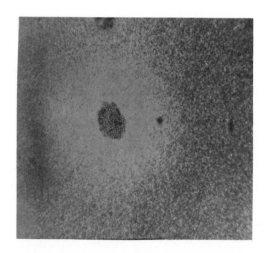

FIGURE 7. *Photomicrograph of the first clone of hybrid
myelomas secreting monoclonal antibody against sheep red cells.
The antibody diffuses around the clonal cells in the middle,
and produces a clear area where the sheep red cells have been
lysed. Beyond this area of lysis, the sheep red cells can be
seen as a dotted background.*

I have often been asked why we didn't take out a patent.
I have heard different versions of why, some quoting very
clever remarks, supposedly made by myself. The plain simple
fact is that on entering the MRC, I had to sign away all my
patent rights, and all other rights of an inventor. Still
today I am not sure how strict this ruling is, and neither am
I interested in knowing, because as a matter of fact I am in
favour of that ruling. Another question which has been asked
is did we realise the economic importance of the fusion
procedure. The answer is yes, but I do not think we ever
guessed at the number of zeros involved in the estimate.

I became more interested in possible applications of the
hybridoma technology in other fields of fundamental biological
research, apart from immunology itself, and in more practical
applications in clinical medicine. And that is why I started
a series of collaborations with colleagues outside the
Laboratory of Molecular Biology, to explore and demonstrate
those possibilities (27-31).

WHO OWNS SCIENTIFIC DISCOVERY, AND WHO SHOULD BENEFIT?

Monoclonal antibodies were an unexpected by-product of
pure, basic, research. It is not the first time that
fundamental technological developments are by-products of
basic research. Sometimes the contribution of the fundamental
research is not so obvious, and a good example is the
automation of biochemical analysis in hospitals. A quick
glance at the historical development of the machines takes you
directly to the first automatic amino acid analyser, developed
by Spackman, Stein and Moore (32), to determine the amino acid
sequence of proteins. Sometimes I wonder whether I am dazzled
by the proximity of the scientific point of view, and I fail
to realise that things are not so obvious as I believe they
are. Science and technology have worked hand in hand for a
long time, but it is the incentive of scientific curiosity, I
submit, that is the main stimulus for most quantum jumps in
technology.

Basic science pays, but in a funny sort of way. How on
earth could the manufacturers of insulin or of growth hormone
have ever imagined that to revolutionise their production, the
best thing was to support research connected with the
restriction phenomena in bacteria. With the emergence of
multi-national companies and huge industrial complexes, the
research and development component in industry has spread
into more general aspects of science, less constrained by the
immediate application. Perhaps more than anything else, the
emergence of high technology areas is eroding the boundaries

between applied and basic research. Still, industry's
interest in science remains narrow-minded, and while industry
is motivated by profit rather than by social benefit, it will
remain narrow-minded. A good example is the commercial
secrecy requirements, which are ever more bureaucratic and
sillier as the company is larger.

It is interesting to recall that less than 10 years ago,
there was mounting criticism, even among scientists
themselves, of the lack of practical achievement in modern
biology. One argument went that, in spite of large amounts of
money being poured into molecular biology, nothing practical
came out, despite the spectacular theoretical successes. And
it is worth recalling those days at this particular moment,
when spectacular successes in practical terms are being
obtained, and are undisputed by the most stringent of
customers, namely the market shares.

After so many years of taxpayer support for basic science,
who benefits from the practical achievements? When a new
discovery is made, it is often considered stupid for
scientists not to take out patents. And yet discovery is the
result of years and years of painful elaboration and of the
co-operation of many, many people. To develop the hybrid
myeloma technique, ideas, methods and materials were
necessary, and yet were not patentable. Ideas like the clonal
selection, with all the experimental facts which made that
idea almost a dogma in immunology. Tissue culture techniques,
and the critical methods of cloning and fusion at high
efficiency. The understanding of myelomas and their
experimental induction, and so on and so on. And in addition
there are the countless little tricks, assay methods etc.which
we take for granted, and which are essential to almost any
modern manipulation. And then there is the atmosphere of the
scientific community, at large and in the micro-environment of
each laboratory. How many times have brilliant ideas risen
prompted by appropriate remarks at tea-time in the laboratory
canteen? And what about the enormous task performed by
collaborators and assistants, and the usually forgotten
secretaries and administrators, who we usually regard more as
a hindrance than as a help, only because we are not in their
shoes. And what about the taxpayer?

The practical achievements of science are part of science
itself. They are advances in public knowledge, and therefore
don't belong to anybody, but to society. The question that
remains with us is how are we going to preserve the atmosphere
which encourages excellence in fundamental research, thereby
producing the most important advances; most important
because they are the ones on which practical advances are
based.

In the past, support for science was largely given in

recognition of its cultural value. The creation of the
research councils and similar organisations, which at the
national and international level encouraged the acquisition of
new scientific knowledge, is an event rather recent in
history, and implied a recognition of practical utility, or at
least, of potential practical value.

The support which society gave to basic science in the
hope of a practical return has, to my mind, been a spectacular
success, materialised in biology and medicine, not only by the
advent of biotechnology, but by modern health care. The
logical continuation is further investment and a more secure
basis for that type of system. The support for basic science
should not drift into a profit-seeking enterprise.
Furthermore, scientists need a good economic basis and
commitment to reseach, and not a commitment to more profitable
accomplishments. The present trend of scientists becoming
entrepreneurs would not be objectionable, provided their
entrepreneurial activities are not based on the academic
resources at their disposal. Mixing of activities seems to me
both objectionable and dangerous in the long run.

There is another trend. Granting bodies themselves, as
well as laboratories and scientific institutions, under the
pressure of the economic recession, are running the risk of
undermining the basis of their existence by trying to compete
with industrial enterprise, public or private. Of course, it
is true that administrators in charge of negotiating with the
Government for funds may find commercial success a way out, or
at least a persuasive argument. I believe what is necessary
is a firmer basis for the support of basic science, especially
because the past has shown that this is money well invested.
Once Bernardo Houssay was asked by a press interviewer whether
he didn't think that it was too much of a luxury to have basic
science in Argentina. He replied "Sir, we are an
under-developed country. We cannot afford to be without basic
science".

The past and present achievements of science deserve at
least a fraction of patent royalties. One of the arguments
used to justify the present patent laws is that they protect
the inventor. I think that there is a misunderstanding about
what precisely an inventor is, and what the role of an
inventor is. The patent laws may encourage people to become
inventors, which usually means to apply the present state of
public knowledge to the solution of certain practical
problems, with a view to making a profit. But patent laws
fail to recognise the importance that institutions engaged in
basic research have and will increasingly have, to provide the
most fundamental advances in technological development. That
also needs encouragement.

It is perhaps time for serious consideration of the

modification of the patent laws. The science councils and
granting bodies should receive a fragment of the royalties
from patents, which at the moment go to enterprises or
individuals without due consideration of the past history of
the invention. But do not misunderstand me: such a
contribution should not go to the particular laboratory from
which the invention originates, but to the organisations which
support basic research in the widest sense.

In this recollection I have tried to use a radioactive
trail, leading to monoclonal antibodies. There came a moment,
somewhere, where messing about with enzymes, isotopes and
antibodies became the origin of the monoclonal antibody
technique. When was that moment? The use of myeloma x
myeloma hybrids to understand allelic exclusion and the
integration of V and C genes, as well as our search for
somatic mutants in tissue culture were obviously critical.
But surely, if that is the case, then we had first to develop
the idea of correlations between structure and somatic
mutations and the technology of tissue culture. In which
case, perhaps we should go back earlier, to the stage where we
started messing about with tissue culture, which in turn was
inspired by the previous work on myeloma proteins. My own
involvement in this subject originated from the idea of using
radioactive iodoacetate to compare differences in the
sequences around the cysteine residues. Messing about with
radioactivity came from phosphoglucomutase labelling of the
active centre. And phosphoglucomutase labelling arose because
the only decent pH meter in the Department of Biochemistry was
in Sanger's laboratory...... and so on.
So, which is the point at which messing about stops, and
the jumping board to big business like the production of
monoclonal antibodies starts? Which brings me to the point of
my plea; to the right and the need for motivated scientists to
mess about with whatever they feel like.

ACKNOWLEDGEMENTS

P.S: It seems appropriate to point out to those who do
not know me personally that the difference between some of my
written and my spoken English is called Peggy Dowding, to whom
I am deeply grateful.

REFERENCES

1. Stoppani, A. O. M., and Milstein, C., *Biochem. J. 67*, 406 (1957).
2. Milstein, C., and Stoppani, A. O. M., *Biochim. Biophys. Acta 28*, 218 (1958).
3. Milstein, C., *Biochem. J. 79*, 591 (1961).
4. Koshland, D. E., and Erwin, M. J., *J. Am. Chem. Soc. 79*, 2657 (1957).
5. Milstein, C., and Sanger, F., *Biochem. J. 79*, 574 (1961).
6. Milstein, C., *Biochem. J. 92*, 410 (1964).
7. Pigretti, M. M., and Milstein, C., *Biochem. J. 94*, 106 (1965).
8. Zwaig, N., and Milstein, C., *Biochem. J. 98*, 360 (1966).
9. Milstein, C., *J. Mol. Biol. 9*, 836 (1964).
10. Hilschmann, N. and Craig, L. C., *Proc. Natl. Acad. Sci. U.S.A. 53*, 1403 (1965).
11. Milstein, C., *Nature 216*, 330 (1967).
12. Milstein, C., *F.E.B.S. Letts. 2*, 301 (1969).
13. Dreyer, W. J., and Bennett, C. J., *Proc. Natl. Acad. Sci. U.S.A. 54*, 864 (1965).
14. Milstein, C., and Pink, J. R. L., *in* "Progress in Biophysical and Molecular Biology" (J. A. V. Bulter and D. Noble, eds.), *21*, 209. Pergammon Press, Oxford and New York, (1970).
15. Milstein, C., Brownlee, G. G., Harrison, T. M., and Matthews, M. B., *Nature New Biol. 239*, 117 (1972).
16. Harrison, T. M., Brownlee, G. G., and Milstein, C., *Eur. J. Biochem. 47*, 613 (1974).
17. Milstein, C., Brownlee, G. G., Cartwright, E. M., Jarvis, J. M., and Proudfoot, N. J., *Nature 252*, 354 (1974).
18. Hamlyn, P. H., Gait, M. J., and Milstein, C., *Nucleic Acids Res. 9*, 4485 (1981).
19. Cotton, R. G. H., Secher, D. S., and Milstein, C., *Eur. J. Immunol. 3*, 136 (1973).
20. Adetugbo, K., Milstein, C., and Secher, D. S., *Nature 265*, 299 (1977).
21. Milstein, C., Cotton, R. G. H., and Secher, D. S., *Ann. Immunol. (Inst. Pasteur) 125C*, 287 (1974).
22. Cotton, R. G. H., and Milstein, C., *Nature 244*, 42 (1973).
23. Kohler, G., Howe, S. C., and Milstein, C., *Eur. J. Immunol. 6*, 292 (1976).
24. Kohler, G., and Milstein, C., *Eur. J. Immunol. 6*, 511 (1976).
25. Galfre, G., Milstein, C., and Wright, B., *Nature 277*, 131 (1979).
26. Galfre, G., and Milstein, C., *Methods in Enzymology, 7B*,

3 (1981).

27.   Williams, A. F., Galfre, G., and Milstein, C., *Cell 12*, 663 (1981).

28.   Howard, J.C., Butcher, G. W., Galfre, G., Milstein, C., and Milstein, C. P., *Immunological Rev. 47*, 139 (1979).

29.   Milstein, C., and Lennox, E., *in* "Current Topics in Dev. Biology" (M. Friedlander, ed.) *14*, 1. Academic Press, New York, (1980).

30.   Voak, D., Sacks, S., Alderson, T., Takei, F., Lennox, E., Jarvis, J. M., Milstein, C., and Darnborough, J., *Vox Sang. 39*, 134 (1980)

31.   Cuello, A. C., Milstein, C., and Priestly, J. V., *Brain Res. Bull. 5* (5), 575 (1980).

32.   Speckman, D. H., Stein, W. H., and Moore, S., *Anal. Chem. 30*, 1190 (1958).

INTRODUCTION

MOVING GENES: PROMISES KEPT AND PENDING

Philip Leder[1]
Philip A. Hieter
Gregory F. Hollis
Aya Leder

Laboratory of Molecular Genetics
National Institute of Child Health and Human Development
National Institutes of Health
Bethesda, Maryland, U.S.A.

I. INTRODUCTION: THE PROMISES KEPT

In the summer of 1976 a particularly gripping article appeared in the *New York Times Magazine* under the following ominous title, "New Strains of Life...or Death" (1). The article, of course, addressed the emerging recombinant DNA technology and its safety, and suggested that many of the promises offered by these revolutionary techniques were in fact unrealistic and, in any case, could be kept by using other--unspecified--techniques. The author acknowledged that these unspecified techniques might require more time but, more importantly, the article then seriously raised the question of whether this sort of research should continue at all.

Six years have passed since that hysterical summer. Many of the overblown issues raised in the course of that debate have been laid to rest. As evidenced by this and many other similar meetings, the research has gone on. It is, therefore, particularly appropriate now--six years later--to review those promises in order to ask which have been kept and which remain to be fulfilled.

From the very outset it was clear that this new approach would allow us to understand the fundamental structure and

---

[1]Department of Genetics, Harvard Medical School, Boston, Mass.

FROM GENE TO PROTEIN:
TRANSLATION INTO BIOTECHNOLOGY

27

organization of genes. It was equally clear that these studies would provide an enormous insight into the molecular mechanisms that operate to regulate the expression of these genes
and assure their orderly modulation. It was further clear
that major practical goals could be reached using recombinant
DNA technology. Production of protein hormones, interferon,
and useful fermentation organisms were among the immediate
short-term goals and talks elsewhere in this Symposium will
chronicle the success of efforts in these directions. Our major purpose will be to detail the way in which a large number
of laboratories using these powerful new techniques have quite
literally changed the way we think about the mammalian genome.

II. A CHANGED PICTURE OF THE MAMMALIAN GENOME

In the summer of 1976, our picture of the mammalian genome
suggested that genes would be encoded in continuous arrays of
nucleotides, organized in simple loci containing only those
genes corresponding to known alleles. Furthermore, our notion
of evolutionary change held that genomic DNA was very stable,
changing slowly one base at a time so as to produce the kind
of single amino acid alterations that distinguish, for example, normal from sickle cell hemoglobin. In keeping with this
notion of the stability of chromosomal DNA was the consequent
notion that genes do not easily move and, in particular, that
they do not move during the lifetime of a somatic cell.
Of course, we now know that all these notions are largely
wrong.
Genes are *not* encoded in continuous sequences. Genes are
not represented in simple loci reflective of their phenotype;
their loci are far more complex, laced with extra copies of
cryptic pseudogenes. Genomic DNA does *not* change slowly one
base at a time during evolution; it changes much more quickly,
by inserting and deleting large chunks of DNA. Moreover, DNA
is *not* stable. Genes move. They move during evolution, as in
the case of a globin and immunoglobulin genes we shall describe below, but they also move during somatic development--
as exemplified by the immune system. In what follows, we
shall recapitulate these new facts and provide some evidence
for them. They represent some of the promises that have been
kept. Subsequently, we intend to provide evidence for and
describe a new class of genetic element that seems to represent a major component of the mammalian genome and to have
been created by a major mechanism of genomic evolution. This
realization provides us with some assurance that our approach
remains promising. And finally, we would like to state the

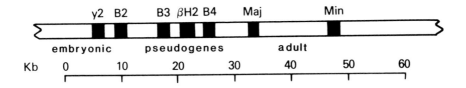

Figure 1.   Diagrammatic representation of the mouse beta
globin gene locus.   The beta gene locus of the mouse is represented on an approximately 50 kilobase long stretch of DNA.
The filled regions represent the positions of beta globin-like
sequences.   The two adult genes, beta globin major and minor,
are indicated.   The embryonic gene, ey, is also indicated.
The beta-like sequences between them are pseudogenes (2,3).

obvious; namely, to set out some of the extremely challenging
problems that are still before us--the promises that have yet
to be kept.

A.   The Complexity of Genes and Genetic Loci

So much evidence has accumulated indicating that most
mammalian genes are encoded in discontinuous pieces of DNA
interrupted by intervening sequences, that this fact hardly
needs mention (see references in review, 2).   Nevertheless,
interrupted genes still provide us with a number of unanswered
questions regarding their role in the physiology of the cell
and in evolution.   Putting aside these questions, however, we
shall take up a less well known feature of the genome, the
fact that genetic loci consist of relatively large arrays of
related gene sequences, some of which encode active genes, but
others of which encode inactive or pseudogene copies.   An excellent example of this is afforded by the beta globin locus

of the mouse in which there are at least seven beta-like genes
spread out over approximately 50 kilobases of genomic DNA
(Figure 1) (2,3). The two 3' most of these are beta globin
major and minor, the genes that are expressed in the adult
red cell. The 5' most gene, y2, is an embryonic gene (4), expressed only in the nucleated red cells that appear in the
yolk sac of the mouse embryo. The remaining four beta-like
sequences are pseudogenes; that is, genes that very closely
resemble beta globin, but have undergone a number of alterations so as to render them incapable of encoding a coherent
globin polypeptide chain. Evidently, these genes have arisen
by some duplication mechanism and, as they likely constituted
redundant gene copies, have begun to drift away from the beta
globin sequence following release of selective pressure. This
complex locus, replete with extra pseudogene copies, is likely
to be a feature of many genetic loci.

B. A Quickened Pace of Genomic Change

The comparison of amino acid sequences of gene products
has led to a view of evolution occurring principally by altering a single nucleotide at a time in an active gene. This
view is deceptive. Obviously, active genes do not tolerate
dramatic changes in their nucleotide sequences that can give
rise to major deletions, insertions, frameshift alterations
or termination codons that can inactivate a crucial sequence.
If a gene is required for the organism's survival, such mutations assure themselves of extinction. Indeed, it is likely
that the last place one should look for the changes that
actually occur in genomic DNA is within the sequence of an
active gene. Selection has assured that such changes will
not often be preserved within it. In contrast, those regions
beyond an active gene, or within it in the form of intervening sequences, that are not subject to the stringent selective pressures, should provide a more authentic picture of the
changes that can occur in genomic DNA. An excellent example
of these alterations is provided by comparing the beta globin
major and minor genes (shown heteroduplexed in the electron
micrograph in Figure 2). As seen in the heteroduplex structure, these genes have conserved their sequences largely within their coding regions, but their flanking regions and the
large intervening sequence that interrupts both genes have
undergone extensive changes so that they no longer anneal to
one another. Extensive sequence studies and comparisons have
shown that these alterations have occurred by either inserting or deleting major blocks of DNA (5).

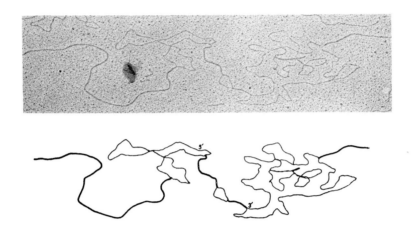

Figure 2. Heteroduplex structure formed between the
mouse beta major and beta minor genes. Cloned fragments encoding the mouse beta globin major and minor genes in the
bacteriophage lambda are heteroduplexed to one another and
are shown in this electronmicrograph. The molecule is shown
diagrammatically below the elctronmicrograph; where thin
lines represent single-stranded DNA and thick lines, double-
stranded or heteroduplexed DNA. The region of homology 5'
consists of approximately 100 bases and the first and second
coding domains as well as the small 5' intervening sequence.
The bubble lacking homology consists largely of the second
or larger intervening sequence and the second region of
homology consists of the third coding domain plus approximately 100 bases to its 5' side. The data is from Tiemeier
et al. (6).

The picture that emerges from an examination of these
altered flanking and intervening sequences suggests that the
rate of evolutionary change is much greater than that suspected by examining phenotypes. Indeed, even the molecular
basis for some of these changes can be suggested by noting
that deletions often take place between two short oligonucleotide sequences that occur several hundred bases from one another in the undeleted gene copy. One can only imagine that,
relative to the single base changes that accumulate within
active genes, these changes occur with much greater frequency.

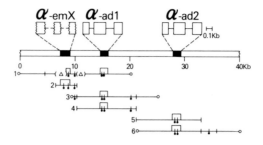

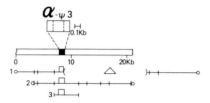

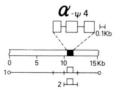

Figure 3. Map of the mouse alpha globin locus and pseudogenes. The alpha globin-like genes of the mouse are shown. The uppermost map represents the active locus consisting (5' to 3') of the alpha embryonic and two adult globin genes. Beneath are the structures of cloned fragments that encode two pseudogenes (αψ3 and αψ4). Below each diagrammatic map are the cloned fragments of DNA that have been isolated from genomic libraries and used to create the physical linkage (9).

C. Dispersed Genes

1. <u>The Alpha Globin System</u>. The alpha globin system of the mouse provides an excellent example of a tightly coordinated gene family whose structure and function are amenable to study. Five closely related members of this gene family have been cloned and sequenced (7-10). Three of these, the embryonic and two adult genes, have been mapped to a locus no more than 20 kilobases in length on mouse chromosome 11 (Figure 3). The remaining two genes are pseudogenes; that is, their structure is very similar to that of alpha globin, but they have undergone sufficient changes so as to no longer encode a coherent alpha globin sequence (2,8,10). One of these genes,

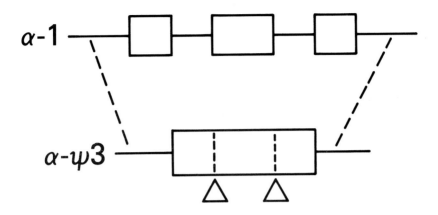

Figure 4.   Diagrammatic representation of the structure
of the αψ3 gene.   The structure of this alpha globin pseudogene was established by determining its nucleotide sequence
(8).   The open boxes represent coding sequences; the lines
connecting them, the two intervening sequences.   As shown in
the diagrammatic representation of the pseudogene, these
intervening sequences have been deleted exactly in accordance
with the structure of a globin-like mRNA.

αψ4, closely resembles the adult alpha globin gene in both its
sequence and organization, but the second pseudogene, αψ3,
differs from the adult gene in a particularly interesting way
(8,10).   It has lost both its intervening sequences in strict
accordance with the GT/AG rule of RNA splicing (Figure 4).
Furthermore, the homology of this gene to the normal adult
globin sequence seems to end very close to the polyA addition
site of a normal globin RNA transcript.
     Neither of these genes could be linked to a major alpha
locus by ordinary molecular gene cloning techniques.   Consequently, an effort was undertaken to map them using mouse-
Chinese hamster cell hybrids that retained varying numbers
of mouse chromosomes (9,11).   These studies quickly indicated
that these two pseudogenes, unlike those of the beta globin
system noted above, had moved from chromosome 11 and now re-
sided on chromosomes 15 and 17, respectively. The genes were
dispersed (Figure 5).

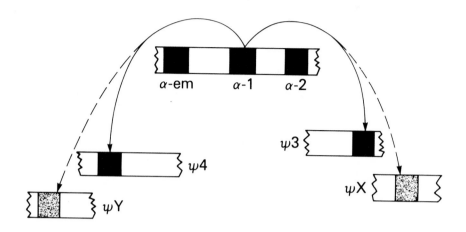

Figure 5. Diagrammatic representation of the major alpha globin locus and its conveyed genes. The major locus is shown in the upper line and the conveyed genes on new chromosomal loci below them. The lowermost, stippled, structures represent additional, postulated, dispersed genes.

Whatever mechanism (see below) was involved in the conveyance and formation of these genes, it seemed likely that if two such sequences existed they might be recent representatives of a phenomenon that was ongoing for millions of years of evolutionary time (Figure 5). Further members of this family should be detectable by reducing the stringency and increasing the sensitivity of the usual blot hybridization assays that are employed to detect DNA fragments that encode genes. An example of such an effort is shown in Figure 6. Here, a probe corresponding to the adult alpha globin sequence detects approximately ten fragments that evidently contained alpha globin-like sequences.

2. The Human Ig Lambda Locus. If the alpha globin gene family of the mouse is complex and contains several dispersed members, it is reasonable to expect that other gene loci will be similarly represented. The human immunoglobulin lambda light chain gene family provides a case in point (12). The human immunoglobulin light chain genes are encoded in several discontinuous segments of DNA, two of which are joined to one another during the somatic differentiation of immunoglobulin-

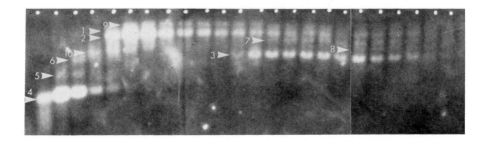

Figure 6. Detection of alpha globin-like fragments of mouse genomic DNA distributed by two-dimensional chromato-graphy and electrophoresis. The fragments were separated, blotted and hybridized on nitrocellulose filters using tech-niques described by Leder et al. (9). Each band represents a DNA fragment that crosshybridizes to·a mouse alpha globin cDNA probe. The numbers indicate discrete bands of which there are at least 10.

producing cells (see references in review, 13). One of these germline segments consists of a constant region gene separated by several thousand nucleotides from a small coding segment called the J (for joining) segment. Formation of an active immunoglobulin gene involves the joining of a germline V re-gion segment encoded some several thousand bases away to this J region segment. The intervening sequence that separates the J and C remains in the active gene and these segments are joined at a subsequent RNA processing step.

The immunoglobulin lambda light chain locus of man con-sists of six separately encoded J constant region sequences that are spread over approximately 40,000 basepairs of genomic DNA on human chromosome 22 (Figure 7) (12,14). If careful *in situ* hybridization experiments are done, additional lambda-like bands can be detected in human genomic DNA. Several of these have been cloned and one characterized further (15). Detailed studies of this gene indicate that it consists of a

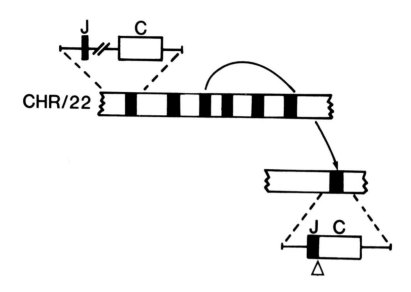

Figure 7. Diagrammatic representation of the human
immunoglobulin lambda genes and their locus. The uppermost
line is the putative diagrammatic arrangement of a germline
human C lambda chain. The middle line represents the six -
gene-containing locus and the bottom line represents the
structure of the pseudogene which is no longer on chromosome
22.

J coding segment covalently joined to a C coding segment,
again in strict accordance with the rules of RNA splicing. In
addition to this interesting feature of its structure, its
homology to a normal lambda C region ends abruptly at a polyA
addition site and in a long sequence of polyA residues. Further chromosomal mapping studies indicate that this pseudogene
is not located on chromosome 22, but in fact is also a dispersed gene (15).

   Both the alpha globin pseudogene noted above and the
lambda immunoglobulin pseudogene referred to here, bear several hallmarks of RNA-like processing. Both have lost intervening sequences in accordance with the rules of RNA splicing,
both share homology with an authentic gene up to the point of
polyA addition and, in the case of the immunoglobulin pseudogene, a sequence of polyA's reminiscent of a polyA tail occurs
precisely at this point. Obviously, the dispersed copies of
these genes resemble an RNA transcript of the authentic gene
more than they do the authentic gene itself. Their structure

and the fact that they have been conveyed to a new locus
strongly suggest that an RNA-like intermediate has played a
role in their formation and conceivably in their conveyance
as well. Inasmuch as these genes bear evidence of RNA processing, we refer to them as *processed genes,* and distinguish
them from the dispersed pseudogenes (such as $\alpha\psi4$) or histone
"orphon"genes (16) which appear to represent large homologous
segments of DNA carrying no evidence of RNA-like processing.

3. <u>Models for the Conveyance of Processed Genes via RNA</u>
The original discovery of these pseudogenes led to the suggestion that RNA, either directly or indirectly, might have
been involved in their formation (8,10). Subsequently, detailed studies of retrovirus structure have indicated that fragments of genomic DNA taken up by retroviruses can undergo
RNA-like splicing (17,18). This observation led Goff et al.
(17) to suggest that the alpha globin processed gene may have
been formed during its conveyance as a retrovirus sequence.
Indeed, retrovirus-like sequences have been found close to
this gene (19).

The structure of the immunoglobulin processed gene is a
little more difficult to reconcile with the retrovirus model.
First of all, its homology with the normal gene ends at a
polyA addition site in a sequence of adenylic acid residues.
If the gene had been taken up from genomic DNA by the retrovirus, there would have been no reason to expect these features. Rather, it seems likely that this gene was taken up
in its processed RNA form, converted, probably via reverse
transcriptase, into a DNA copy, and then incorporated into an
aberrant chromosomal site (Figure 8). While these genes
closely resemble their RNA transcripts with respect to splicing and their 3' ends, they differ from normal transcripts at
their 5' ends. In the case of the alpha globin pseudogene,
homology extends 5' to the site of normal transcription initiation. In the case of the lambda processed pseudogene, its
structure lacks the V region normally associated with the normal immunoglobulin transcript. In each case, if transcription were involved in the formation of these genes, it seems
to have begun at a point 5' to the normal transcriptional initiation site. It is possible to imagine that this aberrant
transcription incorporated new sequences into the globin and
immunoglobulin transcripts. It is further possible to imagine
that these new sequences conferred properties on their RNA
transcripts that allowed them to be mobilized and integrated
into the host chromosome. While transposable elements provide
a biologic precedent for such sequences, it may not even be
necessary to postulate special signals for this purpose. Indeed, the success of DNA transformation experiments using

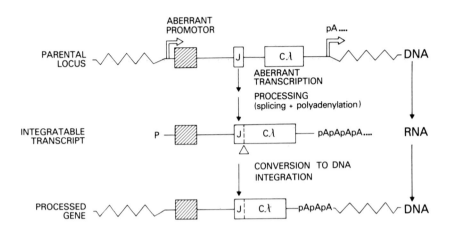

Figure 8. A model for *processed gene* formation and conveyance. The model is described in the text. The hatched
box simply represents a potential nucleotide sequence that
would convey mobility and integration properties to a DNA or
a RNA sequence into which it had been incorporated.

animal cells clearly indicates that illegitimate recombinational events will incorporate DNA sequences into the genome
at a finite rate (20). The fact that these processed genes
appear to have existed once as an RNA probably requires us to
postulate their reverse transcription into a DNA form prior
to integration.

In any case, these findings suggest that such *processed
genes* will be a major element in the genome and that the
examination of features common to their structure will tell
us something more explicit about their mechanism of formation.
For example, if each of these genes does contain an element
that allows them to be easily integrated, this notion has experimentally predictive value. These two initial examples
strongly suggest that genetic information can return to the
genome via an RNA intermediate, and that this may represent a
major mechanism of genomic evolution.

III. EPILOGUE: THE PROMISES TO BE KEPT

We have seen above that recombinant DNA technology has
fulfilled its expectations in terms of dramatically altering
our picture of the mammalian genome. It has already provided
us with so many unexpected features of genomic organization,
that it seems unlikely to have as yet provided our last surprise. But in terms of the promises to be kept, we must turn
to the question of the molecular basis of gene regulation
viewed in the proper context of a developing organism. It is
here that we confront a major challenge. Surely, to understand the nature of control, we must have the ability to manipulate a gene in its natural context. This, in essence,
requires us to be able to remove a gene segment, alter it or
its biochemical environment, and return it to its normal
chromosomal setting in an otherwise normal cellular context.
This form of genetic manipulation, whether invoked to study a
fundamental genetic process, to correct a genetic disorder, or
to manipulate and thereby improve the genetic makeup of a complex organism, presents the most formidable barrier we have
before us. While there are many intermediate steps and much
to be learned from the type of experiment that is feasible
today, it is not unduly optimistic to suggest that an extension of these techniques will eventually allow us to accom-
plish even these more difficult feats.

REFERENCES

1. Cavalieri, L.F., *New York Times Magazine,* Aug. 22 (1976).
2. Leder, P., Hansen, J.N., Konkel, D., Leder, A.,
    Nishioka, Y., and Talkington, C., *Science 209,* 1336-1342
    (1980).
3. Jahn, C., Hutchison, C.A., Phillips, S.J., Waver, S.,
    Haigwood, N.L., Voliva, C.F., and Edgell, M.H., *Cell 21,*
    159-168 (1980).
4. Hansen, J.N., Konkel, D.A. and Leder, P., *J. Biol. Chem.,*
    in press.
5. Konkel, D.A., Maizel, Jr., J.V., and Leder, P., *Cell 18,*
    865-873 (1979).
6. Tiemeier, D.C., Tilghman, S.M., Polsky, F.I., Seidman,
    J.G., Leder, A., Edgell, M.H., and Leder, P., *Cell 14,*
    237-245 (1978).
7. Leder, A., Miller, H.I., Hamer, D.H., Seidman, J.G.,
    Norman, B., Sullivan, M., and Leder, P., *Proc. Natl.
    Acad. Sci. U.S.A. 75,* 6187-6191 (1978).

8.   Nishioka, Y., Leder, A., and Leder, P., *Proc. Natl. Acad. Sci. U.S.A. 77,* 2806-2809 (1980).
9.   Leder, A., Swan, D., Ruddle, F., D'Eustachio, P., and Leder, P., *Nature 293,* 196-200 (1981).
10. Vanin, E.F., Goldberg, G.I., Tucker, P.W., and Smithies, O., *Nature 286,* 222-226 (1980).
11. Popp, R.A., Lalley, P.A., Whitney, J.B., and Anderson, W.F., *Proc. Natl. Acad. Sci. U.S.A. 78,* 6362-6366 (1981).
12. Hieter, P.A., Hollis, G.F., Korsmeyer, S.J., Waldmann, T.A., and Leder, P., *Nature 294,* 536-540 (1981).
13. Leder, P., Max, E.E., and Seidman, J.G., in "Immunology 80" (M. Fougereau and J. Dausset, eds.), p. 34, Academic Press, London (1981).
14. McBride, O.W., Hieter, P.A., Hollis, G.F., Swan, D., Otey, M.C., and Leder, P., *J. Exp. Med.,* submitted.
15. Hollis, G.F., Hieter, P.A., McBride, O.W., Swan, D., and Leder, P., *Nature,* submitted.
16. Childs, G., Maxson, R., Cohn, R.H., and Kedes, L., *Cell 23,* 651-663 (1981).
17. Goff, S.P., Gilboa, E., Witte, O.N., and Baltimore, D., *Cell 22,* 777-785 (1980).
18. DeFeo, D., Gonda, M.A., Young, H.A., Chang, E.H., Lowy, D.R., Scolnick, E.M., and Ellis, R.W., *Proc. Natl. Acad. Sci. U.S.A. 78,* 3328-3332 (1981).
19. Leuders, K., Leder, A., Leder, P., and Kuff, E., *Nature,* in press.
20. Perucho, M., Hanahan, D., and Wigler, M. *Cell 22,* 309-317 (1980).

DISCUSSION

D.C. BURKE: Pseudogenes are not translated.   Are they transcribed?   Once the gene product ceases to be made, do the promoter sites of the pseudogenes start to mutate to become non-functional?

Also, does the mechanism suggested for production of intron-less immunoglobin genes provide an explanation for the existence of active genes which do not contain introns- for example, the histones and the interferons?

P. LEDER: The two examples of pseudogenes that we have evaluated in this regard do not appear to be transcribed. Let me qualify that by saying that they do not appear to be transcribed either in embryonic or adult red blood cells.   In addition, Carol Talkington, whose work is described in a poster session in this meeting, has evidence to indicate that the promoter regions corresponding to these pseudogenes are inactive in in vitro systems.

In answer to your second question, I could really imagine that certain genes, such as those corresponding to the interferons, that lack intervening sequences, could have arisen by a mechanism involving an RNA intermediate. One can imagine several such mechanisms in addition to conveyance by a retrovirus or freely integratable transcript. We have addressed this question elsewhere (see Nishioka, Leder and Leder, PNAS 77: 2806-2809, 1980).

G. HAUGHTON: The observation that genes can jump between chromosomes with great facility has major implications for ideas about evolution. In view of this, what are the implications of larger groups of genes, such as those of the major histocompatibility complex, choosing to remain together through very long evolutionary periods?

P. LEDER: The closely linked nature of the histocompatibility genes may have some implications about the way in which they evolved or the way in which they are regulated. Obviously, we do not know enough about their structure or action to be able to develop a precise mechanism at this point. One might assume, however, that selection has played some role in their grouping. Nevertheless, I would be very surprised to learn that genes closely related to the histocompatibility antigens have not spread to other chromosomal locations. Indeed, I am certain that when this locus is better understood that this will be found to be the case.

G.F. MERRILL: In your model is the original progenitor of a "processed gene" a normal active gene that on rare occasions is abnormally transcribed or is it a pseudogene that undergoes a rare transcription event?

P. LEDER: Our model holds that the two processed genes we have examined are most likely to have arisen from an active gene transcribed from an aberrant promoter site. Obviously we cannot rule out a transcription event having occurred from a pseudogene that is closely related to the active gene structure.

C. SIMONSEN: What is the estimated size of the alpha-globin unit which originally transposed in view of the large duplicated regions observed in murine DHFR or CAD genes. One would expect larger regions of homology at the 5' and 3' ends

CELL CULTURE AND PHYSIOLOGY

Gordon H. Sato

Department of Biology
University of California, San Diego
La Jolla, California
U.S.A.

The first tissue culture experiment of Ross Harrison was greeted with wonder and amazement (1). Some seventy-five years later, it is still useful to ponder the thinking of Harrison's contemporaries. They must have thought along the following lines: the microenvironment of cells within the body of an animal must be so complicated that it should be virtually impossible to recreate it sufficiently well so as to allow life to continue outside the body. I believe that this line of thought was well justified and my talk will deal with the relationship between the requirements of cells in culture to the identification of the relevant elements of their *in vivo* environment.

First, however, I would like to digress and discuss another subject which led to our consideration of the requirements of cells in culture. Ross Harrison's experiment also excited many with the prospect of providing a new and powerful technique for analysing integrated physiology. However, a full fifty years after Harrison's historic experiment, no cell lines existed which were known to express the differentiated functions of the tissue of origin.

Our first efforts in this area were to show that the lack of differentiated function in cell cultures was due to selective overgrowth of fibroblasts (2).

It was now clear to us that the problem was to confer some kind of growth advantage on the specialized cells so that they could effectively compete with fibroblasts. Dr. Vincenzo Buonassisi and I set out to devise an enrichment culture technique analogous to the procedures used so successfully by bacteriologists.

We were helped immeasureably by the generosity of Dr. Jacob Furth, one of the important pioneers in endocrine oncology. Not only did he continue to provide us with valuable tumor material over the next several years but he provided the conceptual impetus for our work over a span of twenty years.

Initially, he provided us with transplantable, murine steroid secreting adrenal cortical tumors and ACTH secreting

FROM GENE TO PROTEIN:
TRANSLATION INTO BIOTECHNOLOGY

pituitary tumors (3,4). We placed these tumors in culture
for short periods of time and injected animals with the surviving cultured cells to obtain culture-derived tumors.

The culture-derived tumors were much easier to grow in
culture than the original tumors and pure strains of hormone-
secreting and hormone-responsive culture lines could be established from them readily (3). In this way, we established
many functional cell lines such as steroid producing adrenal
cells (5), ACTH-secreting pituitary cells (5), growth hormone
and prolactin secreting cells from a Jacob Furth produced
pituitary tumor (6,7), steroid producing Leydig cells (8,9),
pigmented melanoma cells (7), teratoma cells capable of differentiation *in vitro* (10,11), glia (12) and neuroblastoma
cells (13) with distinctive properties of nervous tissue, etc.
We used the somewhat involved technique to give us the best
possible chance for success because it was not known at the
time whether or not it was possible to grow functionally differentiated cells. Today it is commonplace to culture functionally differentiated cells, especially from tumors, without
the elaborate procedure of alternate animal and culture passage.

Although very few people were engaged in this type of
work, we were not alone with this concern for lack of differentiated cell lines. In the early 1920's, Ebeling showed that
pigmented cells could be grown for long periods in culture
(14) and Albert Fischer was able to achieve the same with cartilage cells (15). A long lapse in such activity ensued until
Glenn Fischer established cancerous mast cell lines which produced serotonin and heparin (16). Irving Konigsberg showed
that chick myoblasts could proliferate and fuse to form muscle
straps (17). Later, David Yaffe established rat myoblast lines
which not only fused in culture but could participate in in-
jury repair of crushed muscle (18). Hayden Coon and Robert
Cahn produced clones of normal chick cartilage and pigmented
cells (19,20). The MDCK dog kidney line was established in
1958 (21), but it was not until much later that its functional
role as tubule cells was elucidated (22).

If tissue cultures are to be useful tools for studying
animal physiology, I next reasoned that they should exhibit
growth responses to trophic hormones as do the parental tissues *in vivo*. Again, the impetus for studying this question
came from the work of Dr. Jacob Furth on hormone dependent
tumor growth (23). Accordingly, we set out to develop ovarian
cell cultures whose growth would be dependent on the presence
of gonadotrophins. Dr. Jeffrey Clark initiated this work in
my laboratory and soon obtained the puzzling result that adding gonadotrophins to the culture medium had no effect on the
growth of the cells. This was unexpected because we knew that

if these cells were injected into animals, they would only
grow if the transplant was artificially provided with high
levels of hormones.

We realized that the serum component of the medium was
providing hormones and decided to use serum selectively depleted of hormones. Drs. Katsuzo Nishikawa, Hugo Armelin and
David Sirbasku devised procedures for depleting serum of hormones and these have now become standard practice (24,25).
They extracted serum with charcoal to remove thyroid and
steroid hormones, and passed the serum through a column of
carboxymethyl cellulose to remove basic peptides with growth-
promoting properties. When depleted serum was used, the
ovarian cells would not grow unless hormones were added to the
medium. Optimal growth required the addition of multiple hormones (26).

Another surprise came when we discovered that the crude
luteinizing hormone (one of the gonadotrophins) we were using
was providing novel basic peptides which were the active components of the hormone preparations and that pure luteinizing
hormone was inactive (26,27). This led to our discovery of
ovarian growth factor and fibroblast growth factor, which were
later isolated and purified in the laboratories of Denis Gospodarowicz (28,29) and Ralph Bradshaw (30).

The pattern of results coming from these studies, and a
brief but memorable conversation with Gordon Tomkins, caused
me to wonder if the role of serum in cell culture media might
simply be to provide cells with complexes of hormones. Serum
or its equivalent has been an obligatory component of media
since tissue cultures were first initiated by Ross Harrison
over seventy years ago. No adequate explanation had ever been
given for this requirement. If this surmise about the role of
serum was correct, it should be possible to replace serum with
complexes of hormones. This was a very exciting prospect
because its implications would have far-reaching consequences
for many branches of biology (31).

Dr. Izumi Hayashi, who was a graduate student in the laboratory, quickly obtained evidence in support of the hypothesis (32). She found that the serum in media used to culture
GH3 cells (growth hormone-secreting pituitary cell line that
we had established from a Jacob Furth tumor) could be replaced
with a mixture of insulin (pancreas), transferrin (liver),
triiodothyronine (thyroid), parathyroid hormone (parathyroid),
TSH-releasing hormone (hypothalamus), fibroblast growth factor
(pituitary) and somatomedin C (liver) (Table 1). I have indicated the organ sources of these substances in parentheses.
All of these substances are not hormones. In fact, our hypothesis has now been extended to state that the serum in tissue
culture medium can be replaced by substances from four classes:

TABLE 1.   Hormonal Requirements of GH$_3$ Growth Hormone,
Prolactin, Secreting Pituitary Cells

Substance	Organ Source
1. Insulin	Pancreas
2. Transferrin	Liver
3. Thyroxin	Thyroid
4. Parathyroid hormone	Parathyroid
5. Thyroid stimulating hormone releasing hormone	Hypothalamus (brain)
6. Fibroblast growth factor	Pituitary
7. Somatomedin C	Liver

The hormonal and factor requirement of GH$_3$ pituitary cells
for growth in serum-free medium are presented here with the
organ source of each component. In view of the immense complexity of serum, it is surprising how small a number of defined substances can replace serum. It is also surprising
that the cells give a positive growth response to so many hormones. The serum-free technology is also a powerful tool for
revealing hormonal dependencies. The organs listed are the
organs that would have to be ablated to show these responses
by classical endocrine experiments.

---

(1)   Classical hormones such as insulin and hydrocortisone.
(2)   Growth factors which we consider to be hormones but
whose physiological role has not yet been determined. Substances of this class readily lend themselves to discovery by
cell culture methods and the name "growth factor" is more a
reflection of the assay system for their detection than of
their actual role in physiology.
(3)   Transport proteins. So far only transferrin has
proven useful but it is required by almost all cells.
(4)   Attachment factors, such as fibronectin and collagen.
These substances are often necessary for the proper attachment
of cells to the culture vessels and probably reflects the
usual relationship of epithelial cells to a basement membrane
in vivo.
The data of Dr. Hayashi are remarkable for several reasons.
The first is that serum can be replaced by such a small number
of substances. In view of the immense complexity of serum,
one would not have predicted that serum could be replaced by
seven defined substances for GH$_3$ cells. It is remarkable, on

the other hand, that the cells give a positive growth response
to so many hormones. This surely means that the hormonal regulation of each cell type *in vivo* is much more complex than is
indicated by classical endocrine studies. In fact, the most
challenging aspect of these findings is how the metabolic control by so many hormones of each cell in the body can be
fitted into the pattern of integrated physiology. Finally,
the data demonstrate the immense power of the serum-free technology for uncovering hitherto unsuspected endocrine relationships. Removing serum from cells in culture is the most radical endocrine ablation imaginable. To demonstrate the hormonal dependencies of GH₃ cells revealed by the serum-free technology by classical means, one would have to perform pancrea-
tectomy, hepatectomy, hypophysectomy, thyroidectomy, para-
thyroidectomy and brainectomy. Clearly, such experiments are
impractical and one can predict with fair certainty that in
the future the discovery of physiological hormone responses
will be made with the use of these culture techniques.

The results obtained with GH₃ cells have now been extended
to a sufficiently large number of cells to enable us to say
that all animal cells with the potential for growth can grow
in a serum-free, defined medium containing substances from the
four classes mentioned above (33). Each cell type has a different and unique set of hormonal requirements and in many
cases growth in the defined media is superior to growth in
serum-based media. If a key role of serum in culture is to
provide complexes of hormones, it is understandable why the
defined, hormone-supplemented media can be superior to serum-
based media. Serum is toxic to cells in culture and so is
never used undiluted. It is usually diluted to a concentration of about 10% v/v. This intrinsic toxicity is probably
due to the fact that many potentially cytotoxic substances in
serum do not usually leave the vascular spaces and come into
direct contact with cells.

Cells whose proliferation is driven by hormones usually
require them at concentrations greater than those normally
found in 100% serum. In addition, there are many specialized
areas of the body where certain hormone concentrations are
much higher than that found in the general circulation. For
these reasons, dilute serum cannot be expected to provide the
ideal, specialized environment required by many kinds of cells,
and this probably explains why such cells have never been established in culture using serum-based medium. Use of hormone-
based media should now make it possible to establish cells in
culture that have not been cultured before. A notable example
is the work of Drs. Saverio Ambesi and Hayden Coon at the
National Institutes of Health who have used hormone-based
media to successfully culture normal thyroid cells with their

full array of functional characteristics (34). Such cells
had not been cultured before, despite years of strenuous
effort using serum-based media. In the near future, it
should be possible, by these methods, to establish a great
variety of cell culture lines with potential utility for
studying complex hormonal interactions in animal physiology.

It is now clear that our initial approach for culturing
differentiated cells was not optimal. We took the microbiological approach of forcing diverse cell types to grow in a
standard medium rather than a more physiological approach of
seeking to satisfy their specialized requirements.

Most workers in the field today regard cells in culture
as microorganisms with little regard to their role in the
body. A case in point is the field of growth regulation where
many investigators limit the growth of cells in culture by
physically deleting something from the medium or blocking a
metabolic step by mutation. Such studies can lead to interesting insights into cell biology but usually have nothing to
do with physiologic growth regulation because the substance
deleted is never limiting in the animal. Physiologic growth
regulation deals with questions like how does estrogen cause
uterine endometrial proliferation or how does TSH cause
goiters? Questions such as these are approachable with the
defined media technology.

My main concern over the next few years will be to apply
these ideas to the problem of cancer. Many of the cells we
have studied are tumor cells. When tested in serum-free media,
they have complex hormonal requirements. This means that the
hormonal regulation of tumors may be nearly as complex as the
regulation of normal organ systems. Using serum-free technology, it should be possible to develop a detailed knowledge of
the endocrine physiology of tumors. I feel certain that a
basic approach of this kind to cancer is the best long-term
hope for useful advances in therapy.

Finally, my laboratory is now involved in developing monoclonal antibodies to hormone receptors. The idea is that such
hormones can replace or antagonize hormones required for the
growth of cells in defined media. Recently, the evidence that
antibodies in autoimmune diseases can act as hormones has been
reviewed by Dr. Jesse Roth (35). It is easy to predict that
the general availability of such monoclonal reagents will have
an enormous impact on the physiology and pharmacology of the
next decade.

REFERENCES

1. Harrison, R. G., *Proc. Soc. Exp. Biol. Med. IV*; 140 (1906).
2. Sato, G., Zaroff, L. and Mills, S. E., *Proc. Natl. Acad.*

Sci. 46, 963 (1960).

3. Furth, J., *in* "Recent Progress in Hormone Research, Vol. 2" (G. Pincus, ed.), p. 22. Academic Press, New York, (1955).

4. Furth, J., *in* "Hormone Production in Endocrine Tumors", Ciba Foundation Co-loquia on Endocrinology, vol. 12 (G. E. W. Wolstenholme and M. O'Connor, eds.), p. 3. Little, Brown & Co., Boston, (1958).

5. Buonassis, V., Sato, G. and Cohen, A. I., *Proc. Natl. Acad. Sci. USA, 48*, 1184 (1962).

6. Takemoto, H., Yokoro, K., Furth, J. and Cohen, A. I., *Cancer Res. 22*, 917 (1962).

7. Yasumura, Y., Tashjian, Jr., A. H. and Sato, G. H., *Science 154*, 1186 (1966).

8. Shin, S., Yasumura, Y. and Sato, G. H., *Endocrinology 82*, 614 (1968).

9. Shin, S. and Sato, G., *Biochem. and Biophys. Res. Comm. 45*, 501 (1971).

10. Rosenthal, M. D., Wishnow, R. M. and Sato, G. H., *J. Nat. Cancer Inst. 44*, 1001 (1970).

11. Finch, B. W. and Ephrussi, B., *Proc. Natl. Acad. Sci. USA 57*, 615 (1967).

12. Benda, P., Lightbody, J., Sato, G., Levine, L. and Sweet, W., *Science 161*, 70 (1968).

13. Augusti-Tocco, G. and Sato, G., *Proc. Natl. Acad. Sci. 64*, 311 (1969).

14. Ebeling, A. H., *C. R. Soc. Biol. (Paris) 90*, 562 (1924).

15. Fischer, A. J., *J. Exp. Med. 36*, 379 (1922).

16. Schindler, R., Day, M. and Fischer, G. A., *Cancer Res. 19*, 47 (1959).

17. Konigsberg, I. R., *Science 140*, 1273 (1963).

18. Yaffe, D., *Proc. Natl. Acad. Sci. 61*, 477 (1968).

19. Coon, H. G., *Proc. Natl. Acad. Sci. 55*, 66 (1966).

20. Cahn, R. D. and Cahn, M. B., *Proc. Natl. Acad. Sci. 55*, 106 (1966).

21. Darby, M. D., *T. C. A. Report 14*, 50 (1980).

22. Leighton, J. L., Estes, W., Mansukhani, J. and Brada, Z., *Cancer 26*, 7022 (1970).

23. Furth, J., Ueda, G. and Clifton, K. H., *in* "Methods in Cancer Research" (H. Busch, ed.), p. 202. Academic Press, New York, (1973).

24. Nishikawa, K., Armelin, H. A. and Sato, G., *Proc. Natl. Acad. Sci. USA 72*, 483 (1975).

25. Armelin, H. A., Nishikawa, K. and Sato, G. H., *in* Control of Proliferation and Animal Cells" (Clarkson and Baserga, eds.), p. 97. Cold Spring Harbor Publishers, New York, (1974).

26. Clark, J. L., Jones, K. L., Gospodarowicz, D. and Sato, G.

Nature New Biol. 236, 180 (1972).

27. Armelin, H. A., *Proc. Natl. Acad. Sci. USA, 70,* 2702 (1973).

28. Gospodarowicz, D., Jones, K. L. and Sato, G., *in* Abstracts of 55th Endocrine Society Meetings, Abs. #347, p. A222, (1973).

29. Gospodarowicz, D., *Jrnl. Biol. Chem. 250,* 2515 (1975).

30. Thomas, K. A., Riley, M. C., Lemmon, S. K., Baglan, N. C. and Bradshaw, R. A., *Jrnl. Biol. Chem. 255,* 5517 (1980).

31. Sato, G., *in* "Biochemical Action of Hormones, Vol. III" (G. Litwak, ed.), p. 391. Academic Press, New York (1975).

32. Hayashi, I. and Sato, G., *Nature 259,* 132 (1976).

33. Barnes, D. and Sato, G., *Cell 22,* 649 (1981).

34. Ambesi-Impiombato, F. S., Parks, C. A. M. and Coon, H. G. *Proc. Natl. Acad. Sci. USA 77,* 3455 (1980).

35. Flier, J. S., Kahn, C. R. and Roth, J., *New Engl. Jrnl. of Med. 300,* 413 (1979).

DISCUSSION

D.T. DENHARDT: Are there any advantages to using serum-free media for the purpose of obtaining mammalian cells at defined stages of the cell-cycle?

G. SATO: Yes, because highly selective deletion of hormones and factors can be carried out. It has already been reported, for instance, that deletion of transferrin results in a block in G2.

D. BALINSKY: Other than by trial and error, how do you decide what type and concentration of additives to use for the different cell types?

G. SATO: Except for insulin and transferrin, factor concentrations are physiological. There is now a large body of literature on hormonally defined media. One should start with a medium designed for a related cell type and use whatever clues one has about the hormonal responses of the tissue of origin.

G. KNAPP: Recent literature suggests that studies of promoter (Pol III) mutants in extracts of animal cells and homologous cell culture lines do not give the same results. What are your speculations? Will there be an inherent problem of tissue culture or have we just failed to "tailor the medium"?

G. SATO: I wish I could make some intelligent response to your question but I haven't a clue as to what is going on.

REGULATION OF GROWTH AND DIFFERENTIATION OF MAMMALIAN CELLS BY HORMONES AND EXTRACELLULAR MATRIX

Lola C. M. Reid[1]

Department of Molecular Pharmacology
Albert Einstein College of Medicine
Bronx, New York, U.S.A.

I. INTRODUCTION

For a number of years we have been studying the regulation of growth and differentiation in normal and neoplastic mammalian cells. Our long-range goals are to learn the variables and the mechanisms by which adult tissues maintain specialized functions and how that regulation is lost in neoplastic transformation. A basic assumption in our approach is that we are dealing with"committed" cells, i.e. cells which have already become determined during embryogenesis to be a particular cell type such as lung cells, hepatocytes or prostatic epithelial cells. Once "committed", cells can be influenced only to vary the level of gene expression for the repertoire of genes associated with the specific cell type. We assume this to be true both for normal and neoplastic cells. Our efforts are directed towards identifying those variables which will permit the cells to maintain their respective gene set at levels approximating those _in vivo_. Among these variables are:

1) nutrients including trace elements
2) hormones and growth factors
3) soluble signals from neighboring cell-cell interactions including those transmitted via gap junctions and those secreted by cells
4) substrates of tissue-specific extracellular matrix such as basement membranes

[1] Supported by American Cancer Society grant (BC-301 and PDT-131),and NIH grants AM17702, CA30117, and P30CA1330)

Our focus upon these particular variables is predicated upon
the hypothesis that the community of cells within a tissue
provide cellular interactions which are the primary determinants of the regulation of growth and differentiation in all
metazoan organisms. Among the most important of these cellular interactions is that between the epithelium and the mesen-
chymally-derived cells (1). This relationship was among the
first cell-cell interaction to evolve in metazoans and is a
ubiquitous one in all tissues of higher organisms. Therefore,
we have proposed that the long-known difficulties in culturing
normal cells has been due, in part, to the dissociation of
cells from the tissue matrix and from the community of cellular interactions critical in the regulation of differentiation
(2). If this hypothesis is true, then to observe differentiated expression in vitro requires co-cultivation of the relevant cells or simulation of the communication signals from
those interactions. We have chosen to develop a new cell
culture technology which will mimic these critical cellular
relationships especially that between the epithelial cell
and the mesenchymal cell. The interactions between the epithelium and the mesenchymal cell can be divided into soluble
signals to be found in conditioned medium of cultures of the
cells and insoluble signals to be found in the extracellular
matrix secreted by both epithelial and mesenchymal cells.
In brief, our approach is to plate epitehlial cells on substrates of tissue-specific extracellular matrix and in basal
medium supplemented with hormones, growth factors and the
soluble signals found in conditioned medium.

        Over the past 8-10 years, culture conditions have been
extensively analysed in many labs to ascertain better methods
for the maintenance of cells. Certain variables have been
shown to be critical. Ham (3), McKeehan (4), Waymouth (5)
and others have led the way to the development of better
defined nutritional requirements in the basal medium. Sato
and his associates have demonstrated that the serum supplement
in cell culture medium can be replaced with defined and purified hormones and growth factors (6,7). The use of the serum-
free, hormonally supplemented media has resulted in cell cultures which show more reproducible behavior both in terms of
growth and of differentiation (6,8). The hormonal requirements deduced for neoplastic cells proved to be a subset of
those needed by their normal cellular counterparts (6, 7).
Green and his associates (9,10), Sensenbr et al.(11),   and
Wang et al. (12) have shown that regulatory soluble signals
are present in the conditioned medium of cells and, therefore,
represent the localized signals between cells in the tissue.
Increasingly, though, it has become apparent that maintenance
of differentiative potential of cells is dependent upon com-

plex interactions not only from such soluble signals but also
from insoluble ones present in the extracellular matrix forming the substrate for cells. The mixture of proteins present
in the extracellular matrix is largely insoluble due to
extensive crosslinking (13). Certain types of proteins are
found ubiquitously in the matrix: collagens, glycoproteins
essential for anchorage of cells (e.g. fibronectin), and
proteoglycans (14). Although there is extensive circumstan-
tial evidence for the biological importance of these factors,
the relevance of these proteins as substrates and the mecha-
nism(s) of influencing cells are poorly understood (15,16).

In previous studies we have shown that substrates of
tissue-specific extracellular matrix extracts, referred to as
"biomatrix", are essential for the long-term maintenance of
various types of epithelial cells in culture (17,18). Similarly, others have shown that detergent-treatment of cell
monolayers leaves a residue on tissue culture plates which
proved a more ideal substrate for primary cultures of epithe
lial cells and for other cells with poor affinity for regular
tissue culture dishes (19). The matrix substrates studied,
both the complex mixtures of proteins and individual components from the mixture, have been found to alter cell shape,
growth rates, cellular motility, and the expression of differentiated functions (20,22,23,24).

In this overview of our work will be presented data
from three cellular systems under investigation: normal and
neoplastic rat hepatocytes, normal and neoplastic rat pancreatic islet cells, and normal rat prostatic epithelial cells.
The three systems were chosen as representatives of two of
three hypothetical classes of tissues: proliferative tissues,
differentiative tissues, and regenerative tissues. Each of
the classes is distinct in the cellular interactions and in
the resultant expectations of model systems in culture:
Proliferative tissues =represented by bone marrow, skin or
colon. In these tissues, partially or completely "committed"
stem cells proliferate as long as they are attached to a
form of extracellular matrix known as basement membrane.
Daughter cells of the stem cells become detached from the
matrix (skin, bone marrow) or slide to new regions of presumably chemically distinct matrix (colon) and in so doing lose
their mitotic potential. Thus, the communications between
the epithelium and the mesenchymal cells are concerned primarily with maintenance of the proliferative state. Success in
simulating the proliferative tissues should produce stem cells
anchored to substrates of basement membranes coated on the
cell culture dishes. The attached stem cells should produce
daughter cells which become detached and while floating in the
medium undergo terminal differentiation. Soluble factors in

the medium should affect the kinetics of differentiation (e.g. skin) and/or the direction of differentiation or committment (e.g. bone marrow cells).

Differentiative Tissues=represented by prostate, breast, pancreas, endocrine tissues. In these tissues, a differentiated epithelium is bound to basement membrane, and, in turn, associated with either fibroblasts, or, in tissues with rapid transport or secretion (endocrine tissues, brain), endothelium. Although some proliferation of these cells is possible, it is limited relative to that observed in proliferative tissues and, when it occurs, involves the concomitant growth of all cell types within the tissue. In these tissues, the primary concern of the cellular interactions appears to be the maintenance of specialized functions of the epithelial cells. Simulation of the epithelial-mesenchymal relationship should produce non-growing, differentiated epithelial cells. With a basement membrane from regenerating tissue and with appropriate hormones, the cells should proliferate in a limited fashion.

Regenerative Tissues=a few tissues have an unusual capacity to regenerate either by hyperplasia (liver) or by hypertrophy (kidney). In these tissues, the epithelium is bound to basement membranes and associated with endothelium (in most instances). Under appropriate conditions such as partial hepatectomy or partial nephrectomy, the epithelial-mesenchymal relationship can switch to a proliferative mode and then return to the differentiated condition. Successful simulation of such regenerative tissues should reveal cultures which will either be differentiated and non-growing or rapidly growing (and perhaps with less differentiation) depending upon the composition of soluble signals (hormones, growth factors) added to the medium and upon the chemical composition of the basement membrane.

II. METHODS AND PROCEDURES

Most of the general procedures for these studies are presented in the individual articles on each of the systems (25, 26, 27, 17). However, some of the critical procedures relevant to the new technology are given here.

Animals: Sprague Dawley rats were purchased from Charles River and used immediately for purification of Type I collagen for isolation of biomatrix or for preparation of normal hepatocytes, pancreatic islet cells, or ventral prostatic epithelial cells. Frozen tissues were purchased also from Deutschland Farms (New Jersey). All tissue not used immediately was frozen in liquid nitrogen and stored at -70°C.

Cell Lines: A well differentiated rat Morris hepatoma cell line, $H_4A_zC_2$, was obtained from Dr. Nadel-Ginard (Einstein, New York). It has been found positive for 1)albumin secretion, 2)tyrosine aminotransferase inducibility by dexamethasone, 3)bilirubin metabolism, 4)liver-specific alcohol dehydrogenase production, and 5)ornithine decarboxylase production (28). A well differentiated rat insulinoma cell line was obtained from Oie and Minna (NIH). Clones of it are known to secrete insulin, glucagon and somatostatin (29). GH_3 cells, a rat pituitary carcinoma cell line, were obtained from Sato (UCSD, La Jolla). GH_3 cells secrete growth hormone and prolactin in culture (30). Stock cultures were kept in Falcon 75 cm^2 tissue culture flasks and were passaged and fed once per week. The rat hepatoma cell line and the pituitary cell line were maintained in RPMI 1640 supplemented with 10% fetal bovine serum. The rat insulinoma cell line was maintained in RPMI 1640 supplemented with 10% fetal bovine serum and 20% conditioned medium from the GH_3 cells.

Isolation of Normal Epithelial Cells: The rat hepatocytes were isolated by the procedures of Berry and Friend (31), and the rat islet cells by the procedures of Lacey (32).

Culture Conditions: The cells were cultured on one of three substrates: tissue culture plastic, collagen gels (Type I collagen), or biomatrix and in medium supplemented with 10% fetal bovine serum (SSM) or in medium supplemented with with hormones and growth factors as specified below.

Basal Medium: The cells were grown in RPMI 1640 (GIBCO, New York) suplemented with 10 mM Hepes, 100 units/ml penicillin, and 200 μg /ml streptomycin. When no serum supplement or hormonal additives were present, the medium was referred to as serum-free medium (SFM). With serum supplements, it was referred to as serum-supplemented medium (SSM). If only purified hormones and growth factors were added (described in Table 1), the medium was referred to as defined medium (DM)

Hormones and Trace Elements used in the Defined Medium: The hormones, growth factors, and trace elements utilized for the individual systems are listed in Table 1. For further details on the commercial sources for these factors and for their preparation for use see the reviews by Barnes and Sato (6) and by Bottenstein, et al.(33). It was found crucial to utilize high purity grade reagents and to utilize triple distilled water in preparing these growth factors, since in DM there is no serum to detoxify or mask these contaminants.

Substrates : Cells were plated onto either tissue culture
plastic, Type I collagen gels, or biomatrix prepared either
from normal tissue (rat liver, prostate and pancreas) or from
a transplantable mouse tumor, the EHS tumor line, known to
constitutively secrete large amounts of basement membrane proteins (34). The collagenous and biomatrix substrates were
sterilized by 10,000 rads of gamma irradiation (^{135}Cesium) and
were then stored at 4°C until used. Type I collagen gels
were prepared by the procedures of Pitot, et al (35). Biomatrix (1M) was prepared by our earlier published procedures
(17). Biomatrix (3M) was prepared as follows:

1)Approximately 100 grams of tissue was homogenized
in a polytron homogenizer using 5-10 volumes of ice-cold
3.4 M NaCl and homogenized thoroughly by 10-15 second pulses.

2)The homogenate was filtered through polyester (Tekto,
Inc., Elmsford, New York) of 166-168 mesh per inch and having
a pore diameter of approximately 80 microns.

3)The fibrous residue retained on the polyester was
collected and put again into 3.4 M NaCl, stirred for one
hour at 4°C, and again filtered. This process was repeated
until the extraction of the fibrous residue by the saline solution was complete as determined by the absence of proteins
in the supernatant solution using Lowry or Biorad assays.

4)The fibrous matrix was then rinsed for three, one-
hour rinses in SFM at 4°C.

5)The insoluble residue was added to SFM to which
was added 1 mg DNase and 5 mg RNase (Sigma)per 100 ml of
SFM. For each gram of residue, 100 mls of the nuclease solution was used. The residue was stirred in the nucleases at
37°C for 1 hour. Treatment with the nucleases was continued
until the supernatant contained no evidence of nucleic acids
(absorbance at 260nm). However, in those experiments utilizing cDNA probes assessing specific messenger RNA transcripts,
the biomatrix utilized was not treated with nucleases.

6)The residue was again collected by filtration and put
into 1% sodium deoxycholate prepared in SFM. It was stirred
for 1 hour at room temperature. Extraction of lipids was
judged completed by oil red O staining of samples of the
fibers.

7)The fibers were collected by filtration and rinsed
repeatedly over several hours with SFM.

8)After rinsing, the fibers, referred to as "biomatrix"
were suspended in SFM and sterilized with 10,000 rads of
^{135}cesium. The samples were stored at 4°C until used.
Substrates of the biomatrix were prepared by freezing the
matrix in liquid nitrogen and pulverizing it into a powder
utilizing a liquid nitrogen pulverizer (Freezer-Mill, Spex
Mills, Inc., Metuchen, New Jersey). The powdered sample was

allowed to thaw, and the gelatinous material painted onto
appropriate dishes (petri dishes). It was important to use
the bacteriological dishes, petri dishes, as opposed to tissue
culture dishes, since epithelial cells do not attach well to
the petri dishes and any effects of the substrate on the cells
would, therefore, be due to the biomatrix. The plates so prepared were then irradiated with 10,000 rads of gamma irradiation and stored at 4°C until used. Just prior to use, the
plates were rinsed for one hour in SFM and for one hour in
SSM. The later rinse proved essential for cell survival
for unknown reasons, although we speculate that the serum
supplement may be detoxifying some residues and contaminants
left from the preparative procedures.

Influence of Culture Conditions on the Physiology of Epithelial Cells: The cells were evaluated in a number of ways
to assess the growth parameters under each of the culture
conditions. The attachment efficiency within 24 hours,
the clonal growth efficiency (plating efficiency), the doubling time and the saturation density were measured by methods
described in detail elsewhere (17, 25,25,27).

The cells were studied by both light and phase microscopy
for evidence of cell shape, granularity, vacuolar formation,
organelle content, polarity, and other such parameters. In
some instances, more detailed studies, including ultrastructural analyses, were undertaken to ascertain the influence of
each culture condition on the cellular morphology, and the
correlation of critical aspects of that morphology with the
assessed differentiative potential of the cells.

To ascertain the functional state of the cells, several
tissue-specific markers were evaluated for each systems.
The hepatocyte cultures were studied for tyrosine aminotransferase and its inducibility by dexamethasone, glucuronyl
transferase, glutathione-S-transferase, albumin production,
and other markers (25); the insulinoma and islet cultures
were assessed for insulin production (26); and the rat ventral
prostatic epithelial cells were tested for the production of
an androgen-responsive secretory protein, prostatic binding
protein (27).

III. RESULTS

The technology described incorporates two major changes
with respect to past cell culture methodologies: 1)the elimination or minimization of serum supplementation to the media
and replacement with hormones and growth factors and 2)the
use of substrates of tissue-specific extracellular matrix.

TABLE I. Composition of the Defined Media for the
Cellular Systems under Investigation

Cell Type	Factors Required for Growth
Prostatic Epithelium (27)	insulin (5 µg/ml), transferrin (5 µg/ml), T_3 (10^{-11}M), TRF (10^{-9}M), LH-RF (10 ng/ml) NIH-LH (1 µg/ml), FGF (1 ng/ml), growth hormone (1 ng/ml), hydrocortisone (10^{-6}M), spermine (0.1 µg/ml), and testosterone (10^{-8}M)
Hepatocytes (25)	insulin (1-10 µg/ml), glucagon (10 µg/ml), TRF (10^{-9}M), prolactin (10 µg/ml), growth hormone (10 ng/ml), linoleic acid bound to albumin (5 µg/ml), zinc (10^{-10}M), copper (10^{-7}M), selenium (3 X 10^{-9}M), collagenous substrates (if collagenous substrates are not used, additional factors are required). The hepatocytes require, in addition, EGF (5 µg/ml).
Pancreatic Islets (34)	transferrin (5 µg/ml), T_3 (10^{-10}M), TRF (1 ng/ml), growth hormone (1-5 ng/ml), prolactin (10 ng/ml).

Abbreviations: T_3=tri-iodothyronine; TRF=thyroid hormone releasing factor; LH-RF=luteinizing hormone releasing hormone FGF=fibroblast growth factor; EGF=epidermal growth factor.

The defined media above were developed using malignantly transformed cell lines. The defined medium developed for the neoplastic cells proved to be a subset of those needed for the normal cells. Thus, the defined medium for the rat hepatoma cell line was a subset of the factors needed by the normal rat hepatoctyes. The cell lines could be cultured in the defined medium above with no serum supplementation. Their normal cellular counterpart required approximately 5% serum in addition to the above.

Above is given the composition of the media used for the normal and neoplastic cells. Since the use of serum-free, hormonally supplemented media has been well documented by studies in many laboratories (see the review by Barnes and Sato, 6), it will not be discussed here except where the media show synergy with with the substrates. The emphasis of the work presented here will be on the relevance of substrates especially substrates of extracellular matrix.

Studies of tissue function in normal and diseased states are often complicated by variables which are difficult to control in whole animals. The hope has long been to develop culture systems of normal or diseased cells with retention of cellular phenotypes observed in vivo. However, the culture of normal epithelial cells or of minimally deviant carcinoma cells has proven exceedingly difficult for everyone (36-38). When plated onto tissue culture plastic, the cells undergo rapid morphological and biochemical deterioration resulting in the loss, within two weeks, of all markers of differentiation, and within three weeks, loss of the cultures altogether. Extensive studies of the variables contributing to the inability to culture such cells have revealed the need for careful definition of nutritional requirements (3), proper gas exchange and efficiency of nutrient and waste exchange (3), and requirements for use of serum-free, hormonally supplemented media (6). Innovations in the use of more biologically relevant substrates, in particular the use of floating collagen gels (Type I collagen), were introduced with remarkable enhancement in the attachment, survival, and differentiative state of the cells. The floating collagen gels enabled investigators to culture a wide variety of normal epithelial cells in primary cultures for up to a month and with retention of some differentiated markers for approximately two weeks (23, 24, 39, 40). Morphologically, the cells were distinctly different when plated onto collagen gels in contrast to tissue culture plastic. On plastic, the cells spread to a squamous or "fried egg" shape, appeared increasingly translucent and agranular, and rapidly lost endoplasmic reticulum. On collagen gels, the cells assumed polygonal shapes which became cuboidal or columnar over time in culture. The histology of the cells and ultrastructural studies indicated formation of cellular junctions, the development of polarity of the cells, and well-developed endoplasmic reticulum and Golgi bodies. However, although the cells were more stabilized and retained some functioning, they appeared to undergo a gradual fetalization throughout the time of culture (40). The collagen gels had merely slowed down the deterioration process not eliminated it.

More recently, investigations in a number of labs including my own, have implicated the relevance of utilizing more complex forms of extracellular matrix as substrates for cell cultures. To date, there are four methods by which to make a cell culture substrate from the more complex forms of extracellular matrix. These methods are compared in Table II in which their advantages and disadvantages are discussed.

TABLE II.COMPARISON OF CURRENT METHODS FOR PREPARING CELL
CULTURE SUBSTRATES OF EXTRACELLULAR MATRIX

METHOD: ECM SUBSTRATES=dilute detergent or alkali treatment
            (19,20)   of monolayers of cells (most commonly
                          used are fibroblasts or endothelium)

Advantages: Simplicity in preparation; homogeneous substrates; matrix components have correct stoichiometry and chemical relationships; dramatically improves attachment and plating efficiencies; reduces serum and hormone requirements for
growth; slight augmentation of differentiation (dependent upon
what type of cell used to make the ECM).

Disadvantages: Chemistry difficult due to paucity of material; feeder layer of cells, since under culture conditions,
may produce distinct form of matrix from that which they would
do in vivo where they are interacting with many cell types;
no tissue- or species-specificity (for some studies, this is
an advantage rather than disadvantage); the augmentation of
differentiation by ECM substrates is small relative to that
observed with other forms of matrix.

METHOD: Amniotic Membrane ECM Substrates=dilute alkali treat-
          ment of amniotic membranes to yield basement
          membranes which are then stretched and fastened by
          O-ring gaskets to make substrates.   (41)

Advantages: Preparation is straightforward although some
what more demanding than for ECM substrates; homogeneous substrate; contains only one form of extracellular matrix and its
chemistry is correct for that seen in vivo; dramatically
improves attachment and plating efficiencies; cells survive
longer and are more like they are in vivo (however, its
influence on differentiation, or on serum or hormonal responsiveness of cells is unknown); chemistry studies are possible
since there is plenty of material.

Disadvantages: No tissue- or species-specificity which although a technical advantage means that these basement membranes are distinct from adult basement membranes and means
that this substrate cannot be used to study any biological
phenomenon in which that specificity is relevant (e.g. site
specificity for metastatic potential of tumor cells); the
matrix is an embryonic one and, thus, its influence on cellular physiology may vary from that observed with adult forms
of matrix.

METHOD: <u>Biomatrix</u> (3M)=total extraction of all forms of extracellular matrix from a tissue. Extract is then pulver-
ized at liquid nitrogen temperatures and painted onto
solid-state material to be used as substrates (17,18)

<u>Advantages</u>: Contains all the tissue collagens and many
(most?) of their associated proteins in native form; matrix
chemistry is correct for that seen <u>in vivo</u>; chemistry studies
possible easing the isolation and identification of factors
relevant to biology; dramatically improves attachment and
plating efficiencies, long-term survival, and differentiation;
cells more like <u>in vivo</u> in terms of shape, growth or non-
growth, differentiation, and responsiveness to hormones; for
any given cell (e.g. hepatocyte), can observe either growth
or differentiation depending upon the type of matrix coupled
with the composition of the hormone mixture in the defined
medium.

<u>Disadvantages</u>: Preparation more demanding and requires
carefulness to avoid toxic contaminants; substrates not as
homogeneous as ECM substrates, although recent improvements
(liquid nitrogen pulverization) minimize the problem ; contains all forms of matrix from the tissue used, not just a
form produced by one cell(such as ECM substrate) or by a
pair of interacting cells (such as ECM substrates from amniotic membranes), and therefore, cellular physiology on the
substrates may be somewhat distinct from that seen <u>in vivo</u>.

METHOD: <u>Mixtures of Purified Components</u>=individual factors
from matrix are purified and used either individually
or in combinations as substrates   (15,16)

<u>Advantages</u>: Components are purified, defined and the con-
centrations known; substrates homogeneous; reproducibility of
experiments; analysis of mechanisms of action possible

<u>Disadvantages</u>: Due to insolubility of matrix components,
factors important biologically, have been ignored or remain
unidentified; difficulty in reproducing the stoichiometry
and structural relationships of the components to reflect
that observed <u>in vivo</u>; the biology of the cells on these
substrates gives responses which reflect, in part, those
observed <u>in vivo</u> since the normal response of cells is due
not to one or two factors but to a mixture of components.

<u>CONCLUSIONS</u>: Each of the above methods will contribute infor-
mation enhancing our understanding of matrix biology. The
future will see the use of mixtures of purified components
which include biologically active factors derived from forms
of ECM substrates and biomatrix.

In all the methods, the attachment and plating efficiencies
of the cells can be greater than 90% regardless of the basal
medium or the serum or hormonal supplements. Thus, the ability of cells to attach and survive in culture is determined
strictly by the composition of the substrate. However, if
a substrate is inadequate, then serum or hormonal supplements
must be used to compensate for the deficiencies in the substrate. Long-term survival of cells, i.e. greater than three
weeks, depends upon both substrate and soluble factors. The
composition of substrate and of the soluble factors required
is determined by the cell type. Further clarification of this
point will be given below. The ability of these various forms
of matrix to support either growth and/or differentiation
varied considerably depending upon the matrix type and the
cell type. For all of the forms of matrix, there was usually
an alteration in requirements for serum and/or responsiveness
to hormones. In summary, despite the variability of responses
observed, the relevance of extracellular matrix to anchorage
of cells, to their survival, to growth, motility, and differentiation is clear. The distinctions in the responses of
cells to the various types of matrix indicate that the chemistry of the matrix is variable depending upon its source and
that chemistry is critical in giving a particular response
from cells. To more fully explain these generalizations of
matrix biology, studies are presented from work ongoing in
my own laboratory.

The method utilized for preparing a substrate of extra –
cellular matrix is that described for "biomatrix". The
methodology for preparing the biomatrix has steadily improved
over the last years to optimize for maintaining the cells in
a physiological state approximating that in vivo. Our original method (17), one which is an extensive modification of
that described by Meezan et al.(42), isolated matrix by
exhaustively solubilizing away components presumed undesirable
(e.g. nucleic acids and lipids) to leave an insoluble residue
enriched for components in the extracellular matrix. The
"biomatrix" so prepared permitted much improved attachment
and survival of normal rat hepatocytes with retention of several of the liver-specific markers such as albumin and ligan-
din. However, parallel studies with other systems, pancreatic
islets and normal prostatic epithelial cells, revealed no
significant enhancement in differentiation although the cultures of the cells could be maintained much longer. The new
methods described here and elsewhere (25, 26, 27) now permit
the maintenance of the differentiation indefinitely. The
essential changes are 1)the use of buffer solutions with
high ionic strength to maintain certain components such as
laminins and proteoglycans which are critical for differentia –

tion; 2)the avoidance of nucleases when doing studies on messenger RNA transcript formation since the RNases are difficult to eliminate from the matrix; and 3)the judicious selection of methods for delipidation to eliminate the lipids with out making the matrix toxic (in some instances one can use detergents-e.g. liver biomatrix; in some, one must use organic solvents-mammary or pancreatic biomatrix).

The substrates of biomatrix have proven tissue-specific and, to some extent, species-specific. Thus, for cultures of prostatic cells, we have to use prostatic biomatrix; for hepatocytes, we have to use liver biomatrix; etc. There are important exceptions. If one prepares biomatrix from embryonic tissues (amniotic membranes or placenta) or from certain tumors such as the EHS sarcoma, there is no tissue- or species-specificity observed. Furthermore, embryonic cells whether normal or neoplastic, do not show a tissue specificity. Thus, teratocarcinomas have plated well on every form of biomatrix tested. The tissue-specificity of embryonic cells, however, has not been thoroughly studied; it is likely that some tissue specificity will exist depending upon the age of the embryo and the source of the cells. Lastly, neoplastic cells show tissue-specificity but an altered one from that of their normal cellular counterparts. Interestly, in several preliminary studies, the specificity exhibited suggests a correlation with the metastatic potential of the tumor in vivo. The assay for tissue specificity is critical. As shown in Table 3, the cells are plated on the matrix which is painted onto bacteriological plates, petri dishes. Since petri dishes are not coated or charged with any cations, the cells will not attach. Thus, any attachment and survival of the cells will be to the matrix and not to the plastic plates. One cannot use just attachment of the cells to observe tissue specificity. Many of the cells attached rather well to many types of biomatrix. However, within 3-4 days, the cells died or began to peel off those forms of biomatrix which were in-appropriate. After one week, the cells were growing and/or surviving only on a restricted set of biomatrices or only on one of the types of biomatrix. The relevance of such tissue specificity is important for many known biological phenomena such as the site specificity of metastases, of migrations of cells (for example, haemopoietic cells), of embryonic migra-tion patterns, etc. By using the biomatrices one should be able to characterize these phenomena biologically as well as ascertain the chemical components directing them.

The use of biomatrix substrates facilitated cell cultures with repect to attachment efficiencies, plating efficiencies, long-term survivals of primary cultures, and enhanced differen-tiation. Some of these parameters have been evaluated and are given in succinct form in Table IV.

LOLA C. M. REID

TABLE III. TISSUE SPECIFICITY STUDY OF BIOMATRIX

Substrate	Rat Hepatocytes	Human Prostatic Carcinoma*	Mouse Teratocarcinoma
Petri Dish	−	−	−
Tissue Culture Plastic	++	−	+++
Type I Collagen Gels	+++	−	+++
Biomatrix from			
Liver	+++(rat) ++(bovine)	++(human)	+++(mouse)
Prostate	−	+++(human)	n.t.
Testis	−	++(human)	n.t.
Pancreas	−	−	+++(mouse)
Brain	−(rat) −(bovine)	−(bovine)	+++(mouse)
Kidney	−(rat)	n.t.	+++(mouse)
Lung	−(rat)	++(human)	+++(mouse)

Single cell suspensions were seeded onto the plates at seed
ing densities of 10^5 per 60 mm dish and in medium supplemented
with serum. The cultures were incubated for one week and
then the number of viable cells attached to the plate assessed
The cultures were scored as follows: +=1-15% survival;
++=15-40% survival; +++greater than 40% survival.

*=The autopsy report on the patient from which the transplan-
table tumor line derived indicated metastatic lesions in the
liver, lungs, bones, and throughout the peritoneal cavity.

n.t.=not tested

TABLE IV. GROWTH PARAMETERS OF DIFFERENTIATED EPITHELIAL
CELLS ON VARIOUS SUBSTRATES

Growth Parameter	Cell Type	Medium	Substrates		
			Plastic	Collagen Gels	Biomatrix
A.E.	hepatocyte	SSM	4.7%	68%	83%
	hepatoma	DM	0	88%	95%
	prostate	SSM	1.6%	37%	64%
	insulinoma	SSM	0	3.5%	33%
D.T.	hepatocyte	SSM	no growth	no growth	no growth
	hepatoma	SSM	20 hours	24 hours	24 hours
	hepatoma	DM	no survival	25 hours	40 hours
	insulinoma	SSM	50-55 hours	n.t.	transient growth **
S.D.	hepatoma	SSM	7.8×10^6	5×10^6	just above seeding density
	hepatoma	DM	no survival	5×10^6	just above seeding density

Abbreviations:

A.E. = Attachment Efficiency
D.T. = Doubling Times
S.D. = Saturation Density
SSM = Serum Supplemented Medium
DM = Defined Medium

In Table 4, one can see that attachment efficiencies are
maximized by biomatrix. However, growth of the cells is
actually reduced by biomatrix especially if the cells are
cultured in defined medium (note the data from the hepatoma).
In general, we have observed that the growth and differentiation have been inversely correlated, a phenomenon well known
in studies on embryonic and adult tissues. Thus, those
conditions which have permitted growth have resulted, usually,
in a reduction in expression of a differentiated phenotype,
and conversely, those permitting differentiation are accompanied by slowed growth or no growth. In recent studies on
primary cultures of hepatocytes (43), we have found that
hepatocytes will undergo clonal growth if given the appropriate substrate and hormonal defined medium. The critical
conditions are 1)low seeding densities of less than 10^5 per
60 mm dish, 2)a hormonally defined medium containing various
critical mitogens(insulin, glucagon, and epidermal growth
factor), 3)the avoidance of any serum supplement, and 4)the
use of biomatrix from regenerating rat livers or biomatrix
prepared with low salt extraction (1 M). The addition of
serum to any substrate condition and with or without the
critical mitogens resulted in selection of non-parenchymal
cells and inhibition of growth and survival of the hepatocytes. Although the hepatocytes would undergo division in
the serum-free, hormonally supplemented medium, the cells,
if plated on tissue culture plastic would not reattach; thus,
the collagenous substrates were essential for reattachment.
Simple collagen gels were inadequate, however, since cells
on collagen gels and in serum-free, hormonally supplemented
medium underwent nuclear division and little to no cytoplasmic division. Furthermore, some forms of biomatrix (e.g.
3M biomatrix prepared from normal rat livers) inhibited the
growth. Thus, the hormones and the correct form of biomatrix
worked in synergy to permit (or to stimulate) division.
In summary, our studies on the relevance of extracellular
matrix to attachment, growth and survival of cells indicates
that the matrix is essential for attachment and survival
and that chemical composition of the matrix in combination
with soluble factors (hormones, growth factors)can dictate
whether the cells grow or remain differentiated.

    Studies on the differentiation of all the systems impli-
cates the extracellular matrix as one of the most essential
variables influential to the differentiated phenotype.
Constitutive secretion of insulin by both rat insulinoma
and hamster insulinoma cell lines was augmented more than
ten-fold by plating the cells on an appropriate form of
biomatrix. Cultures of rat ventral prostatic epithelial
cells required biomatrix substrates for maintenance of

androgen-dependent responses such as the secretion of prostatic binding protein. The cultures of normal rat hepatocytes
plated onto biomatrix were assessed for seven separate
liver-specific functions. Six of the seven (tyrosine aminotransferase, glucuronyl transferase, P450, albumin, glutathione-S-transferase, and other markers of bilirubin metabolism) were sensitive to substrate and significantly enhanced
and/or maintained at levels approximating those in vivo.
One marker, ornithine transcarbamylase, proved insensitive
to any matrix tested. The regulation of differentiation
was effected both by the matrix and by hormones acting in
synergy with it. Thus, several markers proved sensitive to
the use of defined medium (TAT, glucuronyl transferase,
glutathione-S-transferase, and albumin). In contrast, ornithine transcarbamylase was not affected by those hormones
present in the defined medium designed for the liver cells.
There are several important points to be made: 1)some markers
will be influence by matrix, some by hormones, some by both,
and, perhaps, some by neither. 2)The presence of serum masked
or eliminated the response of the cells to both matrix and
to the hormones. Thus, the addition of serum to cultures
of hepatoma cells on matrix with or without the necessary
hormones resulted in differentiation which was substantially
less than that with the hormones alone (similar findings were
observed with the insulinoma cells).3)The hormones worked in
synergy with the matrix to effect a particular response by
the cells. Multiple mechanisms are possible to explain this
observation. For example, the matrix might stabilize hormone
receptors and thereby augment or permit greater responsive
ness of the cells to the hormones. 4)The chemistry of the
matrix can dictate whether the cells grow or do not grow and
remain differentiated, a finding correlated with the ability
of the matrix to alter the cells' responsiveness to hormones;
thus, on one matrix, the cells might respond to one set of
hormones; and on another matrix, they might respond to a different set.
     The ability of cells to respond to matrix is not confined
to normal cells. Neoplastic cells also responded to the
matrix and to hormonal stimulation. However, the cells ability to respond was transient. When plated onto biomatrix of
the appropriate type and in defined medium (with no serum
supplement), the cells grew for one to three divisions and
then stopped their growth. The cessation of growth was
accompanied by an augmentation of differentiation. As indicated above and in Table V, the enhancement of differentiation
could be quite significant: a ten-fold increase in insulin
secretion by insulinoma cells for example. The plateau phase
with no growth and increased differentiation persisted for

4 days to three weeks depending upon the cell line. Thereafter, the cells gradually became detached from the biomatrix
and began to grow, losing their differentiation at the same
time. We speculate that since the cells are neoplastic, they
can secrete enzymes which digest away the anchorage proteins
permitting the cells to become unregulated in their growth
with a concomitant loss of differentiative potential. The
length of time the cells stayed plateaued while on the matrix
correlated inversely with the degree of malignancy: thus, the
minimally deviant rat insulinoma cells remained plateaued for
up to three weeks on the biomatrix and in defined medium,
whereas anaplastic carcinomas from the liver remained plateaued for only three to four days. It is likely that this
phenomenon is correlated with the ability of the cells to
secrete enzymes which are degradative for the matrix and/or
hormones.

TABLE V. INFLUENCE OF BIOMATRIX ON THE DIFFERENTIATIVE
POTENTIAL OF NORMAL AND NEOPLASTIC EPITHELIAL
CELLS

CELL	FUNCTION	DAYS CULTURED	TCP	COLLAGEN GELS	3M BIOMATRIX
Rat Insulinoma	insulin[1]	7	0	12-20	60-64
Hamster Insulinoma	insulin[1]	7	0	0	45-50
Rat ventral Prostate	prostatic[2] binding protein	7	0	0	15-20
Rat Hepatocytes	ligandin[3]	35	0	0	39
	glucuronyl[4] transferase	35	0	0	1.2
	albumin[5]	150	0	0	+

1=insulin measured by radioimmunoassay and given in units of
μU insulin/μg DNA/24 hours; 2=prostatic binding protein
measured by radioimmunoassay and given in units of ng protein
secreted/μg DNA/24 hours; 3)ligandin given as μg protein secre
ted/ mg protein; 4)glucuronyl transferase given as ng protein
secreted per mg protein; 5)albumin by immunofluorescence.

In summary, we have found from the various systems studied that extracellular matrix and hormones can interact synergistically to effect either growth or differentiation of epithelial cells. Normal epithelial cells demonstrate a strict tissue specificty for the type of biomatrix required. Neoplastic epithelial cells have an alterred tissue specifi- city with possible correlations to the metastatic potential of the cells. The use of tissue-specific forms of extracel- lular matrix and serum-free, hormonally supplemented media has resulted in cells demonstrating high attachment and plating efficiencies, long-term survival (i.e. months) of even normal cells, and growth and/or differentiated expression depending upon the culture conditions and upon the cell type. Neoplastic cells also respond to matrix and hormonal signals. However, they do so transiently with a length of time which correlates inversely with their malignant potential. A compa- rison of hormonal and matrix requirements of normal and neo- plastic cells indicates that the classical description of neoplastic cells as having "loss growth regulation" can be restated as a reduction qualitatively or quantitatively in their regulation by hormones and/or extracellular matrix.

Our future efforts will continue towards the development of in vitro model systems of normal and minimally deviant neoplastic epithelial cells. Although much of the work remains, the efforts of my own laboratory and that of many others towards the development of appropriate cell culture technologies should enable one in the near future, i.e. within the next five to ten years, to be able to culture almost any type of metazoan cell with significant retention of its chara- cteristic phenotype observed in vivo. The efforts at present are, from necessity, focused upon the technological aspects. Even so, the many studies are revealing fundamental insights into critical regulatory mechanisms controlling normal cells and into mechanisms affected by neoplastic transformation. Thus, the implications of our work may have bearing on many fields in need of cultures of cells which respond in vitro as they do in vivo.

REFERENCES

1. Grobstein, C. Science 118, 128 (1956).
2. Reid, L., and Rojkind, M. , in "Methods in Enzymology" (W. Jakoby and I. Pastan, eds.), 58, 263 (1979).
3. Ham, R. and McKeehan, W. , in "Methods in Enzymology" (W. Jakoby and I. Pastan, eds.), 58, 44 (1979).
4. McKeehan, W. and McKeehan, K. Proc. Nat. Acad. Sci. 77, 3417 (1980).

5. Waymouth, C. J. Natl. Cancer Inst. 22, 1003-1017 (1959)
6. Barnes, D. and Sato, G. Cell 22, 649 (1980)
7. Reid, L. and Sato, G. in "Biochemistry and Mode of Action of Hormones" (H. Rickenberg, ed.) p. 219. University Park Press, Baltimore (1978).
8. Orly, J., Sato, G. and Erickson, G. Cell 20, 817 (1980).
9. Green, H. Harvey Lectures 74, 101 (1979).
10. Green, H., Rheinwald, J. and Sun, T., in "Cell Shape and Surface Architecture", p. 493. Allen R. Liss, Inc., New York (1977).
11. Sensenbr, M., Jaros, G., Moonen, G. and Meyer, B. Experientia 36, 660 (1980).
12. Wang, J., Steck, P. and Kurtz, J. Cold Spring Harbor Symposium 46 (in press).
13. Miller, E. Molec. and Cell. Biochem. 13, 165 (1976).
14. Hassell, J., Robey, P., Barrach, H., Wilczek, J., Rennard, S., and Martin, G. Proc. Natl. Acad. Sci. Biol. 77, 4494 (1980).
15. Kleinman, H., Klebe, R., and Martin, G. J. Cell Biol. 88, 178 (1981).
16. Yamada, K. and Olden, K. Nature 275, 179 (1978).
17. Rojkind, M., Gatmaitan, Z., Mackensen, S., Giambrone, M., Ponce, P., and Reid, L. J. Cell Biol. 87, 255 (1980).
18. Reid, L., Gatmaitan, Z., Arias, and Rojkind, M. New York Acad. Sci. 349:70-76.
19. Gospodarowicz, D., Greenburg, G. and Birdwell, C. Cancer Res. 38, 4155 (1978).
20. Gospodarowicz, D. and Lui, G. J. Cell. Phys. 109, 69(1981)
21. Hume, W. and Potten, C. Brit.J. Derm. 103, 499 (1980).
22. Chambard, M., Gabrion, J. and Mauchamp, J. Cr. Ac. Sci. D. 291, 79 (1980).
23. Emerman, J. and Pitelka, D. In Vitro 13, 346 (1977).
24. Michalopoulos, G. and Pitot, H. Exp. Cell Res. 94, 70 (1975).
25. Gatmaitan, Z., Biempica, L., Arias, W. and Reid, L. (Submitted).
26. Reid, L., Mackensen, S., and Oie, H.(In Preparation).
27. Reid, L. and Morrow, B. (In Preparation).
28. Leinwand, L., Strair, R. and Ruddle, F. Exp. Cell Res. 115, 261 (1978).
29. Gazdar, A., Chick, W., Oie, H., Sims, H., King, D., Weir, G., and Lauris, V. Proc.Natl.Acad.Sci. USA 77, 3519(1980)
30. Tashjian, A., Yasumura, Y., Levine, L., Sato, G., and Parker, M. Endocrinology 82, 342 (1968).
31. Berry, M. and Friend, D. J. Cell Biol. 43, 506 (1969).
32. Lacey, P. and Kostianovsky, M. Diabetes 16, 35 (1969).
33. Bottenstein, J., Hayashi, I., Hutchings, S., et al. Methods in Enzymology 58, 94 (1979).

34. Fong, H., Chick, W., and Sato, G. Cold Spring Harbor Symposia 46 (in press).
35. Michalopoulos, G., Sattler, G., O'Connor, L. and Pitot, H. Cancer Res. 38, 1866 (1978).
36. Buonassisi, V., Sato, G. and Cohen, A. Proc. Natl. Acad. Sci. 48, 1184 (1962).
37. Bissell, D. and Guzelian, P. Ann. New York Acad. Sci. 349, 85 (1980).
38. Jakoby, W. and Pastan, I., eds."Cell Culture". Methods in Enzymology. volume 58 (1979).
39. Murakami, H. and Masui, H. Proc. Natl. Acad. Sci. USA 77, 3464 (1980).
40. Michalopoulos, G., Sattler, G. and Pitot, H. Life Sci. 18, 1139 (1976).
41. Liotta, L., Klerinerman, J., Catanzaro, P. and Rynbrandt, D. J. Natl. Cancer Inst. 58, 1427 (1977).
42. Meezan, E., Hjelle, T., Brendel, K. and Carson, E. Life Sci. 17, 1721 (1977).
43. Reid, L. and Gatmaitan, Z. (in preparation).

DISCUSSION

R.G. SMITH: You made a comparison between cells on type I collagen with cells on a biomatrix with respect to their differentiated function. How do the biomatrices isolated from tissue compare - in terms of differentiated functions - with the results of cells cultured on a synthetic biomatrix consisting of various collagen types and heparin sulfate?

L. REID: Our studies of cells cultured on individual components of extracellular matrix are quite preliminary. We know that use of laminin alone, type IV collagen alone, or the two in combination do not give the differentiated expression seen when cells are plated on biomatrix. We have not tested heparan sulfate at this time.

H. BUSCH: What are the differences in the biomatrix chemistry of benign and malignant tumors? And what are the biomatrix requirements of human malignant tumors?

L. REID: We have not done chemical analyses on the biomatrix prepared from any neoplastic tissue. As indicated in the data presented, many malignant tumors do show requirements for appropriate forms of extracellular matrix. The tumor cells cannot be plated on just any form of biomatrix; they show some tissue specificity. However, in contrast to normal cells which demonstrate a strict tissue specificity in biomatrix requirements, the tumors will grow on several types of matrix.

MICROCARRIERS AND THE PROBLEM OF
HIGH DENSITY CELL CULTURE

William G. Thilly
Debra Barngrover[1]
James N. Thomas[1]

Department of Nutrition and Food Science
Massachusetts Institute of Technology
Cambridge, Massachusetts, U.S.A.

Mammalian cell cultures are needed for production
purposes when (a) proteins require specific post-
translational processing to be functional; (b) viruses,
as opposed to viral antigens, are the products; or
(c) cells themselves are the desired product, as in
tissue reconstruction.
When anchorage-dependent cells are cultured on suitable microcarriers, cell culture is no longer limited
by surface area, and the problems of high density
culture of anchorage-dependent and-independent cells
become the same. These problems include rapid pH
decrease, oxygen and nutrient depletion and accumulation of products such as ammonia to growth-inhibitory
concentrations.
Each of these problems and, in some cases, their solutions will be discussed in terms of the behavior of
microcarrier cultures at cell concentrations at or
above 10^7 cells/ml.

---

[1]Supported as Flow Research Fellows by Flow Laboratories,
Inc., McLean, Virginia.

I.   INTRODUCTION

    Bacteria and yeasts can be grown in large-scale, deep suspension cultures; industrial-scale mammalian cell culture is usually accomplished in multiple low-productivity units (such as roller bottles), which require multiple, labor-intensive manipulations. This is because the mammalian cells used must be attached to a suitable surface in order to grow. Thus, the maximum cell number is limited by the surface area available. A solution to this surface limitation problem was introduced by Anthony van Wezel in 1967 (1). He and his colleagues demonstrated that some mammalian cell types would attach to a beaded commercial anion-exchange resin and grow to confluency in stirred suspension cultures (2,3). However, the system was limited by the "toxicity" of the beads, which at low concentrations (1 mg/ml or 6.0 cm^2/ml) resulted in a significant loss of cell innoculum and at higher levels (> 2 mg/ml or > 12.0 cm^2/ml) in a total inhibition of cell growth. This bead "toxicity" was ascribed to either a toxicity of the beads themselves or to a competitive absorption by the beads of key nutrients (4). Various techniques were proposed to correct this toxicity. These techniques included pretreating the beads with serum or nitrocellulose, increasing the cell innoculum size or adding conditioned media or polycationic macromolecules to the growth medium (5,6). The positive charge density of the microcarriers turned out to be the key parameter; and, when crosslinked dextran beads carried a positive charge of 2 meq/gm, human fibroblasts readily attached and grew (7,8). Cell densities of 1.5 to 5 million cells/ml media (depending on cell type and culture conditions) are now routinely achieved in suspension cultures using these controlled charge microcarriers.

    Microcarrier cultures are no longer surface limited. The cell density achieved (cells/cm^2) at 5 mg/ml (30.0 cm^2/ml) is generally less in microcarrier culture than for the same cells in petri dishes. Addition of more microcarriers above the commonly used 5 mg/ml does not significantly increase cell number.

    Further refinements in the area of microcarrier technology have recently been added (9). For several epithelial cell lines, bead to bead transfer can be achieved without enzyme treatment (trypsin) by simply growing the cells in a low-calcium medium and sequentially diluting the culture with fresh medium and beads. Mitotic cells can be isolated from microcarrier cell suspensions by treating with colcemid and increasing the stirring rate. Cells arrested in mitosis detach from the microcarriers and are easily collected by aspiration of the supernatant (10).

There are still many problems to be solved in microcarrier technology. Effective bead to bead transfer for human fibroblasts has yet to be achieved. So far, over 50 different cell types, including 10 primary cell types, have been shown to grow on microcarriers (9). This work needs to be extended, especially for the more fastidious primary cells. Finally, we earnestly believe that limitation with regard to surface area does not affect the production capacity of microcarrier cell culture systems. Therefore, we are now turning to the identification and elimination of growth limiting factors in high density cell fermentors.

We should, of course, define what we mean by high density culture. In 1975, using commercial anion-exchange resin beads, "high density" in microcarrier cultures meant anything over 10^6 cells/ml. Today, most cells will grow to a density of 1-2 (human fibroblasts), 2-3 (many epithelial cell lines) or 4-5 (chick fibroblasts, CHO cells) x 10^6 cells/ml in stirred batch culture using commercially available microcarriers (Superbeads^{TM}, Flow Laboratories, Inc., McLean, VA) and standard serum-supplemented cell culture media. Using perfusion culture, we have pushed MDCK cells up to 2 x 10^7 cells/ml. Twelve years ago Thayer and Himmelfarb (11) reached densities as high as 6 x 10^7 cells/ml in their spin filter cultures of mouse lymphoma cells. Therefore, "high density" is a relative term which we employ here to mean cell densities which significantly exceed those obtained using present standard media formulations and stirred single-batch suspension or microcarrier cell systems.

We see a role for simple mass balance approaches to fermentor operation. If we provide enough surface area for anchorage dependent cells, then growth of cultures should be limited only by (a) supply of things which cells need to grow, (b) accumulation of things which stop cell growth.

Studies leading to the serum-free growth of cells, such as Prof. Sato has discussed in this volume, and the carefully balanced cell media devised by Ham and his co-workers (13) elicit our admiration. Our group is aiming at bulk production, so what the cells need, we will have to deliver. Sato, Ham and others are expanding our knowledge as to just what cells do require.

In high density single-batch cultures, nutrient supplies are rapidly depleted, especially the nutrients used as energy sources (carbohydrates, glutamine and branched chain amino acids). Nutrient levels in most media were originally chosen as either the minimum that would support rapid growth (12) or else a level midway between the minimum and the toxic level (13). For high density cultures, a level just below the toxic level would seem to be better for rapidly utilized nutrients, but adding high levels of every nutrient leads to problems

with osmolarity. Thus, amino acids, carbohydrates, vitamins
and minerals in the media must be carefully balanced to prevent toxicity on one hand or too rapid depletion on the other.
Lipids need to be added with a suitable carrier such as serum,
liposomes (14) or albumin (15). Oxygen availability is a
problem in deep suspension cultures because less air-liquid
surface area is available per cell, which means less gas exchange. Hormone supply is a problem when hormone-supplemented
serum-free media are used. For example, the serum-free medium
developed by Taub *et al.* (16) for MDCK cells would not support
the growth of MDCK cells in microcarrier culture, unless cAMP
and five-fold higher amounts of insulin and transferrin were
added. With these additions, growth similar to that seen with
serum supplement was observed (9). The amount of insulin necessary was proportional to the number of cells indicating that,
as cell number increases, hormone supply must also increase.
Monitoring the cultures for differential depletion rates of
the nutrients and hormones is our first step in determining
which are growth limiting.

   Toxic metabolite accumulation is another obvious problem
to be anticipated in high density cultures. Lactic acid produced from a high rate of glycolysis causes a rapid drop in
pH to below growth sustaining levels. Ammonia, formed primarily from deamination of glutamine, rapidly accumulates to
levels above 2 mM, which is growth inhibitory to many cells
in microcarrier culture. Specific growth inhibitory peptides
have been shown to be released by some cells (17). Other, as
yet unidentified toxic metabolites may also be accumulating
in high density cultures. Research may be advantageously addressed to prevention of the production of these metabolites,
or failing that, to their effective and inexpensive removal.

   II. PROBLEMS ASSOCIATED WITH HIGH DENSITY CELL CULTURE
            A. Choice of Experimental System

   Both batch and perfusion systems can be used to advantage
in the *in vitro* study of the environmental requirements of
cells grown on microcarriers. In batch systems cells are
incubated for the duration of the experiment without the addition of fresh media or media components. This method is
easy and provides fewer opportunities for contamination, but
cells cultured in this way are subjected to continual changes
in their environment. Some nutrients that are rapidly
utilized must be added initially at very high, unphysiological
concentrations to prevent their complete depletion during the
course of the experiment. If oxygen is not continually introduced into the culture environment, it will be depleted, especially when high cell densities are obtained. In fact,when
cells are cultured in a 10 liter, unstirred fermentor at

densities of 1.0 x 10^6 cells/ml and glucose is used   for
energy, oxygen will be used twice as fast as its capacity to
diffuse into the medium. This increases to 20 times in a
stirred 100 liter fermentor, and oxygen may be used at an even
greater rate if other energy sources are used (18).

Perfusion systems consist of a culture environment in
which fresh medium is continuously added at the same rate as
culture medium is removed. In theory, if media components
could be supplied in quantities matching the rates of their
utilization, then a constant concentration of each nutrient
could be maintained within the culture environment. In practice a true "steady state" of this type would be difficult to
achieve, but a perfusion system provides a much more constant
environment than obtainable in a batch system.

In microcarrier cultures maximal cell concentrations
(cells/ml of culture) are generally higher when cells are
grown in a perfusion system than in a batch system (Figure 1)
(19). Two possibilities exist which may explain the lower
growth observed in the batch system. Cells may stop growing
because a factor such as a nutrient is depleted from the
medium. This process may be expressed in the following way:

$$[V_{in} \, c^i_{in}] < N \, \frac{dc^i}{dt}$$

or

$$[\text{rate of input}] < [\text{rate of use}]$$

where V_{in} is the velocity or rate of input (1/min.), c^i_{in}
is the concentration of each media component supplied
(moles/l), and N represents the cell population size (cell
number).

The accumulation of toxic metabolites may also be growth
inhibitory, as expressed in the following:

$$[V_{out} \, c^t_{out}] < N \, \frac{dc^t_{out}}{dt}$$

or

$$[\text{toxic metabolites removed}] < [\text{toxic metabolites produced}]$$

where V_{out} is velocity or rate of output (1/min.) and c^t_{out} is
concentration of toxic metabolites removed (moles/l).

To determine whether growth inhibition was the result of
nutrient depletion or toxic metabolite accumulation, an experiment was designed in which the concentration of all nutrients
were decreased, but the rate of perfusion remained constant.

Media components, except salts, were diluted to either 25 or 50% of their original concentration. Oxygen was supplied only by diffusion from the head space. The effect of media concentration on cell growth is shown in Figure 2. Diluting all

COMPARISON OF CULTURE METHODS

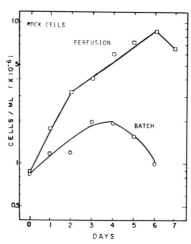

Figure 1.   Growth response curves for MDCK cells grown in batch and perfusion systems using Dulbecco's modification of Eagle's medium (DMEM).

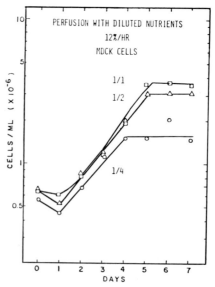

Figure 2.   Growth response of MDCK cells perfused (12%/hr) with DMEM diluted to ¼ or ½ of its original concentration or undiluted DMEM.

nutrients twofold did not significantly change growth rates or final cell concentrations of MDCK cells. This evidence supports the theory that a toxic or inhibitory compound(s) is produced by cells which will retard growth. The experimental evidence, as it stands now, could also be interpreted to mean that oxygen may be the growth-limiting factor missing from the culture environment.

In our next perfusion studies, we plan to dilute the medium components and increase the rate of perfusion by the reciprocal of the dilution factor while monitoring oxygen concentration. Stated another way, media components will be diluted to ½ or ¼ of their normal concentration, but the perfusion rate will be increased 2 or 4-fold, respectively. These experiments should help determine whether or not an inhibitory or toxic compound is produced by cells, which, if allowed to accumulate in the culture environment, will lead to growth inhibition. We also intend to examine the problem of oxygen availability in high density cell culture. An apparatus might be attached to the perfusion system through which incoming medium will be pumped and oxygenated. Oxygen may be effectively introduced into the culture environment through various other means which also will be investigated.

Having introduced our general approach to increasing fermentor productivity, we now can turn to some of our specific observations. First, we will consider the matter of carbohydrate supply and the closely linked problem of lactic acid accumulation. Secondly, we will consider amino acid utilization and the linked problem of ammonia accumulation. We think we have found a generally applicable solution to lactic acid production which eliminates the need for exogenous pH control. We are still working on a solution to the ammonia problem. Explaining the high rate of glutamine utilization and ammonia production may be more difficult than solving the engineering problems, however.

B. Carbohydrate Utilization and the Problem of Lactate Accumulation

In cells cultured in the presence of glucose, there is a high rate of glycolysis and lactate production (20), which, in high density cell cultures, leads to a rapid drop in pH to inhibitory levels. Suggestions for controlling this pH drop have included (1) replacement of the media, either by perfusion or frequent media changes and (2) continual titration of the system with a base to maintain optimum pH. The first suggestion, though useful experimentally, is expensive in large-scale cultures and the second requires elaborate equipment and an engineer to maintain it.

Eagle, in 1958 (20), showed that fructose was more efficiently used than glucose in cell cultures, with less production of lactate. Reinwald and Green (21) demonstrated that replacement of glucose with maltose (which is slowly hydrolyzed to glucose by maltase in the serum) prevented a decrease in pH, while Reitzer et al. (22) showed that HeLa cells grew at an equal rate on either glucose, galactose or fructose. We, therefore, decided to examine the effects of fructose, maltose and galactose in high density microcarrier cultures. The pathways of utilization of these carbohydrates is shown in Figure 3.

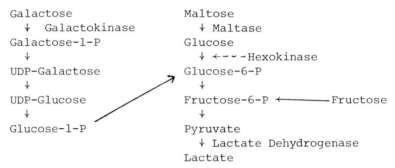

Figure 3.   Pathways of Carbohydrate Utilization

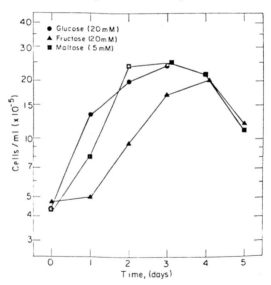

Figure 4.   Growth of MDCK cells in 100 ml Superbead^TM spinner cultures (microcarriers 5 mg/ml). Media (Dulbecco's modification of Eagle's minimum essential medium) was modified to contain the indicated levels of carbohydrates and 4 mM glutamine. Cultures were grown without media replacement in 10% CO_2.

As can be seen in Figure 4, 20 mM glucose, 20 mM fructose and 5 mM maltose all supported growth of MDCK cells equally well in microcarrier cultures. In the absence of any carbohydrate, the cell numbers increased by 20% the first day and decreased thereafter. Decreasing numbers of cells indicates detachment of the cells from the microcarriers.

The time dependence of the pH changes is shown in Figure 5. In the cultures grown in 20 mM glucose, the pH rapidly dropped below 7.0, while in the cultures containing maltose or fructose, the pH remained above 7.0 and, in the case of fructose, above 7.2 during the five days of observation. This work has now been extended to other cell types such as human diploid fibroblasts in larger (4 liter) cultures and higher cell concentrations. In every case the pH change seen in the fructose-containing cultures is less than 0.2 pH units.

As expected, the pH changes reflect lactate production with a lactate production in the glucose-containing cultures which was 3-4 times higher than that seen in the fructose- or maltose-containing cultures (Figure 6).

Comparison of the utilization rates of the carbohydrates (Figure 7) shows that at the same rate of cell growth, fructose is utilized much more slowly than glucose. More than 50% of the initial fructose remains after 5 days, while 5-20 mM glucose is completely depleted from the media in 3 days.

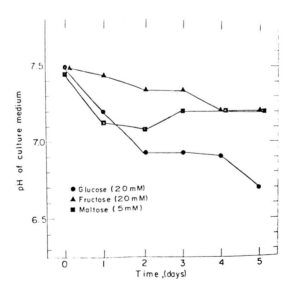

Figure 5.   pH of culture medium in MDCK microcarrier cultures with different carbohydrate supplements. Experimental conditions same as in Figure 4.

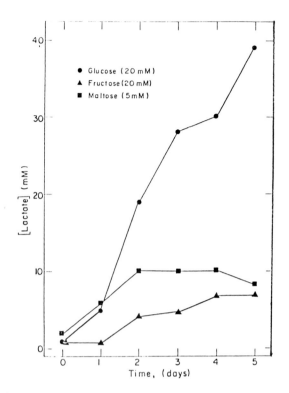

Figure 6.   Lactate levels in culture medium in MDCK
            microcarrier cultures with different carbo-
            hydrate supplements. Experimental conditions
            same as in Figure 4.

     The lactate/pyruvate ratio in the medium reflects the redox
potential of the cytosolic NADH/NAD pool. The physiological
range for the lactate/pyruvate ratio is between 6 and 15 (inferred from 23-25). At the start of the culture, the ratio is
below 6 (Figure 8) because of the formulation of the medium.
This might fruitfully be studied in terms of more physiological
initial conditions. In the glucose-containing medium, the
ratio increased to 30 by day 2 and continued to increase after
that. In the fructose-containing cultures, the ratio stayed
between 6 and 15 during exponential growth (days 1 and 2) and
rose to non-physiological levels only after the growth plateau
was reached (days 4 and 5). In the cultures started on 5 mM
maltose, the ratio remained below 20 during the entire observation period.
     Reitzer et al., recently reported that glutamine is the
major energy source for some cultured cells (22). We, therefore, examined glutamine utilization and ammonia production in
these cultures with various carbohydrate supplements.

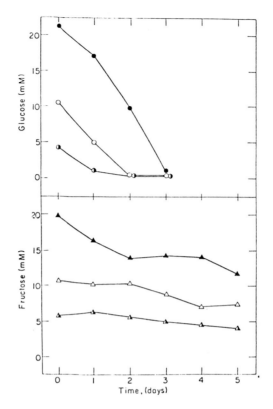

Figure 7.   Glucose (top) and fructose (bottom) re-
            maining in culture medium in MDCK micro-
            carrier cultures with different carbohydrate
            supplements. Experimental conditions same
            as in Figure 4.

Glutamine was rapidly utilized and was depleted by day 2 in
the maltose-containing culture and by day 3 in the fructose-
or glucose-containing cultures (Figure 9). Ammonia was produced at the fastest rate in the maltose-containing culture,
and initially at a slower rate in the glucose and fructose-
containing cultures (Figure 10). In all these cases the
ammonia concentration remained below 2 mM which does not equal
the near disappearance of the initial 4 mM glutamine. Glutamine
utilization and ammonia production are discussed further when we
consider amino acid utilization. Our observations are in accord
with those of Reitzer *et al.* in finding that glutamine utilization does indeed play a role in the cell's energy requirements.
    Cultures grown in 20 mM galactose attain the same density
(2×10^6 cells/ml) seen in glucose-containing cultures, but as
with fructose-containing cultures, the pH remains above 7.3.

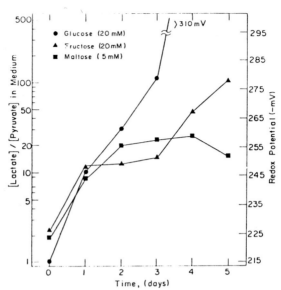

Figure 8. Lactate/pyruvate ratio in culture medium in MDCK microcarrier cultures with different carbohydrate supplements.   Experimental conditions same as in Figure 4.

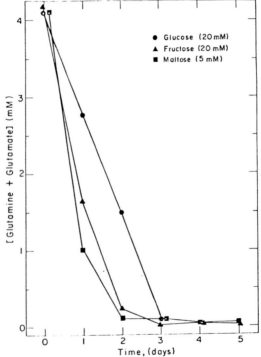

Figure 9.   Glutamine and glutamate remaining in culture medium
in MDCK microcarrier cultures with different carbohydrate supplements. Conditions same as in Figure 4.

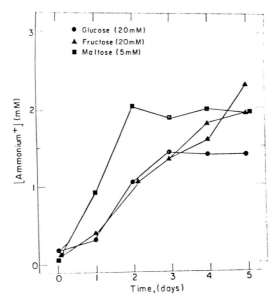

Figure 10.   Ammonia levels in culture medium in MDCK microcarrier cultures with different carbohydrate supplements. Experimental conditions same as in Figure 4.

Thus, pH can be easily controlled in high density cell
culture by the substitution of galactose, maltose or fructose
for glucose. Galactose has not served as a suitable carbohydrate source for some other cell types tested (Victor Edy, Flow
Laboratories, Inc., personal communication). The use of maltose
requires attention to the amounts of maltase present in a
culture medium (21). Fructose has been a suitable carbohydrate
source for all cell lines examined to date, results in no significant pH change in cell cultures approaching 10^7 cells/ml
and requires no other modification of media formulae. One
commercial laboratory, Flow Laboratories, Inc., McLean, VA,
has recently introduced a commonly used culture medium (DMEM)
with fructose substituted for glucose.

C.   Amino Acid Utilization in Perfusion and Batch Cultures

Fisher (26-28), in a series of studies using dialyzed serum,
examined the nutritional requirements of cultured cells and
found glutamine and cystine, along with a few other amino acids,
to be necessary for survival and growth of cells *in vitro*.
Eagle (12) found that 12 amino acids are required for proliferation of strain L mouse fibroblasts in a medium containing 0.25
to 2% dialyzed horse serum. Cells die within 1 to 3 days if

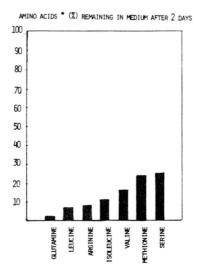

ANTINO ACIDS * (%) REMAINING IN MEDIUM AFTER 2 DAYS

* INITIAL CONCENTRATIONS IN μ M WERE GLUTAMINE- 3188,
LEUCINE- 762, ARGININE- 354, ISOLEUCINE- 785, VALINE-
728, METHIONINE- 158, SERINE- 451

Figure 11.   Heavily utilized amino acids remaining in medium
after 2 days' incubation in batch cultured MDCK
cells.

any one of the 12 amino acids is omitted. Since these early
experiments, many studies have been conducted to determine the
precise amino acid requirements for a variety of cells in
culture. Now, we have examined amino acid utilization for
MDCK cells grown on microcarriers in both batch and perfusion
experiments.

Figure 11 shows the quantity of the most actively utilized
amino acids remaining in the medium after 2 days' incubation in
batch culture. Amino acid utilization is compared on day 2 because this corresponds to the maximum cell concentration (2.8
x 10^6 cells/ml) reached during incubation in the batch system.
Glutamine is almost completely removed from the medium in just
2 days, even though it was initially present in the highest
concentration. Over 75% of serine, methionine, valine, isoleucine, arginine and leucine are utilized during the same
length of time. Figure 12 shows the moderately consumed amino
acids after 2 days of incubation. Only 15 to 40% of these are
removed by the second day. Some amino acids do not change in
concentration or increase in the medium as shown in Figure 13.
Aspartic acid and proline remain constant, but glycine and
alanine are produced by MDCK cells after 2 days of incubation
in batch culture.

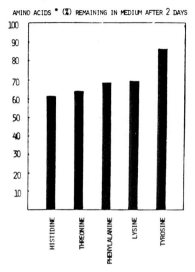

AMINO ACIDS * (%) REMAINING IN MEDIUM AFTER 2 DAYS

* INITIAL CONCENTRATIONS IN μ M WERE HISTIDINE- 165,
THREONINE- 777, PHENYLALANINE- 376, LYSINE- 696,
TYROSINE- 684

Figure 12.   Moderately utilized amino acids remaining in
medium after 2 days' incubation in batch cultured
MDCK cells.

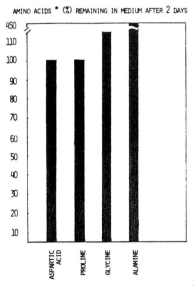

AMINO ACIDS * (%) REMAINING IN MEDIUM AFTER 2 DAYS

* INITIAL CONCENTRATIONS IN μ M WERE ASPARTIC ACID- 22,
PROLINE- 30, GLYCINE- 46, ALANINE- 73

Figure 13.   Slightly utilized or produced amino acids in
medium after 2 days' incubation in batch cultured
MDCK cells.

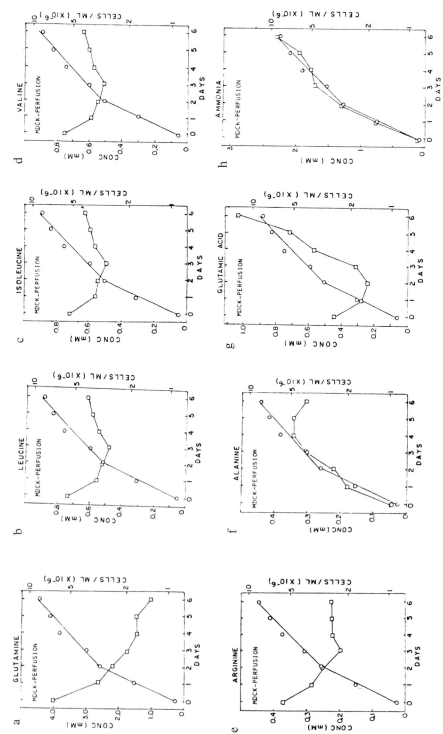

Figure 14.

    In the perfusion study amino acids are consumed from the
medium at the highest rate during the log growth phase. When
cell growth begins to slow, so does amino acid consumption.
Figure 14 compares the changes in concentration of glutamine,
leucine, isoleucine, valine, arginine, alanine, glutamate and
ammonia to the concentration of cells during a 6-day perfusion
study.

The removal of glutamine from the medium (Figure 14a) slows
somewhat as cell growth slows, but glutamine levels continue
to decline through day 6. Leucine, isoleucine, valine and
arginine concentrations initially decline, but then increase
during the period of slow cell growth (Figure 14 b-e).
Levels of glutamic acid in the medium (Figure 14 g) fall for
2 days, but increase sharply for the remainder of the experimental period. Alanine, as shown in Figure 14 f, is produced
in parallel to cell growth for 4 days and then is slightly
consumed. Ammonia is produced in parallel to cell growth
(Figure 14 h). The significance of this observation will be
discussed in detail later in this paper.
    The differential utilization of amino acids in both the
batch and perfusion cultures may be explained through their
metabolic functions. Certainly, all the amino acids are incorporated to one degree or another in cellular protein, but
some may perform more than one metabolic function.

    1. GLUTAMINE. Glutamine serves in the biosynthesis of
aspartic acid, proline (29) and asparagine (30). Its availability may also be rate limiting for *de novo* purine synthesis
in vitro (31). The presence of glutamine is also critical for
DNA replication as measured by autoradiographic determinations
of ^{3}H-thymidine incorporation (32). One report has indicated
that an excess of glutamine overcame the growth inhibitory effect of serum starvation and that quiescent cells were stimulated to undergo DNA synthesis by exposing them to a serum-

Figure 14. A comparison of the utilization and production
            rate of several amino acids and ammonia in per-
            fused (12%/hr) cultures of MDCK cells.

depleted (0.5% serum), but glutamine-rich medium (32). Glutamine may even serve to a limited degree as a lipid precursor
(33). The major function of glutamine in cells grown in culture in terms of mass utilization may be as an energy source
through the donation of its carbon skeleton for oxidation *via*
the tricarboxylic acid (TCA) cycle (34). Several researchers
have found carbon dioxide to be the major end product of glutamine metabolism in cultured cells (35-38). Approximately
40% of the energy requirement of Chinese hamster fibroblasts
(39) and 30% of the energy requirement of human diploid fibroblasts (38) were met by glutamine in a 5.5 mM glucose medium.
Lowering glucose concentration resulted in higher rates of
glutamine utilization. In summary, glutamine is used for many
anabolic processes, but it seems that most glutamine is utilized as an energy source.

2. BRANCHED-CHAIN AMINO ACIDS. The branched-chain amino
acids, particularly leucine and isoleucine showed rapid utilization in both single batch and perfused MDCK microcarrier
cultures. Thus, they also seem to perform more than one
function in cellular metabolism. Utilization of branched-chain
amino acids increase more than 4-fold in primary myoblasts incubated in a medium depleted of glucose (40). Intracellular
ATP, pyruvate, α-ketoglutarate, malate and citrate remain constant, despite the absence of glucose or lactate consumption.
When the utilization of branched-chain amino acids is inhibited
by clofibric acid, alanine and citrate production are blocked,
and intracellular levels of pyruvate, α-ketoglutarate, malate
and citrate are greatly reduced. These data suggest that,
under conditions of glucose depletion, branched-chain amino
acids, as in the case of glutamine, are used as energy sources
in addition to their anabolic roles.

3. GLUTAMATE. In perfused MDCK microcarrier cultures, we
found that glutamate concentration initially declines, but then
increases. This is probably because the rate of glutamine
breakdown exceeds the rate of glutamate utilization at the time
when cells begin to form multilayers on microcarriers. This
suggests that, under slow growth conditions, glutamine was converted to glutamate faster than glutamate could enter the TCA
cycle for oxidation.

4. ALANINE. Alanine, which also increases in both batch
and perfused MDCK microcarrier cultures, is probably formed
through a transamination reaction involving pyruvate and glutamate or a branched-chain amino acid. It is worth noting that,
when pyruvate levels fall toward the end of the experimental
period, the rate of alanine release also declines.

5. AMMONIA. A by-product of amino acid catabolism which accumulates· in the culture medium of growing cells is ammonia. As depicted in Figure 14 h ammonia concentration increases in parallel to cell number. At concentrations above 2 mM, ammonia has been found to be toxic (41-43). Holley et al. (43) found that when 2 mM ammonia is added to the culture medium in combination with 11 mM lactate, growth of BSC-1 cells is reduced by about 30%. The inhibitory effect of the combination is more pronounced when serum concentrations are low. When cell densities of greater than 2.0 x 10^6 cells/ml are obtained, the concentration of ammonia in the culture environment could easily be growth inhibitory.

III. AN ATTEMPT TO INTEGRATE AMINO ACID AND CARBOHYDRATE
     METABOLISM IN CELL CULTURE

Figure 15 depicts the point at which amino acids enter the TCA cycle with the thick arrows showing the probable route during heavy utilization. Glutamine, which has been shown to contribute significantly to the energy production of cells *in vitro*, enters the TCA cycle at the level of α-ketoglutarate. Glutaminase, the first enzyme involved in the utilization of glutamine as an energy source, is elevated during growth of human diploid fibroblasts (44). After 2 days of growth, the maximum activity of the enzyme is reached, but after cells reach confluency, enzyme activity declines. In addition, the activity of glutaminase is independent of glutamine concentration.

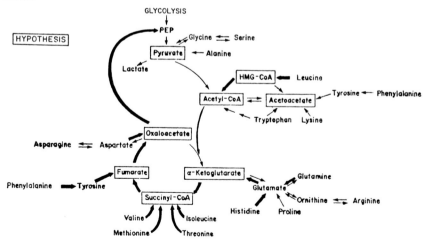

Figure 15.   Primary points of entry of amino acids into the
TCA cycle.

Valine, isoleucine, threonine and methionine enter the TCA cycle at succinyl-CoA while leucine enters through acetyl-CoA. All these amino acids have been shown to be utilized by a variety of cells cultured *in vitro* (19, 45-51). The amino acid that enter the TCA cycle at or before succinyl-CoA tend to be removed from the medium at the highest rate. As a result, GTP should be formed from GDP through substrate phosphorylation when succinyl-CoA is converted to succinate. If more oxaloacetate is formed than acetyl-CoA available to react with it, then oxaloacetate may be converted to phosphoenolpyruvate (PEP) *via* PEP carboxykinase (PEPCK) (Figure 16). This enzyme has a specific requirement for GTP and has been found in cultured human fibroblasts (52). Normal human fibroblast cells contain most of their PEPCK activity associated with their mitochondria (53, 54), and its activity can be induced by dibutyryl cyclic AMP, dexamethasone or hydrocortisone. All of these compounds are stimulatory for cell growth to one degree or another (14, 16, 55, 56). The activity of mitochondrial PEPCK in human diploid fibroblasts increases over 2-fold during cell growth (53). The PEP derived from amino acids may exit the mitochondria or may remain in the mitochondria where it is converted to pyruvate with the release of 1 mole of ATP. The pyruvate so formed may then enter the TCA cycle *via* acetyl-CoA and undergo oxidation. That portion of amino acid derived PEP that moves to the cytosol will also be converted to pyruvate with the formation of ATP. Some of this pyruvate may be converted to lactate (57). In fact, over 10% of the ^{14}C-label given in glutamine is incorporated into lactate by human diploid fibroblasts. A summary of these pathways is shown in Figure 16.

Amino acids may, therefore, contribute significantly to the energy metabolism of cultured cells in two ways. First, they can enter the TCA cycle at various points and be oxidized to CO_2. This would include the conversion of some of the amino

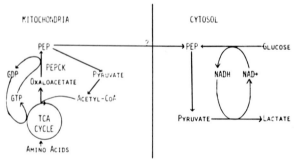

Figure 16. Possible interaction of amino acids in the energy metabolism of cultured cells.

acid carbon into PEP which would eventually enter the TCA cycle after releasing 1 mole of ATP in the conversion of PEP to pyruvate. This step would be especially important when adequate levels of glucose are not present. Secondly, amino acid derived PEP in the cytosol will also form pyruvate with the release of ATP. This would provide additional substrate on which lactate dehydrogenase can act to lower cytoplasmic NADH and thereby prevent NAD^+ from becoming the limiting factor in glycolysis. Of course, amino acid carbon can also exit the mitochondria and be converted to pyruvate in other ways. The best two possibilities besides PEP are through aspartate or malate.

A. Comparison of *In Vivo* and *In Vitro* Amino Acid Metabolism

The metabolism of amino acids in the intact animal is markedly different from amino acid metabolism in continuous cell culture. For that matter, amino acid metabolism varies among cells *in vivo* under various conditions. Glutamine is consumed by cells in the rat gut but is produced by skeletal muscle cells from the same animal (58). Primary liver cells produce or consume glutamine, depending on the pH (58). Primary kidney cells do not normally consume glutamine, but, under acidic conditions, will remove large amounts from the circulation and release ammonia, possibly as a buffering mechanism. In contrast, glutamine is heavily consumed by almost all cells grown continuously in tissue culture. This is even true for myoblasts that produce glutamine under all conditions *in vivo*.

In vivo, both alanine and glutamine are released from muscle to the blood in quantities greater than other amino acids (59, 60). If a rat epitrochlaris muscle preparation is perfused *in vitro* for 6 hours, alanine and glutamine will also be released in the greatest quantities (61). In perfused rat diaphragms, more than 50% of the carbon chains of aspartic acid, asparagine, glutamic acid, isoleucine and valine that enter the TCA cycle are converted to glutamine (62). Less than 20% of these amino acids are converted to carbon dioxide but 30% are converted to lactate.

Alanine and glycine are the amino acids released in the greatest amount from primary myoblasts in culture. Glutamine is not produced, being slightly consumed by these cells. This is in contrast to its production in whole animals and in perfused muscle preparations (51). An established myoblast line consumes even more glutamine, a behavior which, as noted above, is similar to that of many other established cell lines. Table 1 is a summary of the behavior of muscle tissue or myoblasts under a variety of experimental conditions.

TABLE 1.

COMPARISON OF IN VIVO AND IN VITRO GLUTAMINE
METABOLISM IN MYOBLASTS

GROWTH ENVIRONMENT OF CELLS	IN VIVO	IN VITRO		
		PERFUSED TISSUE	PRIMARY CULTURE	ESTABLISHED CULTURE
RESPONSE	HIGHLY PRODUCED	MODERATELY PRODUCED	SLIGHTLY CONSUMED	HIGHLY CONSUMED

B. Comparison of *In Vivo* and *In Vitro* Glucose Metabolism

Glucose, the carbohydrate source presented most commonly
to cells *in vivo,* is also metabolised differently by cells in
culture than in those in the animal. The uptake of glucose by
cells in the animal is limited by transport into cells (63).
Glucose utilization in cultured cells is limited by phosphorylation (51).

A good example of altered glucose metabolism can be found
in hepatocytes. Those that are grown in culture use glucose
differently than those found in the whole animal (64). Hepatocytes' primary energy source *in vivo* are derived from fatty
acid oxidation. Liver cells in intact animals, therefore, consume little glucose; instead, glucose is produced from precursors such as lactic acid. In contrast, established hepatocyte cell lines consume large amounts of glucose (64), and
this seems to be true for a wide variety of established cell
lines (65). When the glucose metabolism of two established
cell lines, one transformed and one non-transformed, are compared with primary hepatocytes, the established cell lines
consume glucose at a 40-fold greater rate and convert 80 to
90% of the carbohydrate to lactate. The primary hepatocytes
convert approximately 50% of the total metabolised glucose to
lactate. It appears that, of the large amount of glucose consumed, only 5% enters the TCA cycle, 80% generates ATP through
glycolysis and the rest is used for anabolic processes (66).

C. Considering the Basis for Alterations in *In Vitro*
Cellular Metabolism

Cells grown *in vitro* generally consume large amounts of both
glucose and glutamine, while cells grown *in vivo* consume smaller amounts of glucose and, in the case of myoblasts, produce
glutamine. The reason for these differences may be environ-
mentally linked. Some nutritional, hormonal or physical
factor may be missing or some toxic or inhibitory product may

be formed which brings about an alteration in cellular metabolism. Several hypotheses have been suggested in the past
which attempt to explain the abnormalities in metabolism associated with cells grown in tissue culture. One of these
involves a change in membrane structure such that cells in
culture lose their capacity to exclude exogenous glucose (34,
51, 67) and produce lactic acid as a means of detoxifying
glucose (68). Another reasonable hypothesis is that cells
grown in tissue culture need more energy to function because
of a defective Na^+-K^+-ATPase (69). Amino acids and glucose
would, therefore, need to be consumed at elevated rates to
meet the higher energy demand of cells in this situation.

   A hypothesis that we would like to suggest is that some
factor or factors are preventing the proper functioning of the
electron transport chain and, therefore, normal oxidative
phosphorylation. This would ultimately result in the incomplete utilization of energy substrates and the accumulation of
toxic metabolites.

   A number of factors may lower the rate of oxidative phosphorylation, but one that needs more investigation is the requirement of cells for oxygen in the culture environment. Much
work has already been devoted to determining acceptable pO_2
levels for cells grown in a variety of culture systems
(70-74), but unfortunately adequate levels of oxygen are still
taken for granted in many studies.

   The oxygen utilization rate (OUR) in FS-4 cells has been
shown to increase as cell numbers increase (Figure 17). Since

OUR vs CELL NUMBER

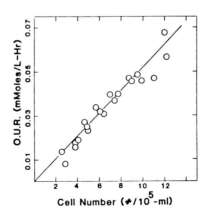

Figure 17.   A comparison of OUR and cell number in FS-4
             cultures containing glucose (18).

a linear relationship is seen, the OUR per cell remains relatively constant, and this is independent of the dissolved
oxygen concentration between 25 and 75% of air saturation (18).
Therefore, as the density of the cell population increases, so
does the rate at which oxygen is removed from the culture medium.
If steps are not taken for its supplementation, then oxygen becomes a limiting factor in high cell density cultures.

A major difference between cells cultured *in vitro* and
those found *in vivo* is the means by which they receive their
oxygen. Hemoglobin functions as an oxygen buffer in the body.
Its effect is to maintain a relatively constant tissue pO_2
level of 40 mm Hg even during wide fluctuations in environmental pO_2. Since cells cultured *in vitro* must rely on gas
diffusion through a liquid medium, often from the surface of
the medium, the pO_2 of the environment surrounding the cells
may vary considerably during prolonged incubation. This
variation becomes particularly acute if cells are grown at
very high densities. Cell populations grown at suboptimal
and/or varying oxygen levels may, over time, alter their
metabolism to cope with their environment. After a period of
time cell populations may become so altered that they lose
their sensitivity to oxygen and continue to use glucose and
amino acids incompletely, even in the presence of adequate
oxygen. Tumor cells seem to be less sensitive to changes in
oxygen tension than non-transformed cell lines (75). Key
enzyme activities involved in oxygen utilization and glycolysis
do not change in tumor cells in the presence of low or high pO_2
levels. To a certain degree, this may also apply to established
cell lines. These lines do alter their profiles of enzyme
activities in response to changes in oxygen tension, but established cells may be less sensitive than primary cells or cells
found in the intact animal to those changes (75).

A lack of a nutritional or hormonal factor could prevent
the proper functioning of the electron transport chain. Adreno-
cortical cells fail to properly oxidize pyruvate in the absence
of antioxidants, such as vitamin E or selenium (76). The absence of these antioxidants may cause a block in the oxidation
of mitochondrial NADH as a result of the inefficient functioning of the ubiquinone cycle.

At this point, our hypotheses about alterations in cellular
metabolism are fairly general, but we feel that they are at
least experimentally testable. It may be instructive to follow
the behavior of primary cells, such as chick fibroblasts or
mouse fibroblasts grown in current uncontrolled oxygen conditions. Under the current culture conditions, these cells enter
a "crisis period" after a few cell divisions in which only a
few cells (or none) emerge capable of continued growth, e.g.,
mouse 3T3 cells. Has this step selected for cells capable of
surviving or adapting to the vicissitudes of uncontrolled
and/or low pO_2?

IV.   SUMMARY

Growth of anchorage-dependent mammalian cells to densities of 1-5 x 10^6 cells/ml is possible using microcarriers and conventional conditions of medium, serum and oxygen supply.

Growth to cell densities of 1-2 x 10^7 cells/ml has been achieved in perfused microcarrier cultures for MDCK cells and chick fibroblasts. Perfusion systems do not increase the actual yield (cells/liter of medium), but can be used in experiments to test for and identify limiting substrates or toxic metabolites.

The minor medium alteration of substituting fructose for glucose eliminates for many cell types the overproduction of lactate responsible for growth-limiting decreases in pH. The rapid use of glutamine and the branched-chain amino acids as energy sources apparently leads to accumulation of ammonia to growth-inhibitory concentrations.

It is clear that patterns of substrate utilization are quite different in cell culture from *in vivo* patterns. We wonder if cells in culture have been selected for their ability to adapt to the non-physiological but conventional conditions? In particular, we wonder about the nature of the metabolic pathways responsible for amino acid catabolism in the presence of apparently sufficient carbohydrate and oxygen?

We conclude that achieving high-density cell culture fermentors of importance in industrial production will depend on our understanding of metabolic limits on cell growth and adapting our approaches to avoid reaching those limits in culture.

REFERENCES

1.   van Wezel, A.L. (1967) *Nature,* 2,64.

2.   van Wezel, A.L. (1972) *Prog. Immunobiol. Stand.* 5,187.

3.   van Wezel, A.L. (1973) "Microcarrier cultures of animal cells", in *Tissue Culture: Methods and Applications,* Kruse, P.F. and Patterson, M.K., Eds., p. 372-377. Academic Press, New York.

4.   Horng, C. and McLimans, W. (1975) *Biotechnol. Bioeng.* 17, 713.

5.   van Hemert, P., Kilburn D.G. and van Wezel, A.L. (1969) *Biotechnol. Bioeng.* 11, 875.

6.   Spier, R.E. and Whiteside, J.P. (1976) *Biotechnol. Bioeng.* 18, 659.

7.   Levine, D.W., Wong, J.S., Wang, D.I.C. and Thilly, W.G. (1977) *Somat. Cell Genet.* 3, 149.

8. Levine, D.W., Wang, D.I.C. and Thilly, W.G. (1979) *Biotechnol. Bioeng.* 21, 821.

9. Crespi, C.L., Imamura, T., Leong, P.-M., Fleischaker,R.J. Brunengraber, H. Thilly,W.G. and Giard, D.J. (1981) *Biotechnol. Bioeng.* 23, 2673.

10. Ng, J.J.Y., Crespi, C.L. and Thilly, W.G. (1980) *Anal. Biochem.* 109, 231.

11. Himmelfarb, P. Thayer, P.S. and Martin, H.E. (1969) *Science,* 164, 555.

12. Eagle, H. (1955) *J. Biol. Chem.* 214, 839.

13. Ham, R.G. and McKeehan, W.L. (1979) "Media and Growth Requirements" in *Methods in Enzymology,* 58, 44, Colowick, S.P. and Kaplan, N.O., Eds. Academic Press, New York.

14. Bettger, W.J., Boyce S.T. Walthall, B.G. and Ham, R.G. (1981) *Proc.Nat.Acad.Sci. USA,* 78, 5588.

15. Yamane, I., Kan, M., Hoshi, H. and Minamoto, Y. (1981) *Exp. Cell. Res.,* 134, 470.

16. Taub, M., Chuman, L., Saier, M.H. and Sato, G. (1979) *Proc. Nat. Acad. Sci. USA,* 76, 3338.

17. Holley, R.W., Böhlen, P., Fava, R., Baldwin, J.H., Kleeman, G. and Armour, R. (1980) *Proc.Nat.Acad.Sci. USA* 77, 5989.

18. Fleischaker, R.J. and Sinskey, A.J. (1981) *European J. Appl. Microbiol. Biotechnol.* 12, 193.

19. Butler, M., Imamura, T. and Thilly, W.G. (1982) (in preparation).

20. Eagle, H., Barban, S., Levy, M. and Schulze, H.O. (1958) *J. Biol. Chem.,* 233, 551.

21. Rheinwald, J.G. and Green, H. (1974) *Cell,* 2, 287.

22. Reitzer, L.J., Will, B.M. and Kennell, D. (1979) *J. Biol. Chem.,* 254, 2669.

23. Bucher, T., (1970) *Pyridine Nucleotide-dependent Dehydrogenase,* Springer Verlag, Berlin, p. 439.

24. Williamson, D.H., Lund, P. and Krebs, H.A. (1967) *Biochem. J.,*103, 514.

25. Brunengraber, H. Boutry, M. and Lowenstein, J.M. (1973) *J.Biol. Chem.,* 248, 2656.

26. Fisher, A. (1947) *Biol. Rev.,*22, 178.

27. Fisher, A. (1948) *Biochem. J.,* 43, 491.

28. Fisher, A. (1953) *J. Nat. Cancer Inst.,* 13, 1399.

29. Levintow, L., Eagle, H. and Piez, K.A. (1957) *J. Biol. Chem.,* 227, 929.

30. Levintow, L. (1957) *Science,* 126, 611.

31. Raivio, K.O. and Seegmiller, J.E. (1973) *Biochemica et Biophysica Acta,* 299, 283.

32. Zetterberg, A. and Engstrom, W. (1981) *J. Cell. Physiol.,* 108, 365.

33. Reed, W.D., Zielke, H.R., Baab, P.J. and Ozand, P.T. (1981) *Lipids,* 16, 677.

34. Pardridge, W.M., Davidson, M.B. and Casanello-Ertl, D. (1978) *J. Cell. Physiol.*, 96, 309.
35. Lavietes, B.B., Regan, D.H. and Demopoulas, H.B. (1974) *Proc. Natl. Acad. Sci. USA,* 71, 3993.
36. Stoner, G.D. and Merchant, D.J. (1972) *In Vitro,* 5, 330.
37. Kovaceric, Z. and Morris, H.P. (1972) *Cancer Res. 32, 326.*
38. Zielke, H.R., Ozand, P.T., Tildon, J.T., Sevdalian, D.A. and Cornblath, M. (1978) *J. Cell. Physiol.*, 95, 41.
39. Donnelly, M. and Scheffler, I.E. (1976) *J. Cell. Physiol.* 89, 39.
40. Pardridge, W.M., Duducgian-Vartavarian, L., Casanello-Ertl, D., Jones, M.R. and Kopple, J.D. (1981) *Am. J. Physiol.,* 240, E203.
41. Visck, W.J., Kolodry, G.M. and Gross, P.R. (1972) *J. Cell. Physiol.*, 80, 373.
42. van Wezel, A.L. (1981) (personal communication).
43. Holley, R.W., Armour, R. and Baldwin, J.H. (1978) *Proc. Natl. Acad. Sci. USA,* 75, 1864.
44. Sevdalian, D.A., Ozand, P.T. and Zielke, H.R. (1980) *Enzyme,* 25, 142.
45. Butler, M. and Thilly, W.G. (1982) *In Vitro* (submitted).
46. Lucy, J.A. and Rinaldini, L.M. (1959) *Exp. Cell. Res.,* 17, 385.
47. McCarty, K. (1962) *Exp. Cell. Res.,* 27, 230.
48. Mohberg, J. and Johnson, M.J. (1963) *J. Natl. Cancer Inst.,* 31, 611.
49. Griffiths, J.B. and Pirt, S.J. (1967) *Proc. Roy. Soc. B.,* 168, 421.
50. Lambert, K. and Pirt, S.J. (1975) *J. Cell Sci.,* 17, 397.
51. Pardridge, W.M., Duducgian-Vartavarian, L., Casanello-Ertl, D., Jones, M.R. and Kopple, J.D. (1980) *J. Cell. Physiol.,* 102, 91.
52. Raghunathan, R., Russell, J.D. and Arinze, I.J. (1977) *J. Cell. Physiol.,* 92, 285.
53. Sumbilla, C.M., Ozand, P.T. and Zielke, H.R. (1981) *Enzyme,* 26, 201.
54. Arinze, I.J., Raghunathan, R. and Russell, J.D. (1978) *Biochimica et Biophysica Acta,* 521, 792.
55. Barnes, D. and Sato, G. (1980) *Cell,* 22, 649.
56. Taub, M. and Sato, G. (1980) *J. Cell. Physiol.,* 105, 369.
57. Zielke, H.R., Sumbilla, C.M., Serdalian, D.A., Hawkins, R.L. and Ozand, P.T. (1980) *J. Cell. Physiol.,* 104, 433.
58. Schrock, H. and Goldstein, L. (1981)*Am. J. Physiol.,* 240, E519.
59. London, D.R., Foley, T.H. and Webb, C.G. (1965) *Nature,* 208, 588.
60. Marliss, E.B., Aoki, T.T., Pozefsky, T., Most, A.S. and Cahill, G.F., Jr. (1971) *J. Clin. Invest.,* 50, 814.
61. Garber, A.J., Karl, I.E. and Kipnis, D.M. (1976) *J. Biol. Chem.,* 251, 826.

62. Chang, T.W. and Goldberg, A.L. (1978) *J. Biol. Chem.*, 253, 3685.
63. Goodman, M.N., Berger, M. and Ruderman, N.B. (1974) *Diabetes*, 23, 881.
64. Bissell, D.M., Levine, G.A. and Bissell, M.J. (1978) *Am. J. Physiol.*, 234, C122.
65. Paul, J., in *Carbohydrate and Energy Metatolism*, 239, (E.N. Willmes, Ed.) Academic Press, New York (1965).
66. Reitzer, L.J., Wice, B.M. and Kennell, D. (1979) *J. Biol. Chem.*, 254, 2669.
67. Hatanaka, M. Todaro, G.J. and Gilden, R.V. (1970) *Int. J. Cancer*, 5, 224.
68. Gregg, C.T., in *Growth, Nutrition and Metabolism of Cells in Culture, I.* (G. Rothblat and V. J. Cristofalo, Eds.) Academic Press, New York (1972).
69. Racker, E. (1976) *J. Cell. Physiol.*, 89, 697.
70. McLimans, W.F., Blumenson, L.E. and Tunnah, K.V. (1968) *Biotech. Bioeng.*, 10, 741.
71. McLimans, W.F., Crouse, E.J., Tunnah, D.V. and Moore, G.E. (1968) *Biotech. Bioeng.*, 10, 725.
72. Werrlein, R.J., Glinos, A.D. (1974) *Nature*, 251, 317.
73. Richter, A., Sanford, K.K. and Evans, V.J. (1972) *J. Natl. Cancer Inst.*, 49, 1705.
74. Balin, A.K., Goodman, D.B.P., Rasmussen, H. and Cristafalo, V.J. (1976) *J. Cell. Physiol.*, 89, 235.
75. Simon, L.W., Robin, E.D. and Theodore, J. (1981) *J. of Cell. Physiol.*, 108, 393.
76. Gill, G.N., Crivello, J.F., Hornsby, P.J. and Simion, M.H., in "Growth of Cells in Hormonally Defined Media" (Sirbasku, D.A., Sato, G.H. and Pardee, A.B., Eds.)(An Abstract of a paper given at the Ninth Cold Spring Harbor Conference on Cell Proliferation).

ACKNOWLEDGEMENTS

The research reported here was supported by the generous gifts of Flow General, Inc., MacLean, VA. We gratefully acknowledge the generosity of Dr. Michael Butler, Manchester Polytechnic Institute, U.K., Prof. Toshiko Imamura, University of California at Riverside and Messrs. Robert Fleischaker and Charles L. Crespi of MIT for sharing their experimental results for discussion at this conference prior to publication.

DISCUSSION

*G. SATO: A few years ago, Robert Cruth designed the farm of the future, and the farm of the future would have cow cells growing in culture for hamburger, and plant tissue culture cells were growing to provide the fire. If you mix them together of course the plant cells could soak up ammonia and CO_2 and produce oxygen. But another way to do it might be to use branched-chain keto acids to form citrulline and arginine as sinks for ammonia. Kidney cells are somewhat like liver cells. They should have the capacity to synthesize glutamine.

*W.G. THILLY: We have actually tried the alpha keto acids and found that they did not serve as substrates. If I could end with a moment of levity I have been thinking of a new cell culture medium, one that has fructose so that lactic acid is not produced. Then we only need something where ammonia would either be re-utilized or trapped in some way. Then it occured to us that if the pH isn't changing, we really don't need the CO_2 incubators to be able to set the pH properly. Some of this glucose utilization reminds one of the futile metabolic cycles where you really don't store the energy. If the cycle is futile the system has got to be exothermic. Then we could get rid of 37^0 incubator too.

*D.H. GELFAND: It is hard to follow your last comment but your abstract points out that for certain applications the cells themselves are the desired product. In certain cases some of these cellular products may be labile and susceptible to degradation if one uses trypsin to dislodge the cells from the microcarrier. Are you working on any microcarriers, perhaps gelatin or something, that could themselves be dissolved so that cells could be harvested without the use of trypsin?

*W.G. THILLY: Yes, that is a very interesting question which we are working on.

---

*Most of the discussion sections were contributed in written form by the participants, after each session. Some contributions, however, are taken from the tape recording of the proceedings. These are marked with an asterisk. In such cases, the editors have often condensed the statement. The words used should not, therefore, be taken to be the literal question or answer. While every effort has been made to paraphrase the wording accurately, the editors apologize to the participants or the reader if there have been unintentional errors or omissions that render the contribution inaccurate or unclear.

CROWN GALL - NATURE'S GENETIC ENGINEER

Milton Gordon
David Garfinkel
Harry Klee
Vic Knauf
William Kwok
Conrad Lichtenstein
Joan McPherson
Alice Montoya
Eugene Nester
Patrick O'Hara
Ann Powell
Lloyd W. Ream
Robert Simpson
Brian Taylor
Frank White

Departments of Biochemistry and
Microbiology and Immunology
University of Washington
Seattle, Washington
U.S.A.

I. GENERAL BACKGROUND

The soil organism, Agrobacterium tumefaciens, is able to transform host plant cells into rapidly growing tissues that synthesize a number of unusual compounds called opines. The causative bacteria are able to utilize the tumor-produced opines as sources of carbon and nitrogen. Thus, the Agrobacteria have converted the plant cells into a unique ecological niche that provides the organism with a competitive advantage. This transformation can be incited on most dicotyledenous plants including many species of great importance to world agriculture. This transformation has been studied extensively since the discovery of the causative organism in 1907 by Smith and Townsend. The discovery

FROM GENE TO PROTEIN:
TRANSLATION INTO BIOTECHNOLOGY

105

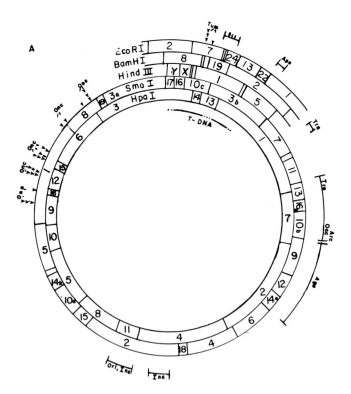

FIGURE 1.   Restriction endonuclease digest
map of pTi-B$_6$806.   The genetic loci are placed
according to the discussion in the text and are
described more completely in a review by Nester
and Kosuge (41).

by A. Braun (1) that crown gall tissues can be grown in the
absence of phytohormones, in contrast to normal tissues
which require phytohormones for growth in vitro, provided
a good selective marker for transformed tissues.

II. MOLECULAR BASIS OF CROWN GALL

The modern molecular basis of studies of the crown gall
transformation was ushered in by the finding that virulent
strains of A. tumefaciens all contain a large plasmid
(Ti-plasmid) of ca 100-150 megadaltons (2,3) and that part
of the plasmid, the T-DNA, is found in the transformed
plant (4). A restriction endonuclease map of an octopine
plasmid is shown in Fig. 1. It has recently been established that the plasmid DNA is covalently attached to high
molecular weight nuclear DNA (5,6,7) and is distributed in
a Mendelian manner upon meiosis (8). The molecular basis
for the uncontrolled growth of the transformed tissues
in planta is unknown at present although crown gall
tissues have a higher phytohormone content than most normal
tissues.

III. FUNCTIONS OF THE TI-PLASMID IN BACTERIA

A. Catabolic Nature

The Ti-plasmid in the bacterial cell codes for catabolic
enzymes. It has been found that the ability of an Agrobacterium strain to utilize opines is usually dependent
upon the presence of a Ti-plasmid. Studies by Gelvin et al.
(9) have indicated that whereas there is a low level of transcription of the Ti-plasmid, the regions from about 3:00 to
6:30 o'clock show high levels of transcription when induced
by various opines. Determination of the functions of
various regions of the Ti-plasmid have involved deletion
mutants or transposon insertions. The studies of Ooms (10)
have indicated that sequences from the catabolic region of
the Ti-plasmid from about 1:00 to 5:00 o'clock can be
deleted with no loss of the virulence. Figure 2 shows that
as the deletion extends into the T-DNA region, the ability
of the resulting tumors to synthesize octopine is lost, and
as the deletions extend further counterclockwise, the virulence of the plasmid is lost.

B. Origin of Replication

The region of the plasmid at ca 7:00 o'clock, contained in SmaI fragment 2, is the origin of replication. Plasmids which contain transposon insertions in this region have reduced virulence and are difficult to isolate (10). Recently, Vic Knauf in our laboratories has found that the introduction of HpaI fragment 11, which contains the putative site of replication, into A. tumefaciens resulted in the elimination of the virulence plasmid. This incompatibility with the Ti-plasmid is in agreement with the assignment of the site of replication to HpaI fragment 11.

C. Virulence Region

The region of the Ti-plasmid from ca 9:00 to 11:30 is very important for the virulence of the Ti-plasmid. It has been found in a number of laboratories that transposon insertion into this region abolishes virulence (11,12,13, 14). Many, but not all, of these avirulent mutants can be complemented by the presence of a compatible plasmid bearing the wild-type fragment of the Ti-plasmid (15). This trans-complementation suggests that the formation of a definite transcription or translation product is required for virulence.

This region of the Ti-plasmid is the most highly conserved of all regions examined. A series of unusual strains of A. tumefaciens of widely separated geographical origin have been examined for DNA homology. These studies showed that a number of Ti-plasmids showed little DNA homology except for this 9 to 11:30 o'clock region, and with some strains this region only showed homology when examined using conditions that allowed up to 30% base mismatching (16; V. Knauf, unpublished observations). The virulence plasmid found in A. rhizogenes has virtually no homology with the Ti-plasmid of A. tumefaciens except for this region (17). There is currently no good experimental evidence concerning the possible functions of the information encoded by this region of the Ti-plasmid. Two possibilities or combinations of possibilities come to mind. One, the region may code for a system that is responsible for the transfer of the Ti-plasmid or at least portions of it into the plant cell. A second possibility is that this region of the plasmid encodes an "insertion system" that inserts the T-DNA (transferred DNA) portion of the plasmid into the plant genome.

The possibility of a definite transfer mechanism can
be tested by techniques which obviate the need for the use
of intact bacterial cells. A recently developed technique
whereby bacterial spheroplasts are fused with plant protoplasts to give a relatively high percentage of transformation (18) appears to be an ideal procedure to test the
hypothesis, and collaborative studies are in progress.
The possibility that the host and/or the Ti-plasmid codes
for an insertion system can also be examined experimentally.

D. Tumor DNA Region

The region of the plasmid from about 12:00 to 1:30
o'clock as indicated in Figure 1 is the T-DNA or transferred DNA. This region of the Ti-plasmid is transferred
to the host cell during the crown gall transformation.
The T-DNA region shows only a low level of transcriptional
activity when present in the Ti-plasmid within cells of A.
tumefaciens (9). There is no known function for this
region in the bacterial cell, although this region may
be responsible for the higher levels of cytokinins found
in the filtrates of nopaline Ti-plasmid containing strains
of A. tumefaciens (19).

E. Lack of Homology of Virulence Plasmids

The most striking fact which has emerged from studies
of a number of virulence plasmids from A. tumefaciens and
A. rhizogenes is that there are no single regions of the
plasmids that are closely homologous. The known functions
that appear to be universal are the elicitation of uncontrolled growth in the host plant and opine metabolism.

IV. FUNCTIONS OF THE T-DNA IN PLANTA

A. General Considerations

Plant cells which have been transformed are distin-
tuished from their normal counterparts by three important
criteria: They escape from the normal morphological
constraints. They grow in vitro in the absence of phytohormones, and they produce opines. It appears certain
that the information for opine synthesis is carried by the

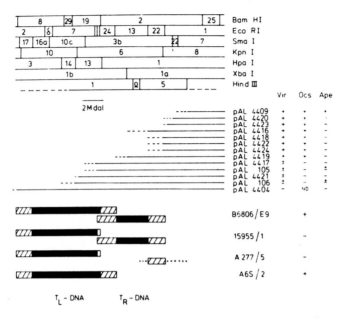

FIGURE 2. Map positions of deletions extending into T-region DNA. Deleted DNA is indicated by a solid line, while the extremities are indicated by a dotted line. This diagram is taken from Ooms (10).

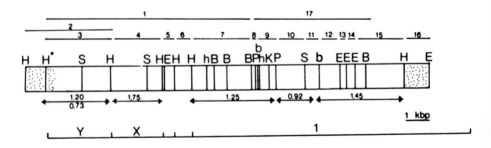

FIGURE 3. Polyadenylated mRNA transcripts of T-DNA in A_6 tumor. The bottom line indicates HindIII fragments.

T-DNA (20). However, the basis for the rapid uncontrolled growth in planta and the growth in vitro in the absence of hormones is not known. The T-DNA may code for the synthesis of phytohormones. Alternatively, more subtle phenomena may be occurring such as derepression of host hormone synthesis, activation of host genes by promoter insertion, or repression of phytohormone catabolism. It should be noted that the T-DNA is an unusual DNA in that it is metabolically quiescent in the bacterial cell but is active in the plant cell. The possible evolutionary mechanism which can give rise to a procaryotic gene that is only active in a eucaryotic cell should be fascinating. There are no obvious sources of opine synthesis genes in plants.

B. Transcriptional Activity

Gelvin et al. (21) have analyzed the polyadenylated transcripts derived from the T-DNA contained in a number of separate tumor lines. The results obtained with the tumor line A6/S2 are depicted in Figure 3. Essentially identical results were found with all tumor lines examined. These mRNA molecules account for about 50% of the T-DNA. The pattern obtained suggests that the transcripts are initiated entirely within the T-DNA. The finding of an identical pattern of mRNA transcripts from the T-DNA integrated into different sites in independently initiated tumors on different tobacco cultivars and different plant species suggests that the transcription of the T-DNA is independent of the flanking host DNA sequences. Similar results have been reported by Willmitzer et al. (see note in proof in 22).

C. Translational Activity of T-DNA Transcripts

At the present time, the only definite products coded for by the T-DNA are the enzymes which catalyze the synthesis of opines. The mRNA coding for the synthesis of octopine synthetase (also known as lysopine dehydrogenase) was isolated from an octopine producing tumor by hybridization to a specific fragment from the T-DNA. In vitro translation of the mRNA with a wheat germ system gave rise to a protein of the same size as octopine synthetase, Mr 39,000, which was precipitated by antiserum specific for purified octopine synthetase. No such material was produced by

extracts from normal cells (20). Similar studies were
undertaken by McPherson et al. (23).

D. Production of Tumor Morphology Mutants by
Alterations of the T-DNA

The T-DNA region in the plant cell is necessary for the
formation of tumors. As indicated above, deletion mutants
which lack large portions of the right end of the T-DNA
are non-virulent. Deletions of the left end of Hind fragment
Y do not destroy virulence (L. W. Ream and F. White, unpublished). A number of laboratories including those of
Schell and Van Montagu, Schilperoort, and our own have
utilized the technique of transposon insertion to generate
mutants in the T-DNA (10,12,14,24,25).

The technique used in our laboratory is illustrated in
Figure 4. This procedure results in Ti-plasmids in which
only the T-DNA region has been mutated by transposons.
A map showing the locations of 75 Tn5 and 3 Tn3 insertions
is shown in Figure 5. In no case was a mutant containing
a single Tn5 or Tn3 insertion in the T-DNA avirulent, and
in many locations the insertion of a transposon had no
effect. This suggests that there is no region in the T-DNA
which is required for the transfer and integration of T-DNA
into the plant. Three loci controlling tumor morphology
could be clearly distinguished. The first group of
mutants contained Tn5 insertions within a 1.25 kb region
and produced tumors that were significantly larger than
normal (tml, see Figure 6). The second group of 9 mutants
incited tumors with massive outcropping of roots on
Kalanchoë stems and contained the Tn5 insertions within a
1 kb cluster (tmr, Figure 7). The third group of nine
mutants incited tumors with shoots growing from the tumor
callus and contained 3 Tn3 and 6 Tn5 insertions over a 3.1 kb
region (tms, Figure 8). These morphological changes may be
a reflection of the altered ratios of cytokinins to auxins
in the tumors as it is known that auxins promote root growth
while cytokinins promote shoot growth (10,26).

In collaboration with the laboratories of Dr. Roy O.
Morris and Dr. T. Kosuge, we have undertaken the analyses
of the trans-zeatinriboside content of these various types
of tumors. The trans-zeatinriboside contents in nanograms/gm
of tissue are as follows: normal Xanthi tobacco tissue 1,
parental A6 tumor 40, tms 400, tmr 1, and tml 40. The
results seem to indicate that the tmr locus either makes or
modulates the synthesis of the cytokinin. The levels of

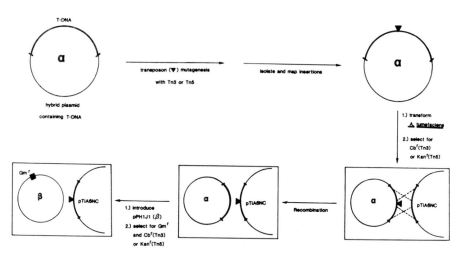

FIGURE 4. Experimental strategy used to induce specific
Tn3 and Tn5 insertions into the T-DNA of pTiA6NC. Plasmids
(α) containing different T-DNA fragments cloned into either
pRK290 or pHK17 were mutagenized by either Tn3 or Tn5. The
mutagenized plasmids were isolated, and the locations of the
transposons were determined by restriction enzyme mapping.
Hybrid plasmids containing insertions distributed throughout
the T-DNA were transformed into A. tumefaciens, which harbors
pTiA6NC. A. tumefaciens strains in which the T-DNA::Tn3 or Tn5
mutations had recombined into pTiA6NC were selected by mating
pPH1J1 (β) into each strain. The resulting transconjugants
were analyzed for the introduction of Tn3 or Tn5 into pTiA6NC
by homologous recombination. T-DNA is shown in heavy lines.
Boxes denote steps of the procedure that occur in A. tumefaciens, while the other steps occur in E. coli.

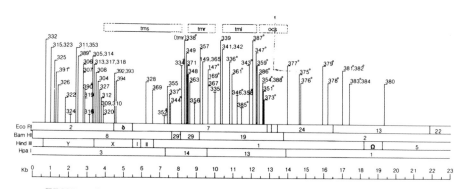

FIGURE 5. Restriction endonuclease map of A₆ T-DNA
showing the location of 75 Tn5 insertions and three Tn3
insertions (392, 293, and 394).

FIGURE 6. A tml tumor on Kalanchoe. The tml tumors are the smooth ones at the top of the leaf. The normal A6 control tumors are the smaller rough ones at the bottom of the photo.

FIGURE 7. A tmr tumor on Kalanchoe.

FIGURE 8. A tms tumor on a Xanthi tobacco stem.

indole acetic acid in these tumors are currently under
investigation as are the copy numbers of various mRNA molecules. Mixed infection with tmr and tms mutants results
in the formation of tumors with relatively normal morphology.
Alternatively, exogenous auxins or kinetins stimulated
normal tumor formation with tms and tmr mutants, respectively. These results again suggest that the basis for
the morphological changes observed are altered phytohormone
levels (10). Work in our laboratory (L. W. Ream, unpublished observations) with double mutants containing transposons in the tmr and tms loci give strains that show only
slight virulence. A tumor, 159550/1 which has at least 10
copies of the T-DNA, has a very high level of trans-
zeatinriboside, 400 ng/gm (27).

All of these results strongly imply a direct or modula-
ting role of the T-DNA in the synthesis of cytokinins in
tumor cells. These speculations should be tempered by the
fact that a number of virulence plasmids have recently been
discovered that have little homology with the above plasmids
whose T-DNA has been extensively studied (see Sections IIIC
and IIIE). The functions of T-DNA may be similar even
though the DNA of the plasmids are not homologous.

V.   ORGANIZATION OF T-DNA

The sites of integration of the T-DNA and the extent of
the T-DNA have been determined by restriction endonuclease
mapping and DNA sequencing. Some generalities concerning
tumors incited by octopine and nopaline Ti-plasmids are
possible. It should be borne in mind that these concepts
may change as further examples are studied: (a) The
variable nature of junction fragments generated upon the
digestion of a number of T-DNA's indicates that there are
no unique sites of integration into the plant DNA. (b) A
given Ti-plasmid does not always give rise to the same
integration pattern. (c) The T-DNA contains a segment that
is colinear with the Ti-plasmid. This so-called "core-DNA"
(28) has a high degree of homology in many octopine and
nopaline plasmids. (d) Certain regions of the T-DNA are
preferred as sites of attachment; however, in the case of
octopine tumors, the ends are not absolutely fixed and the
copy number of different segments of the T-DNA can vary.
(e) The size of the T-DNA may vary as illustrated by the
insertion of sequences containing transposons. The
increase in the nopaline type T-DNA from 21 to 34 kb by

Tn7 insertion illustrates this point (29). These results indicate that T-DNA–containing foreign genes will still be transferred to plant cells.

The organization of a number of octopine tumors was recently examined by Thomashow et al., (30). The results are shown in Figure 9. Each of the tumors contains one copy of the left end of the T-DNA. The left ends of the tumors all fall within the same region of the Ti-plasmid; however, the right ends are quite variable. Tumor 15955/1 does not contain the complete region coding for octopine synthetase. A very interesting feature of B6 806/E9 and 15955/1 is the presence of the so-called satellite fragment of T-DNA which is present in multiple copies. This fragment is not necessary for the transformation of the host cell, and may be inserted as an independent event. The use of this fragment as a vehicle for the insertion of foreign genes into plants is attractive since we do not expect the insertion of the satellite DNA to transform the host cell. Multiple copies of the left end of the T-DNA have also been seen. The tumor 159550/1 contains 10 or more copies of the left T-DNA region (W. Kwok, unpublished).

The tumor which appears to have the simplest organization of T-DNA is A6 S/2, and we have undertaken the detailed analysis of this tumor. As of now, we have sequenced about 1100 bases of the A6 tumor left junction region and the corresponding Ti-plasmid DNA. The left hand junction contains a large inverted repeat, 228 bp long separated by a 300 bp AT rich region. The left hand region of the T-DNA is characterized by a large number of direct and inverted repeats (Figure 10), however, portions of the T-DNA that are colinear with the plasmid are identical to the base. This scrambling of plasmid DNA sequences at the host/T-DNA junction is very similar to that observed at the junction of adenovirus 12 and cell DNA (31). The left terminus of the T-DNA is scrambled Ti sequences and at position 1 the sequences are a direct copy of the Ti-plasmid sequences.

The organization of the T-DNA of nopaline type tumors has been extensively studied (6,32). There are two head to tail tandem repeats in the T-DNA separated by what appear to be scrambled sequences of either plasmid of plant origin. The right end of nopaline type T-DNA is exactly determined, and it appears that the T-DNA of various nopaline tumors is more uniform. A 23 base pair sequence is found in the Ti-plasmid near the terminal regions of nopaline T-DNA (Yadav, N., Barnes, W., and Chilton, M.-D., personal communication). A similar sequence was found in the A6 T-DNA

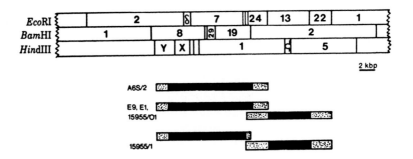

FIGURE 9.   T–DNA maintained in various octopine tumor
lines.   The crosshatched areas indicate extremes of the
T–DNA.

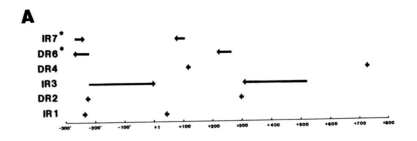

FIGURE 10.   Repeated sequences near the left end of
A6S/2 T–DNA.   The arrows indicate the extent, location and
relative orientation of repeated sequences.

```
A6 Left      GGAATGGCGAAATCAAGGCAGGATATATTCAATTGTAAATGGCTTCATGTCCGGGA

T37 Left     CTGTTGGCTGGCTGGTGGCAGGATATATTGTGGTGTAAACAAATTGACGCTTAGAC

T37 Right    TAAACTATCAGTGTTTGACAGGATATATTGGCGGGTAAACCTAAGAGAAAAGAGCG


Consensus    ---------------G-CAGGATATATT-----GTAAA-------------G--
```

FIGURE 11. Termini sequences from A6 and T37 Ti plasmids. The top line represents the DNA sequence near the left end of the A6 T-region. The vertical arrow indicates the point of divergence between this Ti plasmid sequence and the A6S/2 tumor T-DNA sequence. The middle and bottom lines are sequences near the left and right ends, respectively, of the T37 T-region (Yadav, N., Barnes, W., and Chilton, M.-D., personal communication). The arrows indicate the points of divergence between the Ti and the T-DNA sequences from BT37 tumors.

(Figure 11) at ca position 1 which is at the junction of
the scrambled T-DNA sequences and the sequences which are
direct copies of the Ti-plasmid. This 23 base sequence
may possibly play a role in the insertion and excision of
the T-DNA.

At this juncture, several laboratories are sequencing
the T-DNA region of a number of Ti-plasmids. We can confidently expect that the complete sequences of both octopine
and nopaline type T-DNA will be determined within the next
year. These results should give us a better idea of the
mechanisms of insertion and excision of the T-DNA. A very
useful outcome of the sequencing determinations will be
the availability of clones containing promoter sequences
for possible use in genetic engineering of plants. The
location and amino acid composition of proteins may be
inferred from regions of open reading frames.

VI. GENERATION OF PLANTS WITH FOREIGN DNA

A. In Planta

A prime requisite in any program of agronomically
significant genetic engineering is the ability to form
intact plants. If possible, the newly introduced genetic
traits should be seed transmitted; however, this trait is
not absolutely necessary since many important plants may be
vegetatively propagated. Up to now the techniques used
in these investigations have involved most of the following
steps: A tumor is incited in planta and the tumorous growth
is axenically propagated in vitro in the absence of
hormones. A phytohormone-independent clone of the tumor
is then checked for the presence of T-DNA, and attempts are
made to generate plants from this cloned tumor.

The first reversal of tumor formation was obtained by
A. Braun (33). A cloned nopaline teratoma was treated with
a cytokinin to induce shoot formation, and shoots were
grafted onto a normal plant. Some of the grafted shoots
showed relatively normal morphology even though they retained tumorous traits such as nopaline synthesis and the
ability to grow in vitro in the absence of phytohormones.
These shoots also retained T-DNA (34). The T-DNA, as well
as tumorous traits were lost upon meiosis.

In a similar type of experiment the same nopaline tumor
was treated with cytokinins and the resulting shoots were
induced to root by transfer to a medium which contained no

phytohormones. The resulting plants were fertile. Both the
original reverted plants and the progeny obtained by selfing
contained identical small end fragments of the T-DNA
sequences that were present in the parental tumor (35).
Thus, this procedure either results in the selection of
cells that have lost a portion of the nopaline T-DNA or
this loss was incited by the treatment. The frequency by
which plants can be regenerated from teratoma tumors by
this treatment is under investigation; however, the treatment does not succeed with all tumors. The plants with
T-DNA may be superinfected with octopine or nopaline strains
of A. tumefaciens. The resulting tumors contain normal
T-DNA sequences (A. Powell, B. Taylor, unpublished).

Otten et al. (8) have utilized a "shooter" mutant of
an octopine plasmid containing the bacterial transposon,
Tn7, to obtain a plant that synthesized octopine. The T-DNA
of the resulting plant was most unusual in that the T-DNA
had undergone an internal deletion which eliminated Tn7 and
adjacent T-DNA sequences. The ability to synthesize octopine dehydrogenase was transmitted through pollen and eggs
as a single dominant factor with Mendelian segregation
ratios. The frequency with which this very interesting
event occurs has not been reported. Similar results have
been obtained with potatoes.

A procedure that promises to be generally applicable
has been reported by Wullems and coworkers at Leiden (36).
Shoots were obtained by the fusion of normal protoplasts
of a streptomycin-resistant tobacco with protoplasts of
octopine or nopaline tumors. When grafted onto normal
tobacco plants, the shoots flowered and set seed. In nine
cases, the Fl plants retained the ability to synthesize
nopaline. The results of further applications of this
interesting technique are awaited.

A. rhizogenes has been reported to yield tumorous
tissues from which plants can be readily regenerated (37,38).

B. Transformation of Single Cells

There are now available a number of procedures which
transform single plant cells: (a) The introduction of the
Ti-plasmid encapsulated in liposomes into protoplasts (39)
which occurs at a frequency of less than 10^{-6}. (b) The
fusion of bacterial spheroplasts with plant cell protoplasts,
frequencies of ca. 10^{-3} (18). (c) The infection of 72 hr-
old protoplasts with partially regenerated cell walls with
intact cells of A. tumefaciens, frequencies of 10^{-2} (40).

The latter of these techniques has been used to generate shoots that express opine synthesis (36). The interest in these single cell techniques is that they promise to obviate the need to form tumors. It appears that with suitable selection techniques, clones of plant cells containing foreign DNA can be obtained by utilizing bacteria containing nonvirulent Ti-plasmids bearing suitable foreign genes with active promoters.

C. Transformation Techniques Which Do Not
     Utilize the Ti-Plasmid

We do not know of any fundamental differences between animal and plant cells which indicate that plant cells or plant protoplasts should not be amenable to transformation by many of the techniques which have been so brilliantly successful with mammalian cells. A number of laboratories are currently attempting to apply these techniques to plant cells.

VII. THE VISTA AS SEEN IN JANUARY, 1982

Although in the opinion of this reviewer, the introduction and expression of foreign genes other than opines has yet to be demonstrated, we can confidently expect this to be achieved in the near future. It is probable that the first practical uses of the transformation of plant cells will involve traits which will confer desirable traits when expressed in every cell. The introduction of genes whose expression can be satisfactorily controlled represents a great challenge. We must also remember that extensive field testing is required of newly "engineered" plants. A petri dish is not a field and a laboratory is not a kitchen! There are many pitfalls. Engineered foods may not taste good or bees may not pollinate engineered flowers. Nevertheless, I think there is reason to be optimistic. The progress in understanding the crown gall transformation of plants over the past decade has been breathtaking even to those of us involved in the day-to-day vexations of working in this area.

ACKNOWLEDGMENTS

    We wish to thank our colleagues for permission to quote
unpublished material. The support of the National Cancer
Institute, the American Cancer Society, the National
Science Foundation, and Standard Oil Company (Indiana)
is gratefully acknowledged.

REFERENCES

1. Braun, A. C., Cancer Res. 16, 53-56 (1956).
2. Zaenen, I., Van Larabeke, N., Teuchy, H., Van Montagu,
    M., and Schell, J., J. Mol. Biol. 86, 109-27 (1974).
3. Watson, B., Currier, T. C., Gordon, M. P., Chilton,
    M.-D., and Nester, E. W., J. Bacteriol. 123, 255-64
    (1975).
4. Chilton, M.-D., Drummond, M. H., Merlo, D. J., Sciaky,
    D., Montoya, A. L., Gordon, M. P., and Nester, E. W.,
    Cell 11, 263-71 (1977).
5. Chilton, M.-D., Saiki, R., Yadav, N., Gordon, M. P.,
    and Quertier, F., Proc. Natl. Acad. Sci. U.S.A. 77,
    4060-64 (1980).
6. Lemmers, M., Debeuckeleer, M., Holsters, M., Zambryski,
    P., Depicker, A., Hernalsteens, J. P., Van Montagu, M.,
    and Schell, J., J. Mol. Biol. 144, 355-78 (1980).
7. Thomashow, M. F., Nutter, R. C., Postle, K., Chilton,
    M.-D., Blattner, F. R., Powell, A., Gordon, M. P.,
    and Nester, E. W., Proc. Natl. Acad. Sci. U.S.A. 77,
    6448-52 (1980).
8. Otten, L., DeGreve, H., Hernalsteens, J. P., Van
    Montagu, M., Schieder, O., Straub, J., and Schell, J.,
    Mol. Gen. Genet. 183, 209-13 (1981).
9. Gelvin, S. B., Gordon, M. P., Nester, E. W., and
    Aronson, A. I., Plasmid 6, 17-29 (1981).
10. Ooms, G. Ph.D. Thesis, University of Leiden, The
    Netherlands (1981).
11. Ooms, G., Klapwijk, P. M., Poulis, J. A., and
    Schilperoort, R. A., J. Bacteriol. 144, 82-91 (1980).
12. De Greve, H., Decraemer, H., Seurinck, J., Van Montagu,
    M., and Schell, J., Plasmid 6, 235-48 (1981).
13. Garfinkel, D. J., and Nester, E. W., J. Bacteriol.
    144, 732-43 (1980).

14. Garfinkel, D. Ph.D. Thesis, University of Washington, Seattle, Washington (1981).

15. Klee, H., Gordon, M. P., and Nester, E. W., J. Bacteriol., in press (1982).

16. Thomashow, M. F., Knauf, V. C., and Nester, E. W., J. Bacteriol. 146, 484-93 (1981).

17. White, F. Ph.D. Thesis, University of Washington, Seattle, Washington (1981).

18. Hasezawa, S., Nagata, T., and Syono, K., Mol. Gen. Genet. 182, 206-10 (1981).

19. Regier, D., and Morris, R. O., Biochem. Biophys. Res. Comm., in press (1981).

20. Schröder, J., Schröder, G., Huisman, H., Schilperoort, R. A., and Schell, J., FEBS Letters 129, 166-68 (1981).

21. Gelvin, S. B., Thomashow, M. F., McPherson, J. C., Gordon, M. P., and Nester, E. W., Proc. Natl. Acad. Sci. U.S.A., in press (1982).

22. Willmitzer, L., Otten, L., Simons, G., Schmalenbach, W., Schröder, J., Schröder, G., Van Montagu, M., de Vos, G., and Schell, J. Mol. Gen. Genet., 262, 255-62 (1981).

23. McPherson, J. C., Nester, E. W., and Gordon, M. P., Proc. Natl. Acad. Sci. U.S.A. 77, 2666-70 (1980).

24. Ooms, G., Hooykaas, P. J. J., Moolenaar, G., and Schilperoort, R. A., Gene 14, 33-50 (1981).

25. Garfinkel, D. J., Simpson, R. B., Ream, L. W., White, F. F., Gordon, M. P., and Nester, E. W., Cell 27, 143-53 (1981).

26. Skoog, F., and Miller, C. O., Symp. Soc. Exp. Biol. 11, 118-31 (1957).

27. Akiyoshi, D. E., McDonald, E., Morris, R. O., Kwok, W., Nester, E. W., and Gordon, M. P., in preparation.

28. Depicker, A., Van Montagu, M., and Schell, J., Nature 275, 150-53 (1978).

29. Hernalsteens, J. P., Van Vliet, F., Genetello, C., DeBlock, M., Dhaese, P., Villarroel, R., Van Montagu, M., and Schell, J., Nature 287, 654-56 (1980).

30. Thomashow, M. F., Nutter, R. C., Montoya, A. L., Gordon, M. P., and Nester, E. W., Cell 19, 729-39 (1980).

31. Deuring, R., Winterhoff, V., Tamanoi, F., Stabel, S., and Doerfler, W., Nature 293, 81-84 (1981).

32. Zambryski, P., Holsters, M., Kruger, K., Depicker, A., Schell, J., Van Montagu, M., and Goodman, H., Science 209, 1385-91 (1980).

33. Braun, A. C., and Wood, H., Proc. Natl. Acad. Sci. U.S.A. 73, 496-500 (1976).
34. Yang, F.-M., Montoya, A. L., Nester, E. W., and Gordon, M. P., In Vitro 16, 87-92 (1980).
35. Yang, F. M., and Simpson, R. B., Proc. Natl. Acad. Sci. U.S.A. 78, 4151-55 (1981).
36. Wullems, G. J., Molendijk, L., Ooms, G., and Schilperoort, R. A., Cell 24, 719-27 (1981).
37. Tempe, J., Chilton, M.-D., and Tepfer, D., in press (1982).
38. White, F., Ghidossi, G., Gordon, M. P., and Nester, E. W., in preparation (1982).
39. Dellaporta, S. Ph.D. Thesis. Worchester, Massachusetts (1981).
40. Marton, L., Wullems, G. J., Molendijk, L., and Schilperoort, R. A., Nature 277, 129-31 (1979).
41. Nester, E. W., and Kosuge, T., Ann. Rev. Microbiol. 35, 531-65 (1981).

DISCUSSION

J.R. GEIGER: If the T-DNA is acting like a transposon, is anyone looking for a simpler system for the insertion of foreign DNA onto plant chromosomes?

M.P. GORDON: Many features of the T DNA indicate that it is not acting like a transposon. Many laboratories are looking for simpler systems to insert foreign DNA into plants. These approaches range from attempts to decrease the size and complexity of the Ti-plasmid, injection of DNA into plant cell nucleii, and direct uptake of foreign DNA by protoplasts.

K. SIROTKIN: If I recall correctly that the DNA responsible for tumorigenesis (9:30-11:00 o'clock region) is not found in the plant tumor, then is T-DNA responsible for tumor maintenance?

M.P. GORDON: The 9:30-11:00 o'clock region of the tumor inciting plasmid is necessary for virulence. The T-DNA region is always found in tumors, hence, we can infer that this region is necessary for tumor maintenance.

MONOCLONAL ANTIBODIES—
PRODUCTION AND USES

MONOCLONAL ANTIBODIES: THE PRODUCTION OF TAILOR-MADE SEROLOGICAL REAGENTS

Dale E. Yelton
Pallaiah Thammana
Catherine Desaymard
Susan B. Roberts
Sau-Ping Kwan
Angela Giusti
Donald J. Zack
Roberta R. Pollock
Matthew D. Scharff

Department of Cell Biology
Albert Einstein College of Medicine
Bronx, New York
U.S.A.

I. INTRODUCTION

Immunoassays have been used successfully for years to detect, quantitate, and localize small amounts of macromolecules in complex biological mixtures. While such assays can be made both sensitive and specific, certain properties of conventional antiserums have limited their usefulness, especially for routine diagnosis and therapy. Perhaps the most important of these limitations has been the limited supply of useful antiserum against weak immunogens. This is in part due to the heterogeneity of the immune response which

Supported by grants from the NIH (AI5231) (AI 10702), NSF (PCM8108642) and ACS (NP-317). DEY and DJZ are Medical Scientist Trainees supported by NIGMS grant 5T32GM7288. CD is a Chargee de Recherche (Inserm) on leave and supported in part by a Fogarty International Fellowship. RRP is supported by a postdoctoral fellowship (DRG-502F) from the Damon Runyon-Walter Winchell Cancer Fund.

FROM GENE TO PROTEIN:
TRANSLATION INTO BIOTECHNOLOGY

129

results in each antiserum being a mixture of antibodies with varying affinity, cross reactivities, and effector functions. The particular mix of antibodies or predominance of a subset of antibodies produced by an animal at a certain time in its immune response may be useful for a specified purpose. However, as that mix changes with time or from animal to animal, the nature of the antiserum also changes and it may be impossible to recapture the specificity or other properties again. A second major problem with conventional immunization is that many of the most biologically interesting molecules, such as tumor or differentiation antigens, are small parts of complex mixtures. While specific antiserums can sometimes be generated by extensive and repeated absorption, the heterogeneity and unpredictability of the immune response makes it difficult to repeatedly generate large amounts of antiserum with the same properties. These and other problems with conventional immunization have discouraged and hindered the production of antiserum against many important and useful antigens and have certainly limited the use of immunoassays in the routine diagnostic laboratory to those antigens which can be obtained in relatively pure form and which induce a good immune response.

With the description by Köhler and Milstein of the hybridoma technology for making monoclonal antibodies (1), many of the problems with conventional immunization can now be overcome and the potential of antibodies as reagents in the research and diagnostic laboratory and perhaps as therapeutic agents has been greatly expanded. The hybridoma technique has been described in many places and is summarized in Figure 1. Cultured myeloma cells are fused to spleen cells from an immunized donor. The parental myeloma cells are killed by growing the hybrids in selective (HAT) media and the normal spleen cells die spontaneously in culture. Some of the spleen x myeloma hybrids which survive in culture are making large amounts of an antibody that reacts with the immunizing antigen and was originally being produced by the spleen cell. These hybrids maintain many of the properties of the myeloma parent. They will grow continuously in culture, can be frozen and recovered, and form tumors when injected into animals. Since the hybrid arose through the fusion of a myeloma cell with a single antibody forming spleen cell, it produces a homogeneous antibody with a single amino acid sequence. When the hybrid forms a tumor in a recipient animal, that animal accumulates as much as 10mg/ml of that particular antibody in its serum or ascites fluid.

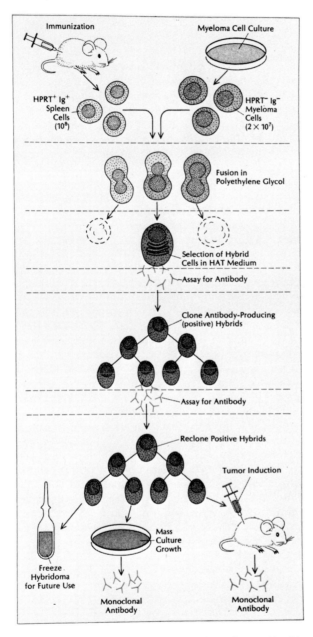

FIGURE 1. The production of monoclonal antibodies (reprinted with permission of Hospital Practice).

II. BENEFITS OF MONOCLONAL ANTIBODIES

The benefits of monoclonal antibodies are enormous. They are a chemically defined reagent. Once a hybridoma producing a desired antibody has been identified, hundreds of milligrams of that antibody can be generated. When the stock of antibody is used up, the cells can be recovered from the freezer and the same exact antibody produced again in large amounts. The amounts of antibody which can be generated and the ability to renew the antibody whenever it is needed make it worthwhile to fully characterize its specificity, affinity and effector functions. A particularly useful hybridoma can be shared with other laboratories or the monoclonal antibody can be produced and distributed commercially providing the opportunity for all investigators carrying out a particular type of study to use the exact same reagent. Because of these benefits, it has become worthwhile to try to generate monoclonal antibodies against antigens such as certain tumor and differentiation antigens whose existence has been suspected but in many cases not proven. The identification of such monoclonal antibodies is greatly facilitated by the nature of the technique which essentially clones the individual antibodies which make up the heterogeneous repertoire of antibodies in conventional antiserum. It is thus possible to immunize with an impure antigen such as a whole cell and screen for a monoclonal antibody that reacts with one of the many determinants on the cell.

These benefits have stimulated investigators in all areas of biology to generate monoclonal antibodies against the particular antigen that they are studying. As the technique was used more widely, it became obvious that even more useful reagents could be generated if one defined the specific assay or task to be accomplished and then picked from amongst a number of monoclonals the one best suited for that purpose. At first this meant merely being more discriminating about the particular hybridoma which one chose to use. Hybridomas which were unstable or produced low amounts of antibody either in tissue culture or ascites were discarded. While at first it was enough to obtain a monoclonal antibody that reacted with a particular antigen, some investigators began to seek hybridomas with exactly the specificity or cross reactivity which was needed. For example, we have chosen to screen for mouse monoclonal antibodies by generating rat monoclonals that react with particular classes or subclasses of mouse immuno-globulins (2). For this purpose we initially used a rat mono-clonal antibody called 187 which binds to mouse κ chains. The monoclonal antibody is biosynthetically labeled by incubating the hybridomas with ^{14}C-amino acids in culture. The medium

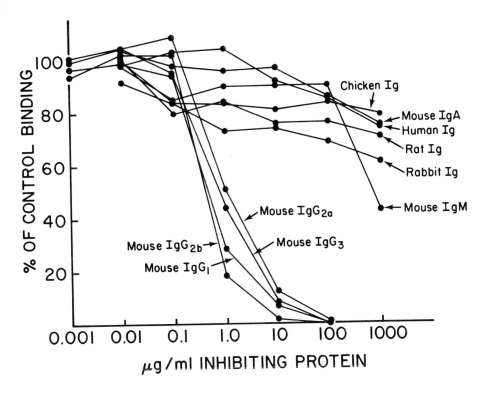

FIGURE 2. Specificity of the 116 rat monoclonal antibody. IgG$_{2b}$ protein was adsorbed to polyvinyl plates. Biosynthetically labeled 116 was incubated in the presence of increasing amounts of the proteins indicated (reprinted from Hybridoma ref 2 with permission).

is then dialyzed to remove unincorporated radioactivity and used without furthur purification (2). The antigen to which we are interested in generating monoclonals is adsorbed to the wells of a polyvinyl microtiter plate. The medium from each hybridoma is incubated in the antigen coated wells and the wells are washed to remove unbound antibody. The radioactive anti κ chain is then added to the wells which after washing are counted in a scintillation counter. We currently use a mixture of two rat monoclonals which react with different sites on mouse κ chains since this improves our specific to non-specific ratio.

Once we have identified hybridomas producing antibodies that react with the immunizing antigen, we often need to determine whether the monoclonal antibody is of the IgM or IgG class. As a first screen, we wanted a monoclonal antibody that reacted equally well with the major subclasses of IgG. One such rat monoclonal is illustrated in figure 2. This antibody reacts with IgG1, IgG$_{2a}$, IgG$_{2b}$, IgG$_3$ and it can be used to simultaneously screen for all of these subclasses. It also illustrates the usefulness of monoclonals that are cross-reactive.

If a number of monoclonals with a particular specificity are available, it is possible to select the one with the optimum affinity or avidity for a particular task. For example, if the monoclonal is to be used for a radioimmunoassay or for targeting a cytotoxic drug to a particular cell type, high affinity antibodies will be selected. On the other hand, if the monoclonal will be used to purify an antigen on an antibody affinity column, a monoclonal with a lower affinity might be used so that antigen can be eluted under mild non-denaturing conditions.

From the very beginning of the application of the hybridoma technology, investigators have selected monoclonal antibodies which could carry out certain effector functions. For example, if the goal is to kill a certain subpopulation of cells with antibody and complement, then monoclonals of the IgM class are selected. On the other hand, if the monoclonal is to be used as a fluorescent tag to select a subpopulation of cells which had to be kept alive, then an IgG$_1$ antibody which will not fix complement is more useful.

III. PRODUCTION OF MUTANT MONOCLONAL ANTIBODIES

Even the most carefully selected hybridoma will not have

TABLE I. Phenotypes of Mouse Myeloma and Hybridoma Variants

	Myeloma	Hybridoma
A. Changes in gene expression		
1. Loss Variants (H^-L^+, H^+L^-, H^-L^-)	+	+
2. Quantitative variants (decreased H and/or L synthesis)	+	+
B. Changes in immunoglobulin structure		
1. Deletions C-terminal	+	+
internal	+	+
2. Class switch	+	+
3. Long Chains	+	
4. Amino acid substitutions	+	
5. Carbohydrate changes	+	+

all of the properties that are desired for a particular task.
With weak immunogens, it is often difficult to generate a
large enough battery of monoclonals to be able to choose
hybridomas with the desired properties. Furthermore there are
some properties that cannot be achieved with any intact immunoglobulins. An alternative tactic for improving the usefulness
of a particular monoclonal is to modify its structure by obtaining somatic mutants with the desired properties. The conceptual and methodological bases for this approach were developed in earlier studies which revealed that immunoglobulin
genes in cultured mouse myeloma cells were extremely unstable
(3). Spontaneous mutations in the expression or structure of
the immunoglobulin genes arise at rates of $10^{-3}-10^{-4}$/cell/
generation and the frequency of such mutants can be increased
to as high as 1% with mutagenesis (4). This somatic instability appears to be limited to the immunoglobulin genes since
mutations in the enzymes hypoxanthine phosphoribosyl transferase, thymidine kinase, and $Na^+K^{\pm}ATPase$ all occur at rates which
are many orders of magnitude lower (4,5). As can be seen from

Table 1, mutants with most of the many possible phenotypes have been isolated from mouse myeloma cells (6). More recently, many similar sorts of mutants have been identified in hybridomas (Table 1) (6,7,8,9,10) and it is likely that mutants of every phenotype have already been isolated and are now being characterized.

If one has obtained a monoclonal antibody with the desired specificity and affinity but the wrong effector functions, it is possible to identify rare subclones of that hybridoma which have rearranged or recombined the variable region of the heavy chain to another constant region. We originally found that class switch variants occurred frequently in an IgG_{2b} producing myeloma (11) and similar events have now been identified by Rajewsky and his colleagues in a number of myeloma lines (12) and in one hybridoma (9) by use of the fluorescence activated cell sorter. We have identified such a class switch in a hybridoma which is making an IgM antibody that binds to the hapten phosphocholine. A subclone of this hybridoma was isolated which was no longer producing IgM but rather was synthesizing and secreting IgG_1 (10). This IgG_1 had the same idiotype and affinity for phosphocholine as the parent. It is not clear whether the class switches which occur in myelomas and hybridomas arise through the same mechanisms used by normal B cells as they differentiate. However, Birshtein and her colleagues have shown that some class switch variants have hybrid chains that contain the C-terminal portion of one class and the N-terminal portion of another (13).

Monoclonal antibodies can also be tailor made by identifying subclones containing deletions in one or another domain. We have generated such deletion variants from an IgG_{2b} hybridoma producing antibody that reacts with the hapten p-azophenylarsonate. Cells were mutagenized with ICR-191 as described for mouse myeloma cells (4). The cells were then cloned in soft agar and overlaid with antibody that was specific for the C_H3 domain of the IgG_{2b} molecule (3). Most of the clones continue to secrete intact antibody and are surrounded by an antigen-antibody precipitate which can be seen with low power magnification. The clones not surrounded by such a precipitate are presumptive variants and are recovered from the agar and grown up. Most of these variants have either lost the ability to synthesize or secrete one or both of their immunoglobulin polypeptide chains or are secreting less antibody than the parent. However, some do produce large amounts of immunoglobulin that continue to bind antigen. These mutants do not react with the anti C_H3 domain antibody which was originally used to overlay the clones and are presumed to have undergone a change in the structure of

TABLE 2. Characterization of Mutants

Rat Monoclonal	MPC-11 Mutants B50-10.1 $(-\frac{2}{3}C_H1)$	M3.11 $(-C_H3)$		Ar 13.4 mutants 16.1	1.2
	% inhibition (RIA)			% binding (RIA)	
116	8	94	C_H1	99	106
180	98	90	C_H2	0	93
196	93	9		1	0
178	99	0	↑	10	0
31	91	0	C_H3	96	0
168	100	0	↓	128	2
			M.W.	39K	39K
			C'	−	+
			Staph A	−	−
			Fc	−	−

their constant region (14).

The location of the change can be identified serologically using monoclonal antibodies that react with sites on the different constant region domains (2,14). This is illustrated for two variants in Table 2. Monoclonals have been generated by immunizing rats with mouse IgG_{2b} (2). These were then screened for their ability to react with the different classes of mouse immunoglobulin and with deletion mutants of the MPC-11 IgG_{2b} mouse myeloma cell line. For example, in Table 2, 116 is not inhibited in its binding to intact IgG_{2b} by the B50-10.1 variant which lacks the C-terminal 2/3 of its C_H1 domain but is inhibited by M3.11 which lacks all of its C_H3 domain. From this it is deduced that 116 recognizes a determinant in or near the C_H1 domain. Monoclonal 180 is inhibited by both deletion variants and is provisionally assigned as reactive with the C_H2 domain. Monoclonals 196, 178, 31 and 168 are not inhibited by M3.11 and are believed to react with the C_H3 domain. These latter four monoclonals do not inhibit each others binding in cross inhibition experiments and therefore react with different determinants of the C_H3 domain.

Having at least in a preliminary way established the specificity of these rat monoclonal antibodies, the antibody produced by each was biosynthetically labelled with ^{14}C-amino acids (2) and examined for its reactivity with the variants derived from the phenylarsonate binding hybridoma Ar13.4. Two such mutants are shown in Table 2. ArM16.1 is missing deter-

minants in both its C_H2 and C_H3 domain but still reacts fully
with two out of the four anti-C_H3 monoclonals. ArM1.2 does
not react with any of the anti-C_H3 monoclonals. The molecular
weight of both of the mutant heavy chains is approximately
39,000 daltons, suggesting that variant 16.1 has undergone an
internal deletion involving both its C_H2 and C_H3 domain while
variant 1.2 has lost all or most of its C_H3 domain. This is
consistent with the defects in effector functions which we
have demonstrated for these two variants (14) (Table 2).
ArM16.1 has lost the ability to lyse sheep red blood cells in
the presence of complement while ArM1.2 is as effective in
complement fixation as the parent. Since at least the binding
of Clq is mediated by the C_H2 domain (17), this is consistent
with 16.1 having changes in that part of the molecule while
1.2 does not. Staph A binding is reportedly mediated by
sites on both C_H2 and C_H3 (18).

These studies and those of others (7,8,9) illustrate that
mutants with changes in the structure of their constant
region can be easily isolated from hybridoma cells and that
such mutants have changes in their effector functions. However these changes may be associated with undesirable as well
as desirable changes. For example, both ArM16.1 and ArM1.2
disappear rapidly from the blood of animals. It would therefore be useful to identify mutants with changes in effector
functions that are associated with small changes in sequence
rather than large deletions. This should be possible using
monoclonal antibodies such as those depicted in Table 2. However, to do this it will be necessary to identify those monoclonals that are reactive with the sites on the constant
region that mediate a particular function. It is possible
that this can be done by determining if a monoclonal inhibits
a particular effector function or if binding the effector site
inhibits the binding of the antibody. This is illustrated in
Table 3 which shows that Staph A inhibits the binding of rat

TABLE 3. Ability of Staph A to Inhibit Binding of Monoclonal

Rat anti-C_H3 Monoclonal	% inhibition of binding
196	11
178	71
31	112
168	86

monoclonal 196 to IgG_{2b}. There are two obvious problems with this approach. First, since Staph A is a large protein, it might have inhibited the binding of all anti-C_H3 monoclonals. However, this is clearly not true since the other three anti-C_H3 monoclonals all bind well (Table 3). Secondly, a loss of binding could be due to conformational changes at some distance from the critical amino acid sequences. This is quite possible but can be tested by isolating mutants that do not react with 196 and examining their structure and function.

IV. DISCUSSION AND CONCLUSION

Mutants such as those described above could provide improved reagents for many immunological assays. For example, fluorescence assays on mixtures of cells are often complicated by the binding of antibodies to Fc receptors on phagocytic or other cells. The usual solution to this problem is to treat antibodies with papain so as to remove the Fc region and thus eliminate non-specific binding. An alternative is to isolate variants such as those described in Table 2 which also do not bind to Fc receptors and provide a permanent reagent that can be used in fluorescent assays. Such mutant antibodies might carry a cytotoxic agent to a target cell without binding to and killing normal cells that bear Fc receptors.

We have not yet isolated any somatic variants of hybridomas which have changes in affinity or specificity. This could be because we have been examining IgM producing hybridomas and such cell lines may have fewer somatic mutations than IgG producing hybridomas (19). Antigen binding mutants of an IgA producing myeloma have been found to arise spontaneously at a high frequency and to be associated with single amino acid substitutions in the heavy chain variable regions (20,21,22). We expect that further studies will also reveal such mutants in IgG and IgA producing hybridomas.

In conclusion, monoclonal antibodies have many benefits over conventional antisera. In addition to providing homogenous reagents which can be generated in large amounts and replenished whenever needed, particular monoclonals can be selected which will be superior reagents for certain tasks. Useful monoclonals can be made even more effective by isolating subclones which have undergone mutations or rearrangements.

REFERENCES

1. Köhler, G., and Milstein, C., Nature 256, 495 (1975).
2. Yelton, D. E., Desaymard, C., and Scharff, M. D., Hybridoma 1, 5 (1981).
3. Coffino, P., Baumal, R., Laskov, R., and Scharff, M. D., J. Cell Physiol. 79, 429 (1972).
4. Baumal, R., Birshtein, B. K., Coffino, P., and Scharff, M. D., Science 182, 164 (1973).
5. Margulies, D. H., Kuehl, W. M., and Scharff, M. D., Cell 8, 405 (1976).
6. Morrison, S. L., and Scharff, M. D., CRC Critcial Rev. in Immunol. 3, 1 (1981).
7. Köhler, G., and Shulman, M. J., Eur. J. Immunol. 10, 467 (1980).
8. Yelton, D. E., Cook, W. D., and Scharff, M. D., Transpl. Proc. 12, 439 (1980).
9. Neuberger, M. S., and Rajewsky, K., Proc. Natl. Acad. Sci. (USA) 78, 1138 (1981).
10. Thammana, P., and Scharff, M. D., in preparation.
11. Preud'homme, J.-L., Birshtein, B. K., and Scharff, M. D., Proc. Natl. Acad. Sci. (USA) 72, 1427 (1975).
12. Radbruch, A., Liesegang, B., and Rajewsky, K., Proc. Natl. Acad. Sci. (USA) 77, 2909 (1980).
13. Birshtein, B., Campbell, R., and Greenberg, M. L., Biochemistry 19, 1730 (1980).
14. Yelton, D. E., and Scharff, M. D., in preparation.
15. Monk, R. J., Morrison, S. L., and Milcarek, C., Biochemistry 20, 2330 (1981).
16. Kenter, A. L., and Birshtein, B. K., Science 206, 1307 (1979).
17. Winkelhake, J. L., Immunochemistry 15, 695 (1978).
18. Lancet , D., Isenman, D., Sjödahl, J., Sjöquist, J., Pecht, I., Biochem. Biophys. Res. Comm. 85, 608 (1978).
19. Gearhart, P. J., Johnson, N. D., Douglas, R., and Hood, L., Nature 291, 29 (1981).
20. Cook, W. D., and Scharff, M. D., Proc. Natl. Acad. Sci. (USA) 74, 5687 (1977).
21. Cook, W. D., Rudikoff, S., Giusti, A., and Scharff, M. D., Proc. Natl. Acad. Sci. (USA), in press.
22. Rudikoff, S., Giusti, A., Cook, W. D., and Scharff, M. D., Proc. Natl. Acad. Sci. (USA), in press.

DISCUSSION

P.A. LIBERTI: You indicated that you had used mutagens to
increase the frequency OF hybrid clone mutation.   Have you
used fetal thymus (extracts) in an attempt to (1) increase
mutation frequency and (2) test Jerne's ideas on the role of
thymus in generation diversity via somatin mutation?

M.D. SCHARFF: Donald Zack is trying to do such experiments.
We have one suggestive result but much more must be done.   It
is worth noting that Lynch and his colleagues and Abbas have
shown that T cells can regulate antibody production by
myelomas both <u>in vivo</u> and in tissue culture.

*N.L. LEVY: I was confused by the slide that you showed on the
deletion mutants where reactions with your monoclonal reagents
indicated that in one variant hybridoma the C_H2 region and
part of the C_H3 region were missing and IN the other only the
C_H3 region was missing.   How do you explain that both variants
have a mol. wt. of 39,000 daltons?

*M.D. SCHARFF: The one lacking the C_H2 region has an internal
deletion equivalent to a whole domain.   Apparently part of the
C_H2 region and part of the C_H3 region (the C-terminal end of
C_H2 an the N-terminal of C_H3) are deleted.   The size of this
internal deletion is equivalent to the whole C_H3 domain that
is deleted from the C-terminal of the other hybridoma.

A.A. ANSARI: Theoretically there is the potential for making
low affinity as well as high affinity monoclonal antibodies.
However, most of the monoclonal antibodies whose affinity
constants have been reported, happen to fall in the category
of low affinity antibodies.   I wonder if you would comment on
that.   Also, do you think one stands a better chance of getting
high affinity hybridoma antibodies if spleen cells from hyper-
immunized animals are used rather than from one- or two-time
immunized animals?   My last question is: to your knowledge,
what is the affinity constant reported for a monoclonal
antibody?

M.D. SCHARFF: It is true that most monoclonal antibodies that
people are using have affinities that are in the lower ranges.
However, some monoclonals with high affinities are available.
In fact some of the commercially available ones are extremely
high.   My feeling is that most conventinal antibodies do not
have very high affinities but that in the mixture that makes up
an antiserum, there are some high affinity antibodies which
play a major role in the usefulness of the antiserum and make

us forget the lower affinity antibodies present. In any case, if you generate a large enough battery of monoclonals, for most antigens some will have high affinities. I do not know if the percentage will increase with hyperimmunization although this is certainly possible. At least in some systems, repeated immunization results in increased suppression and increased anti-idiotype which would produce problems.

W.I. SCHAEFFER: What is the rate of back mutation in the mutants which you develop.

M.D. SCHARFF: I do not have much information on that. According to Rajewsky and his colleagues some class switch mutants revert at frequencies of 10^{-5}-10^{-6}. We have not found true revertants of short chain mutants but would only have detected them if they occurred at frequencies of 10^{-3}-10^{-4}. Donald Zack and Wendy Cook in our laboratory have examined the rate at which one low antigen binding mutant with a single amino acid substitution in the S107 heavy chain reverted, and found it to be 2 x 10^{--4}/cell/generation. Another mutant with a different substitution did not revert at all. I think the answer will depend on the nature of the change in each particular mutant.

MONOCLONAL ANTIBODIES IN THE ANALYSIS OF THE MOLECULAR BASIS FOR HUMAN GENETIC DISEASES

Roger H. Kennett[1]
Kendra B. Eager
Barbara Meyer
Virginia Braman
Suzanne Newberry
David W. Buck
Department of Human Genetics
University of Pennsylvania
School of Medicine
Philadelphia, Pennsylvania 19104 U.S.A.

I. INTRODUCTION

When Köhler and Milstein first reported the production of monoclonal antibodies, we were making antisera against cell-surface antigens on human tumors (1,2). Seeing this new technology as a possible way out of the "jungle" of classical serology with its endless cross reactions and limited antisera availability, we soon began to make mono-clonal antibodies against human tumors in an attempt to an-swer the questions: "Do human tumor-specific antigens really exist on neuroblastomas and leukemias?" and "What is the genetic basis for these antigens?"

Since that time, a large number of antibodies against a variety of human tumors have been reported. Since progress in this area has recently been reviewed by us and by others (3,4,5), we will concentrate here on another application of monoclonal antibodies about which there has been little or

---

[1]This research was supported by the Wills Foundation, and by NSF Grant PCM-26757 and NIH Grant CA-14489.

FROM GENE TO PROTEIN:
TRANSLATION INTO BIOTECHNOLOGY

143

no discussion: The use of monoclonal antibodies for analysis of fine structure variation of proteins and other molecules and for detection of polymorphic variation in gene products produced from loci linked to genes responsible for human genetic diseases.

A.   Detection of Variation in Molecular Structures by Conventional Means

Most of the methods used to detect polymorphic variation, such as starch gel or polyacrylamide electrophoresis (PAGE), depend on differences in charge on the molecules or significant differences in size (6,7). The changes, in fact, comprise only a small proportion of the molecular changes that are possible results of genetic variation in a molecule (8). Many of the methods also depend upon the detection of specific enzymatic activity and thus are not applicable to non-enzymatic proteins or variants that have lost enzyme activity.

On the other hand, it has been shown that antibodies can detect even a single amino acid change (9,10). Since such changes produce neither a change in size nor necessarily a change in charge, antibodies are potentially a more sensitive and discriminating means of detecting polymorphic variation in macromolecules. Antisera have been used to define alleles in blood-group systems and in antigens of the major histocompatibility complex. In these cases, the antisera used are usually derived from the same species as the antigen being tested and so usually detect polymorphic differences without troublesome absorptions. In humans, such sera are often in limited supply because of a dependence on such things as chance immunization of a mother with fetal cells as in the case for most anti-HLA sera. In cases where useful human polymorphic markers have been found by xenogeneic immunization, the specific antisera require extensive absorption, are available in relatively small amounts, and sometimes cannot be reproduced after exhaustion of the original supply used to identify the polymorphic variant (11).

B.   Detection of Structural Variation with Monoclonal Antibodies

While there have been a wide variety of monoclonal antibodies produced in the past several years, a limited number of these have been developed with the detection of polymorphic protein variation in mind. Various monoclonal antibodies that detect polymorphic variation in antigens of the major histocompatibility complex (MHC) in mouse, rat and man (12,13,14) have been described. Some of these even detect allelic products which are present as polymorphisms in more than a single species (15,16).

Monoclonal antibodies against the influenza virus hemagglutinin molecule have been used to characterize the various epitopes on the viral particles. Several groups have compared the reactivity of a panel of these antibodies against a panel of viruses and the evidence indicates that there are on the order of 1000 ways in which monoclonal antibodies can react with this protein (17). This implies that there are as many as 1000 different ways of recognizing and discriminating structural variations in this protein which has a molecular weight of approximately 220,000 daltons.

In both the influenza system and the MHC antigens, it has been shown that monoclonal antibodies can be used not only to detect variation but to also select variants in the population that differ from the original antigen by only a single amino acid (18,19).

In the immunoglobulin (Ig) system, monoclonal antibodies against Ig isotypes as well as anti-idiotype antibodies have been described (20,21,22).

Slaughter et al. (23) have shown that antibodies against the human enzyme alkaline phosphatase can distinguish the common allelic forms previously detected by starch gel electrophoresis. Of the first six antibodies developed, four were useful in discriminating between the common allelic forms and two of the others reacted differently with uncommon variant forms of the enzyme.

The results in these systems indicate that monoclonal antibodies can be used to detect structural variation in the protein and glycoprotein gene products in which variation has been previously defined by conventional antisera or electrophoretic methods. In some cases the variation corresponds to that previously detected and in other cases it does not.

C. Monoclonal Antibodies in the Study of Human Genetic Diseases

There are many human genetic diseases described on the basis of symptoms but for which the gene product affected by the mutation producing the symptoms has not been defined, i.e. Huntington's Disease (HD), Neurofibromatosis (NF, von Recklinghausen's Disease), and Cystic Fibrosis (CF).

Identification of these gene products would allow a more detailed analysis of the pathology, provide a means of detecting heterozygotes in recessive diseases, and eventually
possibly even provide a basis for the establishment of a
rational and practical therapy for the disease.

With the fine degree of discrimination available with
monoclonal antibodies in mind, we have begun to approach
the problem of determining the molecular basis for human
genetic diseases in two ways:

    1. Definition of Structural Variations in Proteins Hy-
pothesized to be Responsible for the Disease. In certain
cases, it is possible to define a specific protein, variation in which may cause the symptoms of the disease. In
cases where the protein does not have a detectable enzymatic activity or is not changed in a way that produces a detectable alteration in charge or molecular weight, it may
be particularly difficult to confirm that an alteration
has actually taken place in the molecule.

    a. Alpha-2-macroglobulin. One example of this is the
question of whether or not alpha-2-macroglobulin is altered
in CF. For several years, this has been one of the popular
hypotheses for the molecular mechanism of CF, i.e. an alteration in the structure of alpha-2-macroglobulin results in
a change in its function of inhibiting serum proteases.
This results in some of the proteases producing abnormal
proteolytic digestion products which are in turn seen as
the various "CF factors" postulated to produce the variety
of pathological effects seen in CF (24,25). In spite of
comparisons of the molecular weight, isoelectric focusing,
and protease binding activity of CF and "normal" alpha-2-
macroglobulin, it has still not been shown clearly whether
the two molecules differ significantly in structure and
function.

    Polymorphic variation in alpha-2-macroglobulin has pre-
viously been detected but has apparently not been reproduced after the original antisera became limiting (26). This
emphasizes one of the advantages of monoclonal antibodies -
the continual availability of an antibody of the same specificity.

    In addition to alpha-2-macroglobulin's postulated role
in CF, a second advantage of using alpha-2-macroglobulin as
a model system is that it is a protein in which two distinct
conformational forms can be distinguished. Alpha-2-macro-
globulin exists as a tetramer with a molecular weight of
725,000 daltons. When the protein is run on non-denaturing

PAGE, the form that has not reacted with proteases runs slower than the form that has been interacted with proteases (26). The "fast" form can be generated from the "slow" or native form by treatment with amines (27). Marynen et al. (28) have recently described a monoclonal antibody that reacts with the fast form and not the slower native form. This again supports the idea that these antibodies may be extremely useful in detecting protein conformational changes.

We have made monoclonal antibodies against alpha-2-macroglobulin derived from CF plasma and have compared the reactivity of these antibodies against the antigen purified from CF and from normal plasma. The reaction of the antibodies with the slow and the fast forms of alpha-2 was also tested. Our goal in these experiments is to determine how effective monoclonal antibodies are in detecting allelic differences in this non-enzymatic gene product and to determine whether any of the structural variation in alpha-2 can be related to CF pathology.

b. Fibronectin. We have made antibodies against a second human protein, plasma fibronectin, to determine whether polymorphic variation can be detected with monoclonal antibodies and whether any of the structural variation can be related to the pathology of neurofibromatosis (NF). This is an autosomal dominant genetic disease in which a variety of neural crest derived cells are affected in various ways. These symptoms appear to involve abnormal migration or division of these neural crest derived cells (29).

Based on observations that fibronectin has an effect on the migration and differentiation of neural crest cells in birds (30), we have hypothesized that a structural change in fibronectin may account for the various and variable symptoms in NF. On that basis, we have made antibodies against normal plasma fibronectin and tested the binding to both normal and NF fibronectin molecules. In this case, we must take into consideration that since NF is a dominant genetic disease, even if fibronectin were affected the NF patients would have some normal fibronectin also present. Another possibility more difficult to analyze is that the mutation affects the amount of normal fibronectin produced and that the NF patient has, for example, less of the molecule present in the plasma or on cell surfaces.

2. Detection of Polymorphic Variation in Loci Linked to the Genes Producing Genetic Diseases - "Jumping onto the Genome". A second method for using monoclonal antibodies to detect the genetic loci and corresponding gene products responsible for genetic diseases involves screening for polymorphic variation in any gene products that can be easily obtained from a large number of individuals and asking whether these polymorphic markers co-segregate with the disease phenotype. This is analogous and in fact complementary to the procedure recently described by Botstein et al. (31) in which they propose to screen for linkage, not to polymorphic gene products, but to restriction fragment polymorphisms (RFP's) detectable at the DNA level with cloned probes hybridized to DNA electrophoresed after treatment with restriction enzymes.

Each of the approaches has the advantage that the material tested does not have to be in a cell type in which the gene product responsible for the disease phenotype is actually expressed - a factor which is often undefined and a matter of speculation. One can simply detect a polymorphic gene product from a gene in the tissue or cell type chosen because of its ready availability and ask whether the genes co-segregate in an informative family with the disease. By screening for this variant and others found in the same molecule or DNA region in other families, this linkage can be confirmed. In the case of a linked polymorphic molecule detected by monoclonal antibodies, the antibodies can be used as a tool for isolating the product or identifying the corresponding mRNA. Current methods can then be used to isolate the cDNA clone and genomic clone corresponding to the protein. Both methods of screening thus allow one to "jump onto the genome" at a locus linked to the "disease gene" so that detection, isolation and characterization of that gene and its product can be approached more specifically.

Once linkage is established by either method, one can look for other linked markers and gradually work down the DNA until the actual gene and gene product involved in the disease is defined. Meanwhile, the linked marker can be used as a diagnostic test for the presence of the linked "disease gene" for families in which two co-segregate. The reliability of the test would, of course, depend on the amount of recombination between the two loci.

To screen for polymorphic variation with monoclonal antibodies, we have chosen to use B-lymphoblastoid lines from members of several families with genetic diseases -

specifically HD and NF. We chose those cell types because they provide a relatively homogeneous population of cells that are easily obtainable and can be easily grown up in culture to large amounts for screening and eventual isolation of antigens or nucleic acids. They are also generally quite stable karyotypically. These lines also express a variety of well-defined polymorphic gene products, i.e. HLA, Ia antigens, Ig heavy and light chains. The basic steps in the procedure will be:

1. Screen our hundreds of monoclonal antibodies against lymphoblastoid lines from sets of parents.

2. Those showing differential binding to two parents' cells will be tested to see if they segregate in the progeny.

3. If segregation of the "alleles" is detected, determine whether there is co-segregation with other known markers including the disease phenotype.

4. Use antibody to isolate protein so that other monoclonal antibodies can be made with the intent of defining other polymorphic determinants on the same molecule so segregation can be analyzed in other families.

5. For markers which appear to co-segregate with the disease, use the antibody and isolated antigen as tools for isolation of the corresponding cDNA clone and use this for screening a genomic DNA library for the genomic clone.

6. Screen for restriction fragment polymorphisms linked to the gene.

7. Detect other genes in the same genomic region until genes more closely linked are defined and ultimately until the structural gene for the "disease gene" is defined.

8. Search for mRNA from this region and characterize the translation product, i.e. look for message present in various tissues that will hybridize to the genomic clone that apparently represents the disease "gene"; translate it in vitro and make antibodies against the translation product so that the molecule may be detected and isolated from normal and diseased tissues.

We describe here the progress we have made in these directions toward using the discriminating power of monoclonal antibodies to detect and analyze the molecular basis of human genetic diseases.

II.   METHODS AND PROCEDURES

A.   Production and Characterization of Monoclonal Antibodies

The procedures for the production of hybridomas and screening for antibodies were previously described in detail (32-34). The plasmacytoma line SP2/0-Ag14 (35) was fused with spleen cells from immunized female BALB/c mice. Immunizations of alpha-2-macroglobulin or fibronectin were given weekly for three weeks (100 ug I.P.) followed by three daily injections (50 ug I.P. or I.V.) and the fusion done the following day. This generally follows the procedure for soluble antigens described by Stahli et al. (36). The fibronectin was plasma fibronectin obtained from BRL. Two antifibronectin antibodies were also obtained from BRL. The alpha-2-macroglobulin was injected as an immunoprecipitate formed between goat anti-alpha-2-macroglobulin and alpha-2-macroglobulin from CF serum.

Screening was done with the enzyme linked antibody assay (ELISA) using peroxidase conjugated goat anti-mouse immunoglobulin (Cappel). Antigen was bound to polystyrene wells at a concentration of 0.5 µg per well. The anti-fibronectin antibodies were tested against plasma fibronectin and the anti-alpha-2-macroglobulin antibodies against the immunoprecipitate as well as control wells containing only goat immunoglobulin bound to the plastic. Those binding to the precipitate and not to goat Ig were considered to be anti-alpha-2-macroglobulin antibodies. This was later confirmed by immunoprecipitation with the monoclonal antibodies.

B.   Purification of Alpha-2-Macroglobulin and Fibronectin for Quantitative Binding Assays

Alpha-2-macroglobulin was purified on a zinc chelate column (37) provided by Dr. J. Kaplan, University of Utah. The purified alpha-2 was characterized as predominantly the slow form by PAGE (26) and conversion to the fast form by treatment with methylamine (26,28) confirmed by the same procedure.

Fibronectin was purified on a column of gelatin bound to cyanogen bromide-activated sepharose (38).

C. Assay of Quantitative Binding to the Proteins

To test the binding of the antibodies to a series of the corresponding proteins from different individuals, 0.5 µg of the purified protein was bound to each well of a polystyrene plate. The binding of each antibody was tested at least four times in duplicate or quadruplicate with the antibody in excess. The amount of binding was normalized by taking the ratio of binding of each of the monoclonal antibodies to the binding of a standard mouse antiserum against the protein to correct for any variation between experiments or due to the variation in protein from sample to sample.

D. Assay of Peroxidase –Conjugated Antibody Against Lymphoblastoid Cell Lines

The ELISA assay was performed as described previously (33). Detection of the HD antigen in the supernatant of lymphoblastoid line was done by growing the cells in culture for five days and collecting the supernatant. An equal volume of supernatant and antibody dilutions were placed in the assay well so that any antigen present in the supernatant would inhibit the binding of antibody to the cells bound to the plastic surface.

E. Immunofluorescence with Anti-Fibronectin Antibodies

Indirect immunofluorescence was done as described by D. Lane and W.K. Hoeffler (39).

III. RESULTS

A. Antibodies to Alpha-2-Macroglobulin

Nineteen of 313 clones from a mouse anti-alpha-2 fusion were selected and reaction with alpha-2 was confirmed. Thirteen of these were chosen because of their degree of binding and stability. Competition assays indicate that they detect at least ten independent binding sites. Six rat spleen cell x SP2/OAg14 antibodies were chosen from a second fusion.

At this point, these antibodies have been tested for their binding to both the slow and fast forms of alpha-2-macroglobulin from five purified samples, two from CF patients and three from non-CF patients. Figure 1 shows the normalized values for the binding.

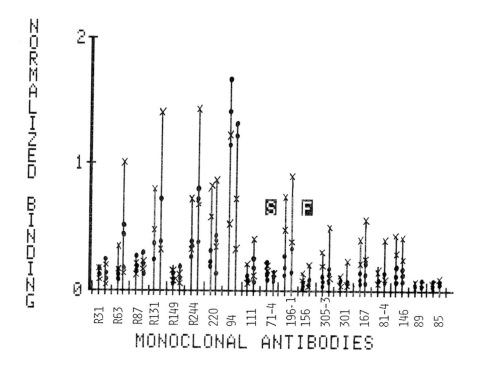

FIGURE 1. Binding of monoclonal antibodies to the slow form and fast form of alpha-2-macroglobulin. Alpha-2 was purified from 2 CF patients (X) and 3 normal patients(O). Part of the purified sample was shifted to the fast form with methylamine. The two forms were bound to plastic wells and each antibody tested for its binding to each of the samples. In each case, antibodies are in excess. Each was tested 2-4 times in duplicate or quadruplicate. To control for variation from experiment to experiment, each sample was also tested with a standard mouse anti-alpha-2 antiserum. The values for each antibody were normalized by taking the ratio of binding of the monoclonal antibody to the binding of the antiserum. The bars indicate the upper limit for the binding of each antibody. Each point represents the binding to a specific alpha-2-macroglobulin from a CF (X) or normal individual (O). Each pair of bars represents the binding to the slow form on the left and the fast form on the right for each monoclonal antibody.

It is clear that some of these antibodies show binding
to some of the alpha-2-macroglobulins tested that differs
significantly from their binding to the same molecule from
other individuals. Whether there are polymorphic differences will have to be determined by testing more samples and
by testing the binding to alpha-2 from appropriate families.
The possibility that the relatively high binding of some
of the antibodies with CF alpha-2-macroglobulins is really
detecting conformational difference will be tested by looking at further samples including individuals who are obli-
gate heterozygotes for the CF gene. Figure 2 shows the
binding of each antibody to the fast form compared to the
binding to the slow form. The data indicate that the conformational change in the molecule does affect the binding
of several of the antibodies.

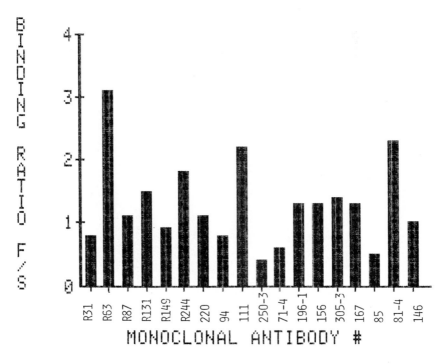

FIGURE 2. Ratio of binding of each monoclonal antibody
to the slow form and fast forms of alpha-2-macroglobulin.
This data includes the binding to each of the five alpha-2-
samples shown in Figure 1. Some antibodies bind better to
one form than the other indicating their binding to the determinant is affected by the protein conformation.

B.   Antibodies to Fibronectin

Figure 3 shows the binding of eight anti-fibronectin antibodies to fibronectin from five normal and three NF individuals. Although there is some degree of variation in binding to these purified proteins, it is clear that there is no obvious distinction between binding to NF and to normal fibronectin with the eight antibodies. Using the antibodies to detect fibronectin on fibroblasts by binding or by immunofluorescence (Figure 4) also shows no distinction between the two types of fibroblasts. Using the antibodies to detect the amount of fibronectin produced by NF and normal fibroblasts in culture using an inhibition assay (40) shows, at this point, no significant difference in the amount produced by the two types of cells in culture (R. Polin, personal communication). The amount of fibronectin in plasma and the specific structure of the cell surface fibronectin still remain as a possible difference between NF and normal.

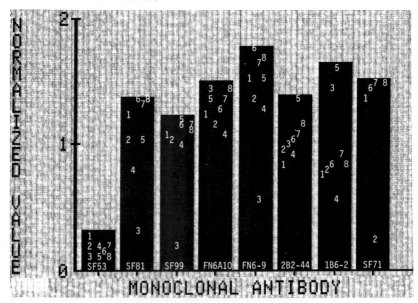

FIGURE 3. Binding of eight monoclonal antibodies to human plasma fibronectins from NF patients (#4,5,6) and normal individuals (#1,2,3,7,8). Fibronectins were purified on a gelatin column and tested following the procedure described in Figure 1 for alpha-2-macroglobulin. Position of the number for each fibronectin indicates the normalized binding of each antibody to that fibronectin. Individuals 7 and 8 are unaffected relatives of patient #6.

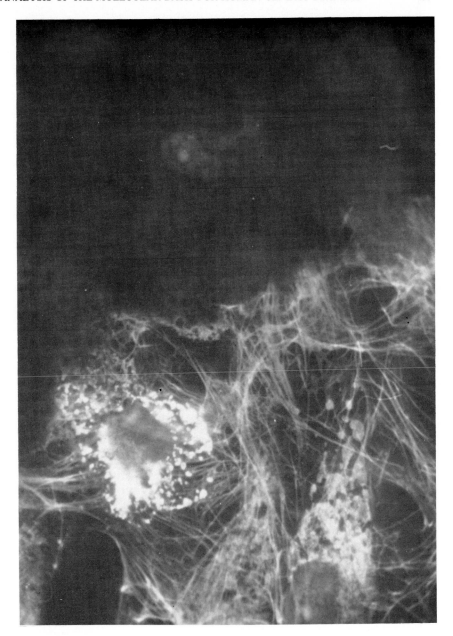

FIGURE 4. Immunofluorescence showing distribution of fibronectin on fibroblasts from an NF patient using monoclonal antibodies. Unstained dark area at top are cells in control area to which monoclonal anti-fibronectin antibody was not applied.

C.   Detection of Antigenic Variation in Lymphoblastoid Cell Lines

Our stimulus for screening lymphoblastoid cell lines for antigens that may be polymorphic came from our work on screening antibodies against lymphoblastoid cell lines and fibroblasts from Huntington's Disease (HD) patients and non-HD individuals.   While screening monoclonal antibodies against HD lymphoid lines, we detected what appears to be a high background against the four HD lines tested.   Further work has shown that the peroxidase conjugated goat anti-mouse immunoglobulin antiserum actually binds to the HD cell line and not to the normal B-lymphoid cell lines (Figure 5).

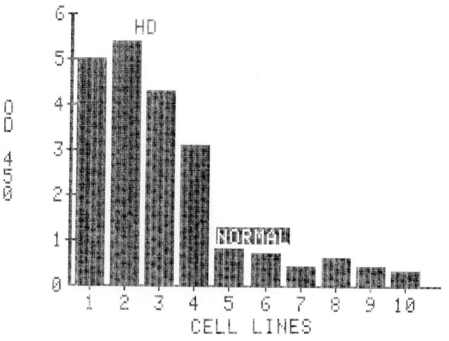

FIGURE 5.   Binding of our peroxidase-conjugated goat anti-mouse Ig antibodies to lymphoblastoid lines from HD patients (1-4) and normal individuals indicating significant binding to the 4 HD lines tested and no significant binding to non-HD cell lines.

We have shown that the molecule detected by the antibody is present in the supernatant of the HD cells grown in culture (Figure 6).

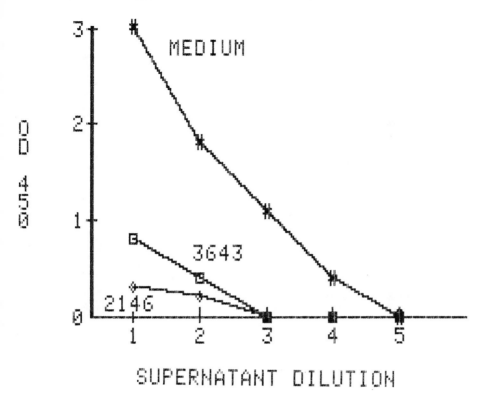

FIGURE 6.   Inhibition of binding to HD cell lines by supernatants from the cell lines.   Supernatant from a culture of the HD cell lines GM-3643 and GM-2146 inhibits the binding of the peroxidase conjugated goat anti-mouse Ig antiserum shown in Figure 5 to the HD cell line GM 3643. This data show the titration of the antibody from 1/1000 (1) to 1/16,000 (5) in the presence of supernatant and control medium in which cells have not been grown.

Precipitation of the antigen from an extract of the HD cell line GM-2080 using Staph. aureus coated with the antiserum followed by sodium dodecyl sulfate polyacrylamide electrophoresis shows two specific bands with molecular weights of approximately 20,000 daltons and 57,000 daltons (Figure 7).

We are presently producing monoclonal antibodies against these precipitated molecules and beginning to screen for the presence of the antigen in HD families to determine whether it segregates with HD.   The similarity of the

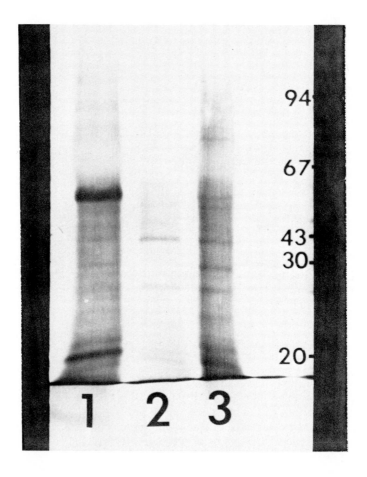

FIGURE 7. Immunoprecipitation of proteins from an extract of HD line GM-2080. Cells were grown in ^{35}S-methio-
nine, extracted in the detergent NP40 and treated with
S. aureus coated with the antiserum that reacts with HD
cell lines and not normal cell lines. Proteins bound by
the antibody were pelleted with the S. aureus and run on
SDS-PAGE and detected by autoradiographs. Lane 1 shows
the 2 bands precipitated from the HD line GM-2080, lane 2
the extract from the non-HD cell line showing that these
bands are not present, and lane 3 the treatment of the HD
line with a control goat antiserum. The position of the
stained molecular weight standards is shown on the right
(X1000).

molecular weight of these bands to immunoglobulin molecules
has led us to hypothesize that the antibodies are detecting
polymorphic variation in an immunoglobulin gene and that
therefore the HD gene may be linked to a gene for one of the
immunoglobulin chains. The fact that immunoprecipitation
from one of the HD lines brings down only a high molecular
weight band suggests that this is probably a heavy chain
gene, and the relatively high molecular weight would lead us
to believe that it may be the mu chain locus. We are beginning to test this hypothesis at both the gene product and
the DNA level.

IV.   GENERAL DISCUSSION

We have considered two different ways in which monoclonal antibodies can be used to help define the molecular
basis for human genetic diseases. The preliminary work
presented here on the binding of these reagents to alpha-2-
macroglobulin and fibronectin and other work in other systems (12-16,17,20-23) support the idea that monoclonal antibodies can play a significant role in discriminating between proteins with different conformations.

The work on the influenza hemagglutinin molecule cited
above suggests that the number of sites· on a protein at
which monoclonal antibodies may be able to detect conformational changes may be even higher than one would expect in
the basis of work with conventional antisera. The various
ways in which these antibodies interact with the protein do
not necessarily represent binding sites that are all spa-
tially distinct but different ways of binding three-dimen-
sional arrays of amino acids on the surface of the protein.
With so many ways to detect the protein surface conformation, it will not be surprising if many of the antibodies
detect differences between proteins from different individuals.

Certainly the most promising and the most generally applicable part of the work described here is the use of monoclonal antibodies to screen for polymorphisms linked to the
genes responsible for human genetic diseases. Screening
with antibodies has several advantages:

1. There are already available hundreds of antibodies
    made against determinants on human gene products
    that can be easily screened against panels of cells.

2. Work with these antibodies in various systems indicates that they frequently are able to detect variation in protein conformation.

3. Once obtained, these antibodies will be in constant supply so that each can be used by many other investigators to look at the segregation of these genes in various families with a variety of other genetic diseases, or families in which other genetic markers of interest are segregating. These antibodies will provide, in effect, a panel of new human polymorphic markers for genetic linkage analysis.

4. If a polymorphism is detected, the antigen can be readily isolated and other polymorphic variations in the protein detected by antibodies against other sites or by conventional methods.

5. Isolation of the protein will also facilitate microsequencing and synthesis of the corresponding polynucleotide probe.

6. The antibodies will be useful for detecting specific mRNA by immunoprecipitation of in vitro translation product or isolation of specific polysomes and therefore facilitate isolation of the corresponding gene.

We can certainly expect, with the combination of monoclonal antibody and molecular genetic techniques now available, that the next few years will provide an increase in our knowledge of the molecular basis for human genetic diseases and even more generally a better understanding of the amount of genetic polymorphism that is present within populations.

ACKNOWLEDGEMENTS

We wish to thank Dr. D. Holsclaw of the Hahnemann Medical College of Philadelphia for his very helpful provision of the samples of CF plasma. We also are thankful for the provision of the NF plasmas by Dr. Anna Meadows of the Children's Hospital of Philadelphia. HD lymphoblastoid lines were obtained from the NIGMS cell bank at the Institute for Medical Research in Camden, New Jersey.

REFERENCES

1. Kohler, G., and Milstein, C., Nature 256, 495 (1975).
2. Kennett, R.H., Jonak, Z., and Bechtol, K., in "Mono-
    clonal Antibodies - Hybridomas: A New Dimension in
    Biological Analyses" (R.H. Kennett, T.J. McKearn, and
    K.B. Bechtol, eds.), p. 155, Plenum Press, New York,
    (1980).
3. Kennett, R.H., Jonak, Z.L., Bechtol, K.B., and Byrd, R.,
    in "Fundamental Mechanisms in Human Center Immunology"
    (J.P. Saunders, J.C. Daniels, B. Serrou, C. Rosenfeld,
    and C.B. Denny, eds.), p. 331, Elsevier/North Holland
    Inc., New York, (1981).
4. Kennett, R.H., In Vitro 17, 1036 (1981).
5. Minna, J.D., Cuttitta, F., Rosen, S., Bunn, P.A., Car-
    ney, D.N., Gazdar, A.F., and Krasnow, S., In Vitro
    17, 1058, (1981).
6. Harris, H., "The Principles of Human Biochemical Gene-
    tics," p. 38, Elsevier/North Holland Inc., New York,
    (1980).
7. Weber, K., and Osborn, M., J. Biol. Chem. 244, 4406
    (1969).
8. Harris, H., "The Principles of Human Biochemical Gene-
    tics," p. 21, Elsevier/North Holland Inc., New York,
    (1980).
9. Crumpton, M.J., in "The Antigens," Vol. 2 (M. Sela,
    ed.), p. 1, (1974).
10. Smith-Gill, S.J., Wilson, A.C., Potter, M., Prager, E.
    M., Feldmann, R.J., and Mainhart, C.R., J. Immunol.
    128, 314, (1982).
11. Berg, K., and Bearn, A.G., J. Exp. Med. 123, 379 (1966).
12. Hammerling, G.J., Hammerling, U., and Lemke, H., Im-
    munogenet. 8, 433, (1979).

13. Galfre, G., Howe, S.C., Milstein, C., Butcher, G.W.,
    and Howard, J.C., Nature 266, 550, (1977).
14. Brodsky, F.M., Parham, P., Barnstable, C.J., Crompton,
    M.J., and Bodmer, W.F., Immunl. Rev. 47, 3, (1979).
15. Smilek, D.E., Boyd, H.C., Wilson, D.B., Zmijewski, C.M.,
    Fitch, F.W., and McKearn, T.J., J. Exp. Med. 151, 1139,
    (1980).
16. Gasser, D.L., Winters, B.A., Haas, J.B., McKearn, T.J.,
    and Kennett, R.H., Proc. Natl. Acad. Sci. USA 76, 4636,
    (1979).
17. Gerhard, W., Yewdell, J., Frankel, M., Lopes, A.D.,
    and Staudt, L., in "Monoclonal Antibodies - Hybridomas:
    A New Dimension in Biological Analyses" (R.H. Kennett,
    T.J. McKearn and K.B. Bechtol, eds.), p 317, Plenum
    Press, New York, (1980).

18. Laver, W.G., Gerhard, W., Webster, R.G., Frankel, M.E., and Aiv, G.M., Proc. Natl. Acad. Sci. USA 76, 1425, (1979).

19. Sherman, L.a., and Randolph, C.P., Immunogenet. 12, 183, (1981).

20. Oi, V.T., and Herzenberg, L.A., Mol. Immunol. 16, 1005, (1979).

21. Reth, M., Kelsoe, G., and Rajewsky, K., Nature 290, 257, (1981).

22. Conley, M.E., Kearney, J.F., Lawton, A.R., and Cooper, M.D., J. Immunol. 125, 2311 (1980).

23. Slaughter, C.A., Coseo, M.C., Cancro, M.P., and Harris, H., Proc. Natl. Acad. Sci. USA 78, 1124, (1981).

24. Wilson, G.B., Pediatric Res. 13, 1079, (1979).

25. Shapira, E., Martin, C.L., and Nadler, H.L., J. Biol. Chem. 252, 7923 (1977).

26. Barrett, A.J., Brown, M.A., and Sayers, C.A., Biochem. J. 181, 401 (1979).

27. Ohlsson, K., and Skude, G., Clin. Chim. Acta 66, 1 (1976).

28. Marynen, P., Von Leuven, F., Cassiman, J.-J., and van den Berghe, H., J. Immunol. 127, 1782 (1981).

29. Riccardi, V.M., N. Eng. Med. 305, 1617 (1981).

30. Sieber-Blum, M., Sieber, F., and Yamada, K.M., Exp. Cell Res. 133, 285, (1981).

31. Botstein, D., White, R.L., Skolnick, M., and Davis, R.W., Am. J. Hum. Genet. 32, 314, (1980).

32. Kennett, R.H., Denis, K.A., Tung, A.S., and Klinman, N.R., Curr. Topics. Microbiol. and Immunol. 81, 77 (1978).

33. Kennett, R.H., McKearn, T.J., and Bechtol, K.B., eds., "Monoclonal Antibodies - Hybridomas: A New Dimension in Biological Analyses," Appendix, p. 361, Plenum Press, New York, (1980).

34. Kennett, R.H., in "Cell Culture" Methods in Fnzymology, (W.B. Jakoby and I. Pastan, eds.), p. 345, Academic Press, New York, (1979).

35. Shulman, M., Wilde, C.D., and Kohler, G., Nature 276, 269 (1978).

36. Stahli, C., Staehelin, T., Miggiano, V., Schmidt, J., and Haring, P., J. Immunol. Met. 32, 297, (1980).

37. Kurecki, T., Kress, L.F., and LasKowski, M., Anal. Biochem. 99, 415, (1979).

38. Engvall, E., and Rouslahti, E., Int. J. Canc. 20, 1, (1977).

39. Lane, D.P., and Hoeffler, W.K., Nature 288, 167, (1980).

40. Polin, R., and Kennett, R.H., J. Pediatr. 97, 540, (1980).

DISCUSSION

J. HARFORD: Are any of the monoclonals to alpha$_2$M directed at the carbohydrate portion of alpha$_2$M?

R.H. KENNETT: We have made no attempt to answer that question. If there is a carbohydrate determinant then it is apparently not present in other serum glycoproteins, because immunoprecipitation followed by SDS-PAGE shows only a single band with the molecular weight of the alpha$_2$ macroglobulin sub-unit (185,000).

R.L. WARD: Your results imply that tumor-specific antigens may not exist. If this is the case, what is your view regarding immuno-surveillance and recognition of transformed cells?

R.H. KENNETT: I have always thought that immuno-surveillance may play more of a role in the case of tumors that do not survive than in the case of those that become actively growing tumors. It may be that most variants that become successful tumors do not express new antigens in such a way that they can be easily recognized as "antigens". The fact that we have immunized across species and have used various antigen preparations makes this explanation less likely in our case. Results that we and others have obtained up to this point would certainly suggest that it is not easy to find a tumor-specific "determinant".   This may mean nothing more than that monoclonal antibodies are so specific that if one screens enough, the same determinant will eventually be found in other molecules expressed on other cell types. This does mean that to define a tumor-specific antigen will take a good deal of biochemical analysis.   Our results with our neuroblastoma antigen show that this is a practical course to pursue but that it takes a good deal of time and effort.

Isolating a monoclonal antibody against a potential "tumor-specific antigen" is only a first step and must be followed by: 1) isolation and biochemical characterization of the antigen including production of antibodies against other determinants, and

2) isolation of the corresponding gene using molecular genetic techniques and analysis at the DNA level.

*J.L. STROMINGER: I am really not surprised that true tumor-specific antigens are not being detected in large numbers because the new protein would mean either a new or an altered gene. If there are a lot of new genes then one would suspect viruses as the prominent carriers of such new genes. So far there isn't a lot of evidence that cancer cells contain new genes. That they have altered genes for some surface protein may be more likely.

*R.H. KENNETT: I am not surprised either, but the use of monoclonal antibodies seemed to me to be the only way to look for them.

R.J. HAY: I would like to inform those of you who have generated hybridomas and are being inundated with requests for them, that the ATCC is soliciting these for deposit for eventual distribution to the scientific community. Hybridomas and other cell lines may be placed with the ATCC either with intent to patent by the originator, or with no restrictions for general distribution.

Also, those who would like to obtain specific hybridomas or fusion lines should request lists of those available from the publications department of the ATCC.

HUMAN CELL SURFACE ANTIGENS STUDIED WITH MONOCLONAL ANTIBODIES *

Jack L. Strominger

Department of Biochemistry & Molecular Biology
Harvard University
Cambridge, Massachusetts U.S.A.

* Supported by research grants from the NIH
(AI-10736, AM-30241, AM-13230 and AI-15669)

This lecture is intended as a tribute to Cèsar Milstein and the revolution in cell biology created by the discovery of the technique of producing monoclonal antibodies. I would like to illustrate the revolution with studies of human histocompatibility antigens. Five topics will be covered briefly. Monoclonal antibodies were employed in each and permitted an advance in a different area relating to histocompatibility antigens. These areas include: 1) the purification of the HLA-A,B,C antigens (Class I MHC products); 2) identification of novel surface proteins, in this case the presumed human analogs of the murine TL and Qa antigens (also Class I products); 3) separation of subsets of HLA-DR antigens (Class II products); 4) cloning a gene for the α chain of HLA-DR antigens and the elucidation of the structure of Class II proteins; and finally 5) identification of subsets of human lymphocytes, in this case a novel subset of allospecific HLA-DR6 cytotoxic T lymphocytes which are OKT 4^+ AND OKT 8^-.

The major histocompatibility complex of the mouse is located on the 17th chromosome, that of man on the 6th chromosome. Three classes of antigens are known to be encoded in this region. The class I antigens in man are called the HLA-A,B,C antigens and in the mouse the H-2K,D,L antigens. In addition to these class I antigens which are expressed on all nucleated cells, the mouse possesses additional class I antigens, called Qa and TL antigens, which are only expressed on some cells and are therefore differentiation antigens. Reagents to detect putative human analogs of mouse Qa and TL have been identified but it is not yet known whether or not they map to chromosome 6. Sometime during evolution one of the class I genes of the mouse (H2-K) appears to have been transposed from the class I cluster to another region in the major histocompatibility complex. The class II antigens are encoded in the I-region in the mouse of which several subregions have been defined, i.e. I-A, I-J, I-E. The HLA-DR antigens in man are the homologs of the murine I-region antigens but so far further subdivision has not been accomplished (although recently serological reagents have begun to define the complexity of these antigens). Finally, the class III antigens are the complement genes which are in the middle of the MHC and beyond the scope of this lecture.

The most interesting feature of this system it its extraordinary polymorphism. It is the most polymorphic system of genes known in the species. In the range of 25 alleles are presently known at the A locus, upwards of 40 alleles at the B locus, 8 alleles at the C locus and in the range of 12 alleles at the DR locus. However, HLA-C and HLA-DR have been studied much more recently than HLA-A and

HLA-B. Since large population blanks at HLA-C and HLA-DR
occur, the number of HLA-C and HLA-DR alleles can be
expected to increase.

   1. Purification of HLA-A,-B and -C antigens. The class
I antigens are composed of two polypeptide chains (1). The
glycosylated heavy chain spans the membrane; it is composed
of an extracellular region, a hydrophobic intramembranous
region and a small intracytoplasmic region. The heavy chain
is firmly but noncovalently attached to a small subunit,
β_2-microglobulin, which is not embedded in the membrane.
The complete or nearly complete primary structures of
several heavy chains and of β_2-microglobulin are now known.
Moreover, HLA-A2 and -A28 have recently been crystallized,
so soon information about the 3-dimensional structure may
also become available. The original procedure for
purification of the intact, detergent-solubilized HLA-A and
-B antigens from the JY cell line was quite difficult.
An immunoaffinity column with rabbit anti-β_2-microglobulin
serum facilitated the purification, but the HLA-A locus and
-B locus allospecificities of the homozygous cell line could
not be separated easily. Monoclonal antibodies which were
allospecific for HLA-A2 and -B7 were developed. Using a
series of immunoaffinity columns containing these monoclonal
antibodies, HLA-A2 and HLA-B7 could readily be purified
(2). Moreover, a small amount of a third Class I antigen,
presumably the HLA-C locus product was picked up finally
on a W6/32 immunoaffinity column. (W6/32 is a monoclonal
antibody which recognizes all the HLA-A,-B and -C antigens.)
The purifications which formerly took 5 days can now be
accomplished in an afternoon, so that has been an enormous
advance.

   2. Identification of novel Class I human surface antigens
with monoclonal antibodies. NA1/34 is one of the first
monoclonal antibodies obtained by Milstein and his
collaborators (3). It was obtained after immunization with
a human thymocyte preparation. It precipitated from human
thymocytes or some T cell lines a complex of proteins of
48,000 and 12,000 daltons. In fact cortical thymocytes
express only the NA1/34 antigen and no HLA-A,-B or -C
antigens. To my mind one of the most interesting
differentiation steps in immunology is the step in which the
cortical thymocytes migrate from the cortex to the medulla,
lose the NA1/34 antigen and acquire the HLA-A,B,C antigens.
Another monoclonal antibody, obtained originally by Kung
et al. also after immunization with human thymocytes, was
called OKT 6 (4). OKT 6 and NA1/34 recognize the same
polypeptide complex as determined by preclearing experiments

although they recognize different epitopes on this complex
(5). Unlabeled NA1/34 competed with itself for binding but
OKT 6 did not compete with the binding of NA1/34; reciprocally
OKT 6 competed with itself but NA1/34 did not compete with OKT
6. Among cell lines derived from individuals with T cell
leukemias, some expressed a very large amount of the NA1/34
antigen, others expressed a small amount and some expressed
none (5,6). The NA1/34 antigen appeared to have a slightly
higher affinity for the antigen than the OKT 6 antibody and
gave slightly higher values than the OKT 6 antibody in
monoclonal antibody binding assays, but otherwise they had the
same cell distribution. In sequential immunoprecipitations
W6/32 removed all material reactive with this monomorphic
HLA-A,B,C monoclonal antibody but material remained which
could be precipitated with NA1/34. Finally after both
W6/32 and NA1/34, an anti-β_2-microglobulin monoclonal
antibody precipitated still another heavy chain which was,
of course, associated with β_2-microglobulin. This heavy
chain band was always a lot broader than the other heavy
chain bands and in fact there is evidence that it contains
more than one polypeptide, i.e. there are at least two
additional polypeptides associated with β_2-microglobulin.

Another monoclonal antibody isolated by the Ortho group
was called OKT 10 (4). OKT 10 is not T lymphocyte specific,
i.e. it recognized a protein on the JY B cell line and other
B cell lines. It recognized a protein on activated, but not
on resting, T lymphocytes and also on a variety of T cell
leukemia lines. Precipitation with OKT 10 following clearing
with W6/32 again revealed two polypeptides, a heavy chain
of around 48,000 daltons and a light chain of around 12,000
daltons (5). SB and HSB are syngeneic lines from the same
individual who had a T cell leukemia; SB is the B cell line
and HSB is the T cell line. OKT 10 precipitated a material
of the same polypeptide composition from each.

The small subunits of the OKT 10 and NA1/34 (OKT 6)
antigens appear to be distinct from β_2-microglobulin (5).
They are distinguished by a small difference in gel mobility
first observed by Ziegler and Milstein (7) who recognized
that the small subunit of the NA1/34 antigen was different
from β_2-microglobulin and called it β_t. Additional evidence
for this difference in the OKT 10 antigen was obtained in the
following way (5). After immunoprecipitation the OKT 10
antigen was denatured by solubilizing the immunoprecipitate
in hot SDS. The heavy chain was not precipitable by anti-HLA
heavy chain serum and similarly the light chain was not
precipitable by goat anti-β_2-microglobulin serum. In our
experiments the heavy chain of the NA1/34 antigen cross-
reacted with the anti-HLA heavy chain serum and was associated

with both β_t and β_2-microglobulin. A great deal of work
remains to be done to characterize these novel class I
antigens, all made possible by the advent of monoclonal
antibodies which recognize them.

    3. <u>Separation of subsets of HLA-DR antigens</u>. HLA-DR
antigens also have a heavy and light chain. Both chains
span the membrane. The polypeptides are 28,000 and 26,000
daltons on SDS gels (plus carbohydrates). In 2-dimensional
gel separations (SDS in one direction and isoelectric
focusing in the other) the heavy chains from different HLA-DR
specificities gave the same pattern. The light chains,
however, differed greatly in isoelectric point and in some
cases the light chain patterns were very complex (8). The
light chain is the polymorphic chain. Minor bands in the
light chain patterns which were ignored at first later
became very important.

    The DC1 antigen was first described using antisera
obtained in Italy after planned immunization of human
volunteers (9). It was found on DR1,2 and 6 cells.
Immunoprecipitations from an HLA-DR6 homozygous cell line,
established from an individual who was the offspring of a
consanguinous mating, showed that the two polypeptides
recognized by DC1 had a slightly different molecular weight
than those recognized by DR6 alloantisera. Two-dimensional
separations established clearly that these two sets of
polypeptides are distinct, i.e. DC1 and DR6 alloantisera
recognize two distinct sets of polypeptides (10). DC1 is
not a "supertypic" determinant on DR antigens, as had been
suggested.

    One of the first monoclonal anti-HLA-DR antibodies
described was Genox 3.53 (11). It had a confusing pattern
of specificity since it reacted with DR1, DR2 and DR6 cells,
and so it was thought to recognize a supertypic determinant.
However, Genox 3.53 is clearly a DC1 monoclonal antibody.
L243, another HLA-DR monoclonal antibody described by Lampson
and Levy (12), immunoprecipitated 2 light chains together
with one heavy chain from the homozygous DR6 cells (13).
The monoclonal antibodies L227 (12) and LKT111 (14) each
recognized only one of these light chains (L1 and L2
respectively) together with a heavy chain (15).

    These monoclonal antibodies, Genox 3.53, L243, L227 and
LKT111 have been used in immunoaffinity columns to separate
3 subsets of DR antigens preparatively (16,17). First, L243
was used to separate the major heavy chain and 2 light chains.
Then a Genox 3.53 immunoaffinity column removed the 2 chains
of the DC1 antigen. The material initially separated by
the L243 column was subsequently further separated into two
subsets of DR antigen making use of the specificities of

L227 and/or LKT111, yielding H1L1 and H1L2. We do not presently know whether or not H1, the heavy chains of the L227 and LKT antigens, are different. They may have a slight difference in gel mobility. N-terminal amino acid sequencing may help in answering this important question.

N-terminal sequencing of the heavy chain of the DC1 antigen has clearly established that it has strong sequence homology to murine I-A antigens (16). No sequence has yet been obtained for the DC1 light chain because it has a blocked N-terminus. The DC1 antigen thus appears to be the long sought human homologue of murine I-A antigens. It has been known for a long time that the HLA-DR antigens (i.e. those proteins recognized by the L243 monoclonal antibody) are the homologues of the murine I-E antigens.

4. <u>Cloning the α chain of the HLA-DR antigen</u>. The α chain of the DR antigens was cloned making use of the specificity of a monoclonal antibody. This work illustrates the extraordinary power of the two revolutionary techniques of modern cell biology, recombinant DNA technology and monoclonal antibody technology. Several years ago we reported detection of the cell-free translation products of the two chains of HLA-DR antigens from 3 different cell lines (18). These products were slightly larger than the <u>in</u> <u>vivo</u> non-glycosylated chains; they contained a leader sequence. Thus, we had developed the background to recognize these cell-free translation products of the DR antigen chains from total mRNA.

Over the years a number of investigators have reported purification or efforts at purification of polysomes by precipitation with heteroantisera (for example, 19). We modified the technique, by using a monoclonal antibody rather than rabbit heterosera (20). A monoclonal antibody prepared against the denatured heavy chain of the HLA-DR antigen was added to polysomes. Those polysomes were then applied to a Staph protein A-Sepharose column which binds antibodies. After washing the column the bound polysomes were dissociated and eluted with Tris-EDTA. The eluted mRNA was recovered on oligo dT-cellulose and analyzed by cell-free translation. Immunoprecipitation with heterosera against the heavy chain of both HLA-DR and HLA-A,B,C antigens indicated that the amount of HLA-DR heavy chain product formed from total lymphocyte mRNA was about 1/5 of the amount of the HLA-A,B,C antigen heavy chain. Cell-free translation of the mRNA recovered from the immunoaffinity purification in the first experiment using the anti-heavy chain monoclonal antibody suggested that the DR heavy chain mRNA was about 30% pure. In another experiment mRNA of

50-90% purity was obtained in a single step. Only one
prominent protein was synthesized by this mRNA, the DR heavy
chain with its leader sequence. We estimated the abundance
of this message at about 0.01-0.02%. The mRNA was pure
enough so that it could be used to probe a cDNA library or
a genomic library directly. This mRNA was also reverse
transcribed and cloned. Essentially all cDNA clones
obtained using message of this purity were HLA-DR heavy
chain clones. mRNA selection with the cloned heavy chain
cDNA selected a message which yielded DR heavy chain in cell
free translation. The cDNA clone was sequenced and contained
cDNA corresponding to half of the hydrophobic region of the
DR antigen and the entire C-terminal hydrophilic region.
The hydrophilic region began with a cluster of basic amino
acids and had a serine which was necessary as a phosphorylation site, since we knew that the heavy chain is
phosphorylated in vivo. Digestion of isolated heavy chain
and its papain cleavage product with carboxypeptidases A and
B confirmed that the cDNA clone was an authentic DR α chain
clone.

The heavy chain cDNA clone was then used to probe a
human placental DNA library in phage λ. Four clones were
obtained which had overlapping restriction sites and yielded
a 3.4 kb Eco R1 subclone containing virtually the entire
gene for DR α chain (with the exception of the DNA encoding
the first two amino acids, the leader sequence and the
5'-untranslated region which are located about 1 kb upstream
in an additional 4 kb Eco R1 fragment)(21). The nucleotide
sequence of the 3.4 kb Eco R1 subclone revealed that it
contained four exons encoding respectively 1) α1, 2) α2,
3) transmembrane, and intracytoplasmic regions and a part of the
3'-untranslated region, and 4) the remainder of the 3'-
untranslated region. The complete amino acid sequence of
the DR heavy chain could be deduced and, although it is
beyond the scope of this report, the fact that α1 and α2
have remarkable similarity to the domains of the HLA-A,B,C
heavy chain is noteworthy. In particular α2 is a highly
conserved Ig-like domain (as is β2 of the light chain (22)).
Thus, this protein may be added to the group of proteins
which have descended from a primitive Ig-like ancestral
domain.

5. Cloned allospecific HLA-DR cytotoxic T lymphocyte
lines are OKT 4^+8^-. In collaboration with Steven Burakoff,
Alan Krensky and Carol Reiss cytotoxic T lymphocytes directed
against DR antigens have been studied. Some time ago we had
shown that CTL could recognize allogeneic HLA-DR antigens as
well as HLA-A,B,C antigens. Using xenogeneic murine CTL in

a secondary CTL response, HLA-A and -B antigens in liposomes
stimulated HLA-A,-B specific CTL. HLA-DR antigens in
liposomes however, stimulated HLA-DR specific CTL (23). This
work has been extended to allogeneic, rather than xenogeneic,
CTL. Three long term human lines were obtained (24) by
stimulating the peripheral blood cells of laboratory donors
with the Daudi cell line which expresses HLA-DR but not
HLA-A,B antigens. Moreover, two separate clones have been
established from these lines by limiting dilution and have
now been in culture for many months (25). These lines and
clones are very potent CTL; they kill appropriate targets
(e.g. Daudi cells) up to 50% at an effector: target ratio
of 0.4. Using a panel of cells, no relationship of killing
to the HLA-A,B antigens was found but specificity for the
DR6 antigens was clearly evident. The stimulator, Daudi, is
an HLA-DR6 cell. This killing was not blocked by HLA-A,B
specific monoclonal antibodies (W6/32) but was blocked by
either heteroantiserum or monoclonal antibody against the
HLA-DR antigens. All these lines and clones were OKT 4
positive and OKT 8 negative and killing was also blocked
by OKT 4 but not by OKT 8. It had previously been concluded
that OKT 8 defines the cytotoxic/supressor subset of
lymphocytes and OKT 4 defines the helper/inducer subset of
lymphocytes (4). However, we suggest that the OKT 4, OKT 8
phenotype is not related to the function of the cell, but
rather to the class of histocompatibility antigens which
is participating in the function manifest by that cell. A
helper cell in which the cell-cell recognition is dependent
on DR antigen is OKT 4 positive but also a killer cell which
recognizes DR antigen is OKT 4 positive. It is not the
function but the class of antigen participating in the
cellular interaction which is related to the OKT 4, OKT 8
phenotype. These data then lead to the interesting
suggestion that the OKT 4 and OKT 8 antigens may be part of
the T cell receptor for histocompatibility antigens.

Thus, progress in these five areas relating to
histocompatibility antigens was made possible by the
discovery of the technique of production of monoclonal
antibodies by Cesar Milstein and his collaborators.

REFERENCES

1. Ploegh, H.L., Orr, H.T. and Strominger, J.L. Cell 24, 287-299 (1981).
2. Parham, P. J. Biol. Chem. 254, 8709-8712 (1979).
3. McMichael, A.J., Pilch, J., Galfre, G., Mason, D.Y., Fabre, J.W. and Milstein, C. Eur. J. Immunol. 9, 205-210 (1979).
4. Kung, P.C., Goldstein, G., Reinherz, E.L. and Schlossman, S.F. Science 206, 347-349 (1979).
5. Cotner, T., Mashimo, H., Kung, P.C., Goldstein, G., Milstein, C. and Strominger, J.L. Proc. Natl. Acad. Sci. U.S.A. 78, 3858-3862 (1981).
6. Cotner, T., Hemler, M. and Strominger, J.L. Int. J. Immunopharm. 3, 255-268 (1981).
7. Ziegler, A. and Milstein, C. Nature (London) 279, 243-244 (1979).
8. Shackelford, D.A. and Strominger, J.L. J. Exp. Med. 151, 144-165 (1980).
9. Tosi, R., Tanigaki, N., Centis, D., Ferrara, G.B. and Pressman, D. J. Exp. Med. 148, 1592-1611 (1978).
10. Shackelford, D.A., Mann, D.L., van Rood, J.J., Ferrara, G.B. and Strominger, J.L. Proc. Natl. Acad. Sci. U.S.A. 78, 4566-4570 (1981).
11. Brodsky, F.M., Parham, P. and Bodmer, W.F. Tissue Antigens 16, 30-48 (1980).
12. Lampson, L.A. and Levy, R. J. Immunol. 125, 292-299 (1980).
13. Shackelford, D.A., Lampson, L.A. and Strominger, J.L. J. Immunol. 127, 1403-1410 (1981).
14. Bono, R., Hyafil, F., Kalil, J., Koblar, V., Wiels, J., Wollman, E., Mawas, C. and Fellous, M. Transplantation and Clinical Immunology 11, 109 (1979).
15. Shackelford, D.A. and Strominger, J.L. Submitted for publication.
16. Bono, R. and Strominger, J.L. Submitted for publication.
17. Bono, R., Andrews, D. and Strominger, J.L. Submitted for publication.
18. Korman, A.J., Ploegh, H.L., Kaufman, J.F., Owen, M.J. and Strominger, J.L. J. Exp. Med. 152, 65s-82s (1980).
19. Shapiro, S.Z. and Young, J.R. J. Biol. Chem. 256, 1495-1498 (1981).
20. Korman, A.J., Knudsen, P.J., Kaufman, J.F. and Strominger, J.L. Proc. Natl. Acad. Sci. U.S.A. in press

21. Korman, A.J., Auffray, C., Schamboeck, A., Kamb, A.,
     Knudsen, P. and Strominger, J.L., in preparation
22. Kaufman, J.F. and Strominger, J.L. Nature, in press.
23. Englehard, V.H., Kaufman, J.F., Strominger, J.L. and
     Burakoff, S.J. J. Exp. Med. 152, 54s-64s (1980).
24. Krensky, A.M., Reiss, C.A., Mier, J.W., Strominger, J.L.
     and Burakoff, S.J. Proc. Natl. Acad. Sci. U.S.A. in
     press.
25. Krensky, A.M., Clayberger, C., Reiss, C.S., Strominger,
     J.L. and Burakoff, S.J. Submitted for publication.

DISCUSSION

*D.H. GELFAND: Have you used your partial HLAD P34 clone as a hybridization probe to look for sequence polymorphism among HLAD type families?

*J.L. STROMINGER: Do you mean do Southern blots on a panel? On a very limited panel we have seen no polymorphism. There are some peculiarities that I can't say are due to polymorphism. But in general that heavy chain clone picks up only one band. Remember that the heavy chain of the HLAD-DR antigen is the constant chain like β-2-microglobulin in the AGC antigen. We see only one gene, that is also matched by the isolation of genomic clones. We have 5 genomic clones from the Maniatis genomic library and they appear to be overlapping clones. So it is possible that this is a single copy gene. If so, and if there are three heavy chains, as seems possible, it also means that the clone doesn't cross-hybridize with the other heavy chains.

IMMUNOCYTOCHEMISTRY WITH MONOCLONAL ANTIBODIES: POTENTIAL APPLICATIONS IN BASIC SCIENCES AND HISTOPATHOLOGY

A. Claudio Cuello

Departments of Pharmacology and Human Anatomy
Oxford, England

César Milstein

MRC Molecular Biology
Cambridge, England

INTRODUCTION

Immunocytochemistry, since its foundation by Coons and collaborators(1), has become one of the most powerful tools to investigate the presence and localization of defined substances in microscopic preparations. This methodology has been of great assistance in solving a number of problems in basic sciences and it is increasingly applied in histopathology. The advent of hybrid myelomas for the continuous supply of monoclonal antibodies with predefined specificities (2) offers new alternatives in immunocytochemistry. Here we summarise our results on the production and application of monoclonal antibodies (McAbs) in immunocytochemistry, both as primary and marker antibodies.

MONOCLONAL ANTIBODIES AS PRIMARY REAGENTS IN IMMUNOCYTOCHEMISTRY

When polyclonal antibodies are applied in immunocytochemistry (ICC) higher concentrations of the antisera are usually required than for radioimmunoassay (RIA), a system conventionally used for sera characterization. As the individual population of antibodies recognize different determinants with different affinities it is frequently the case that the antibodies which are expressed in immunocytochemistry are not the same as those expressed in RIA. This of course does not occur when applying McAbs, as the same

Supported by grants from the MRC (U.K) and The Wellcome Trust.

population of antibodies with a single combining site are
present in the incubation medium, regardless of the concentration of immunoglobulins used in the assay system. The
characterisation of the antibody combining sites done with
RIA will therefore apply to immunocytochemistry when using
the same monoclonal antibody.

Another advantage of McAbs as primary reagents for ICC
is that they immortalize difficult or rare antibodies. An
example of such antibodies is given by the recently developed
rat hybrid line coded YC5/45 which secretes McAb to serotonin.
When this antibody is tested by indirect immunofluorescence
it detects serotonin-containing cell bodies and terminals in
the central(3), peripheral nervous systems(4). The competition experiments demonstrated a strong cross-reactivity with
dopamine, tryptamine and 5-methoxytryptamine when the test
was performed in liquid media(3). This provides an example
of antibody cross-reactivity resulting from multi-recognition
by a single molecular species i.e. the "intrinsic cross-
reactivity" of the monoclonal antibody. Such cross-
reactivity is of a more specific nature than the cross-
reactivity associated with ordinary polyclonal antisera.
Polyclonal antisera are complex mixtures of monoclonal antibodies and their cross-reactivities might result from cross
reaction of individual antibody molecular species. The
intrinsic cross-reactivity to dopamine and other amines of
this McAb is more apparent than real as the antibody shows a
much higher affinity for the serotonin-bovine seroalbumin
conjugate, but does not recognise bovine seroalbumin alone.
Formaldehyde shifts the cross reaction towards serotonin in
hemagglutination tests. When the antibody was tested in
tissue preparations, it completely failed to reveal dopamine
sites in areas of the brain where the presence of this amine
is well documented, namely the zona compacta of the substantia nigra and ventro tegmental groups. It does appear that
with paraformaldehyde fixation, YC5/45 does not reveal the
dopamine antigenic sites in brain tissue preparations. On
further studies using tissue model systems we have established that the formaldehyde fixation is essential for the
generation of "serotonin" immunoreactive sites(5). The
results thus emphasise the relevance of the method used for
detection. When performing immunocytochemical studies,
specificity can be introduced by the antibody interactions
("antibody specificity" of both primary and secondary antibodies) and also by the conditions of the incubations, and
preparation of the tissue ("method specificity"). Here again

McAbs can be invaluable tools to explore "methods specificity" as there is no risk that different antibody molecular
species can be expressed under different experimental
circumstances.

In the case of YC5/45 we have obtained further experimental evidence that the antibody sees only serotonin sites
in fixed tissue by the pretreatment with drugs which affect
the monoamine content in the brain. Thus α-methyl-p-
tyrosine in conditions which produce a very drastic depletion
of the catecholamine content in neurones of the central
nervous system did not affect the immunofluorescence, while
depletors of 5HT such as the tryptophan hydroxylase inhibitor
p-chlorophenyl-alanine results in a marked diminution of the
binding of YC5/45 to neurones in the raphe nuclei system,
an area of the brain known to contain serotonergic neurones,
(Fig. 1).

The use of monoclonal antibodies as primary reagents
tends to improve the quality of the immunostain as unwanted
antibodies or carrier immunoglobulins are eliminated. This
results in a better signal to noise ratio which can be
critical when dealing with small numbers of antigenic sites.
Background is, in these cases, dependent upon the secondary
antibodies used to demonstrate the antibody binding in tissue.
The first demonstration of such applications for an intracellular antigen was provided by the anti-substance P monoclonal antibody coded NC1/34(6)(Fig.2). This McAb is the
product of a rat x mouse fusion. NC1/34 demonstrated also
that McAbs can perform adequately in a radioimmunoassay
system where fentomoles of substance P could be detected(6).
The same system was used to analyse the "intrinsic cross-
reactivity" of this McAb. NC1/34 showed no cross-reactivity
with a number of brain peptides while cross-reacting completely with small C-terminal fragments of the peptide and by 5%
with the related eledoisin. The good correlation of this
particular McAb in RIA and ICC can be explained as follows:
in both cases the NC1/34 recognises the C-terminal portion of
the peptide and this part of the molecule is seemingly not
affected by the aldehyde fixation which very likely affects
the N-terminal portion of the molecule. More recently an
anti-Leu-enkephalin antibody coded NOC1 has been derived from
a mouse x mouse hybridoma which shows greater affinity for
"enkephalines" antigenic sites in fixed tissue preparation
than in radioimmunoassay systems. The full characterisation
of this McAb is in progress.

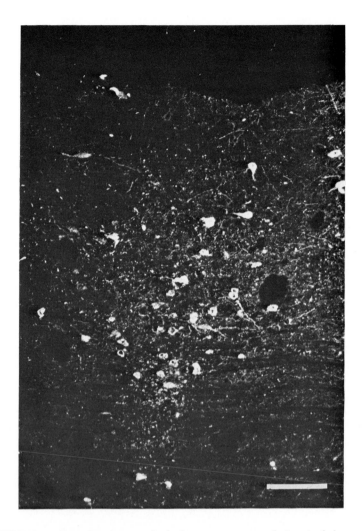

FIGURE 1. Serotonin-containing neurones detected by
immunofluorescence in the rat brain (nucleus raphe dorsalis)
by applying the McAb YC5/45. Scale bar = 100 μm.

   The fact that McAb can be produced in large amounts is
of some relevance as ICC requires larger amounts of antibody
molecules than RIA. Interesting McAbs for research or histopathological diagnoses can be widely distributed to compare
results or to be used as "standards". In this regard we have
made our own McAbs available to other workers and they are

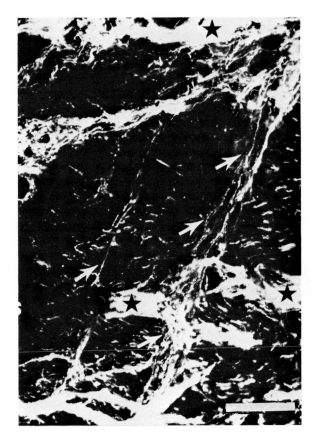

FIGURE 2.   Substance P immunofluorescent fibres as seen
applying the McAb NC1/34 in the rat brain stem.   Individual
axons (arrows) and nerve tracts (stars) are detected at the
entrance of the trigeminal nerve.   Scale bar = 50 μm.

currently distributed by SeraLab (Crawley Down, Sussex, U.K).
It is hoped that other similarly interesting antibodies from
other laboratories will become available.

MONOCLONAL ANTIBODIES AS MARKERS
FOR MICROSCOPIC VISUALISATION

Direct Immunocytochemistry with Monoclonal Antibodies

In the direct immunocytochemical technique the primary
antibody itself constitutes a marker antibody. In this
method the primary antibody is bound to a suitable tag which
allows the microscopical detection of the antigen-antibody
complexes in tissue preparations. The lack of amplification
of this procedure and the loss of immunoglobulins during the
isolation of specific antibodies made the technique a non-
appealing alternative and it was therefore abandoned. The
advent of McAbs can radically change this situation as the
hybridoma secretes only the desired antibody molecules thus
avoiding the use of affinity columns for the purification
of specific antibodies. With McAbs the simple separation of
the immunoglobulins should suffice for an efficient coupling
of specific antibodies to a marker molecule.

In collaboration with Drs Boorsma and Van Leuwen we have
recently explored the use of a direct immunoenzyme histo-
chemical technique by the direct conjugation of the McAb
NC1/34 to the enzyme horseradish peroxidase (HRP)(7). For
this, rabbit immunoglobulin against rat IgG (Dako, Denmark)
was coupled to CNBr-activated Sepharose beads (Pharmacia,
Sweden) and the eluate, presumably pure McAb NC1/34, was
conjugated to horseradish peroxidase by a two-step method(7).
This results in a 400,000 dalton preparation with 1:1 McAb/
HRP and approximately 100% of the McAb were effectively
conjugated in contrast to the usually low yield obtained
when polyclonal antibodies are used.

The large molecular weight of the HRP-McAb conjugate was
similar to that of the PAP (peroxidase-antiperoxidase)
complex. The immunohistochemical application of this enzyme
McAb complex resulted in a very clean immunostain in areas
known to contain the peptide in the rat C.N.S., with long
incubations at low temperature and in the presence of the
detergent Triton X-100 (see Fig. 3). Successful direct
immunostain with NC1/34-HRP could also be obtained with
short incubations (30 minutes) at $37^{o}C$, somewhat sacrificing

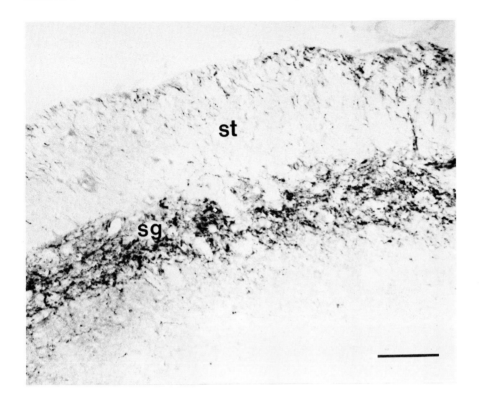

FIGURE 3. Direct immunocytochemistry with HRP conjugated
monoclonal antibodies against the peptide substance P.
Immunolocalization of the peptide on nerve terminals in the
substantia gelatinosa (sg) of the spinal nucleus of the
trigeminal nerve. Individual axons can be seen in the spinal
tract (st). Scale bar = 50 μm. (From Boorsma et al. ref.7.

the excellent signal/noise ratio. This type of fast direct
immunocytochemistry might prove useful when a quick histopathological diagnosis is required.

The introduction of various sizes of gold particles as
antibody markers (8,9) might also offer new possibilities
to direct immunohistochemical techniques with McAbs. It is
conceivable that different McAbs could be tagged to different
sizes of gold particles. This, and the combined use of
immunoenzyme techniques should allow the simultaneous detection of various antigenic sites at light and electron microscopical levels.

Internally Labelled Monoclonal Antibodies

Another radical approach to direct immunocytochemistry with McAbs is the use of internally labelled McAbs. We have developed a procedure for the localisation of tissue antigens which can be referred to as "radioimmunocytochemistry" (see below).

Developing Antibodies

There has been little progress in the use of McAbs as secondary or developing antibodies in ICC. This might be due to the fact that polyclonal serum can more effectively amplify the signal of the primary antibody binding as different antibodies of the antiserum might recognise different determinanats of a single immunoglobulin molecule. This nevertheless can be theoretically overcome by the use of suitable mixtures of McAbs against immunoglobulins of different species Furthermore, McAb can prove to be efficient tools as developing antibodies when they recognise repetitive determinants of a single antigen and this might be the case of antiperoxidase McAbs.

RADIOIMMUNOCYTOCHEMISTRY WITH INTERNALLY LABELLED MONOCLONAL ANTIBODIES

Hybridomas can be cultured in media containing radio-labelled amino acids (10,11). This results in the in vivo biosynthesis of McAbs at high specific activity with radio-active amino acid precursors. These radioactive immuno-logical probes can be effectively used in radioautography of immunoreactive sites (11-13) offering an interesting alternative towards the cellular and subcellular localisation of immunoreactive substances as well as the development of novel immunoradiometric techniques.

The application of internally labelled monoclonal anti-bodies in "radioimmunocytochemistry" is schematically repre-sented in Fig. 4 and can be summarised as follows:

1. Biosynthesis of McAbs in the presence of radioactive amino acids, generally ^{3}H-or ^{14}C-lysine (other isotopes and amino acids can be used). Fig. 4a,b.
2. Separation of medium from cells (hybrid myelomas). Fig. 4c.

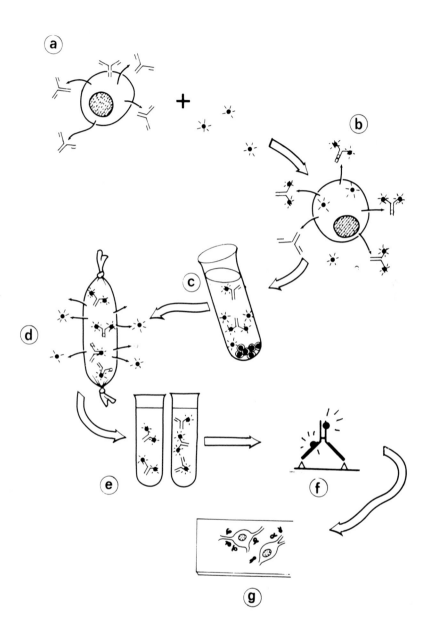

FIGURE 4. (see text).

3. Dialysis of supernatant (or passage through columns) to
    eliminate non-incorporated radioactive amino acids.
    Fig. 4d,e.
4. Incubation of tissue preparations with 3HMcAbs at the
    corresponding dilutions. Fig. 4f.
5. Radioautography for light and electron microscopy using
    conventional procedures. Fig. 4g.

The high specific activity (50-2000 Ci/mmol for ^{3}H-NC1/34
HL) obtained by the internal labelling of monoclonal antibodies results in enhanced sensitivity of this procedure as
compared with conventional ICC. At the light microscopic
level successful radioautographs were obtained 72 hours
following incubation. It is hoped that the use of internally
labelled monoclonal antibodies at light microscope level will
allow the quantification of the immunohistochemical reaction.
Preliminary observations would indicate that this is
feasible. We have shown with optical densitometry that the
immunohistochemical signal is related to the input of radio-
labelled antibodies and discriminate regional localisation
of antibody in subnuclei in the CNS with great accuracy(14).
This method may offer a meaningful alternative for quantification in immunocytochemistry. Previous attempts have been
based on the use of fluorescent markers, or, more recently,
on the intensity of peroxidase reaction following the chain
of antibodies required for the PAP techniques. In all those
cases the final signal of the immune reaction is distant to
the primary antibody binding while with radioimmunocytochemistry the signal should be directly related to the binding of the primary antibody. The validity of these assump-
tions can only be tested experimentally.

Another potential application of "radioimmunocytochemistry" is the detection of minute amounts of antigenic
material which happens to occur in highly localised areas.
This again is possible due to the high specific activity of
internally labelled monoclonal antibodies. An example of
this is provided by the reaction of ^{3}H-YC5/45 on cryostat
sections. Vertebrate retina is suspected of containing only
traces of an indoleamine, probably serotonin(15). Fig. 5
illustrates the binding of ^{3}H-YC5/45 HLK in the frog retina
which has been developed as a radioautograph. This
preparation shows that the radiolabelled antibody binds to
amacrine cell bodies and their processes in the internal
plexiform layer.

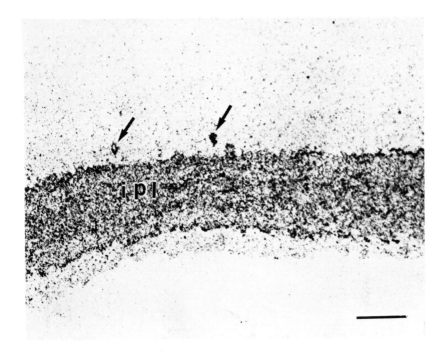

FIGURE 5. Localization of serotonin in cell bodies (arrows) and processes in the internal plexiform layer (ipl) of frog retina as revealed by radioimmunocytochemistry with 3H-YC5/45. Scale bar = 50 μm. (From Osborne and collaborators, in preparation).

Promising applications of this technique are also expected in electron microscopy where the use of conventional immunoenzyme techniques result in a compromise between good subcellular preservation and antibody penetration. Internally labelled monoclonal antibodies do not require developing antibodies for the detection of antigenic sites. Therefore this compromise is less costly in terms of tissue preservation. Another potential advantage of "radioimmunocytochemistry" is the elimination of the electron dense products resulting from the application of immunoenzyme techniques which obscure the fine ultrastructural detail of the immunoreactive sites. The main disadvantage of fine resolution radioautography with monoclonal antibodies resides in the need for high technical expertise and the long exposure times.

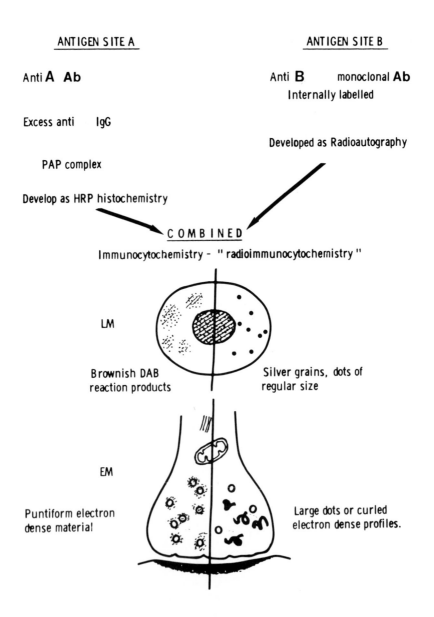

FIGURE 6. Schematic representation of the combined immunocytochemistry for the simultaneous detection of two antigenic
sites by applying internally labelled monoclonal antibodies
(radioimmunocytochemistry) and immunoenzyme techniques (Ab,
antibody; HRP, horseradish peroxidase; DAB, diaminobenzidine,
LM, light microscopy; EM, electron microscopy.

Another application of the radioimmunocytochemical
approach is its combination with immunoenzyme techniques
for the simultaneous detection of two antigenic sites at
light and electron microscopical levels (Fig. 6). In this
connection we have obtained encouraging results for the
simultaneous detection of two intracellular antigens:
substance P and serotonin, in the raphe nuclei, and substance
P (Fig. 7) and Leu-enkephalin in the substantia gelatinosa of
the spinal nucleus of the trigeminal nerve(14). This is an
area of termination of primary sensory fibres, many of which
contain the peptide substance P. This peptide has been
proposed as a transmitter of nociceptive information,while
axo-axonic interaction has been proposed between substance P
and enkephalin containing inhibitory neurones as a basis for
the "gating" of pain messages at the first synapses in the
CNS. The electron microscopical application of the combined
radioimmunocytochemistry with internally labelled McAbs with
conventional immunoenzyme technique showed that the majority
of these peptide-containing nerve terminals are largely
unrelated although they may terminate on a common dendrite
(14).

The combined use of radioimmunocytochemistry and immunoenzyme techniques might find wider applications in fields
unrelated to neurobiology.

MONOCLONAL ANTIBODIES IN HISTOPATHOLOGY

Immunohistochemistry began with research related to
histopathology. Nowadays a number of diagnoses are based on
immunohistochemical analysis of tissue preparations. The
methodology is nevertheless restricted to those researchers
having access to limited sources of relevant antibodies.
Therefore, up to now, the application of this technique has
been restricted to major medical institutions. It is possible that McAbs will change this situation.

An example of the potential use of McAbs in histopathology is given by the application of YC5/45 to detect
serotonin immunoreactive sites in normal and pathological
specimens. At present the microscopic diagnosis of serotonin-
producing carcinoid tumours is based on the morphological
pattern and only occasionally confirmed by silver staining
for the argyrophilic or argentaffinic reaction or the
electron microscopic analysis. All these procedures are

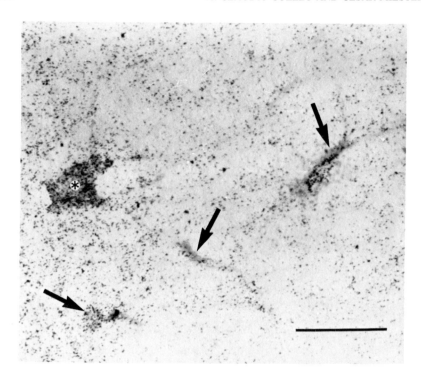

FIGURE 7. Combined radioimmunocytochemistry for substance P
(^{3}H-NC1/34) and immunoperoxidase for serotonin (YC5/45) in
the nucleus raphe magnus of the rat, showing the co-existence
of these two antigens on single neurones (asterisks). Arrows
indicate tangential section of neuronal cell bodies. Silver
grains denote substance P immunoreaction sites. Grey,
homogenous stain immunoperoxidase reaction for serotonin.
Scale bar = 50 μm. (From ref. 14).

either erratic or do not substantially add to the diagnostic
analysis. For example, non-reactive tumours to silver salts
are often found which are otherwise considered carcinoid
tumours by the microscopic pattern and biochemical data.
Immunocytochemistry offers a more reliable approach towards
the microscopic diagnosis of carcinoids. We were able to
demonstrate with the monoclonal antibody YC5/45 that a close
correlation exists for immunohistochemistry of 5HT immunoreactive sites using YC5/45 and the clinical and biochemical
data(16)(see Fig. 8). It would be of interest to expand this
initial experience in order to assess whether this McAb can
be used as a routine diagnostic procedure.

In the past decade advances in neurochemistry have opened new frontiers in our understanding of brain function in health and disease. The discovery by Hornykiewicz(17) that the Parkinson condition is accompanied by a specific deficiency in the dopamine content in the striatum is a dramatic example of this type of accomplishment in neurological research. This discovery was a crucial step in the development of the compensatory therapy with L-dopa in the patients. This also prompted many laboratories to undertake large-scale analysis of chemical correlates in human post-mortem brain in various neurological conditions. This gigantic task is now being carried out with a great deal of international co-operation for which newly developed, sensitive techniques might play a decisive role.

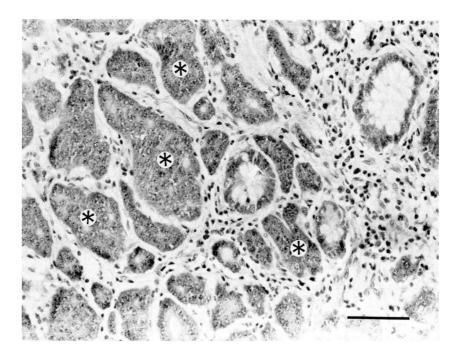

FIGURE 8. Serotonin-producing carcinoid cells (asterisks) stained with the McAb YC5. Immunoperoxidase indirect technique. Human ileum. Scale bar = 100 μm. From ref. 16.

In this regard McAbs can help in the analysis of neuro-transmitter content by RIA of tissue homogenates. These results are very difficult to compare as so much depends on the neuroanatomical sampling. On the other hand immunocyto-chemistry can accurately locate the presence or absence of neural substances in given nuclear structures. As an example of this situation the histopathological application of NC1/34 has shown that while large concentrations of substance P immunoreactive terminals were present in the substantia gelatinosa of the spinal cord of control human nervous systems there was an almost complete absence of such terminals from this region in all patients suffering from familial dysautonomia, a neurological condition associated with specific neuronal losses and with diminished pain sensitivity(18).

It is stimulating to think that, in due time, a number of McAbs to various critical markers will be universally available for histopathological diagnoses. In addition, it is possible that the current or modified version of radioimmunocytochemistry with internally labelled McAbs will allow the quantification of normal and abnormal substances and allow the large-scale scanning of relevant antigens in tissue preparations.

CONCLUDING REMARKS

Monoclonal antibodies have proved to be powerful tools for immunohistochemistry, both as primary and "marker" anti-bodies. The use of internally labelled antibody offers a viable alternative for the simultaneous detection of two antigenic sites at electron microscopical level and for quantitative immunocytochemistry.

It is expected that the wide distribution of relevant McAbs will provide adequate standards for histopathological diagnosis.

ACKNOWLEDGEMENTS

We would like to thank John Priestley, Bruce Wright, Adriana Consolazione and John Jarvis for collaborative efforts in the work mentioned here, as well as the technical assistance of Steve Bramwell, Brian Archer, Judith Lloyd and the secretarial help of Mrs Ella Iles.

REFERENCES

1. Coons, A.H. and Kaplan, M.H. J. exp. Med. 91, 1. 1950.
2. Kohler, G. and Milstein, C. Nature 256, 495. 1975.
3. Consolazione, A., Milstein, C., Wright, B. and Cuello, A.C. J. Histochem. 29, 1425. 1981.
4. Costa, M., Furness, J.B., Cuello, A.C., Verhofstad, A.A. J. and Elde, R.P. Neuroscience (In press). 1982.
5. Milstein, C., Wright, B. and Cuello, A.C. (In preparation).
6. Cuello, A.C., Galfre, G. and Milstein, C. Proc. Natl. Acad. Sci. USA. 76, 3532. 1979.
7. Boorsma, D.M., Cuello, A.C. and Van Leuwen, F.W. J. Histochem. Cytochem. (Submitted) 1982.
8. Slot, J.W. and Geuze, H.J. J. Cell Biol. 90, 533. 1981.
9. De Mey, J., Moeremans, M., Geuens, G., Neydens, R. and De Brabander, M. Cell Biol. Int. 5, 889. 1981.
10. Galfre, G. and Milstein, C. Methods in Enzymology. 73, 46. 1981.
11. Cuello, A.C. and Milstein, C. In: Physiological Peptides and New Trends in Radioimmunology. Ed. Ch. A. Bizollon. Elsevier-North Holland, Amsterdam. p.293. 1981.
12. Cuello, A.C., Galfre, G. and Milstein, C. In: Receptors for Neurotransmitters and Peptide Hormones. (Ed. G.Pepeu, M.J.Kuhar and S.J.Enna). Raven Press, N.Y. p.349. 1980.
13. Cuello, A.C., Milstein, C. and Priestley, J.V. Brain Res. Bull. 5, 575. 1980.
14. Cuello, A.C., Priestley, J.V. and Milstein, C. Proc. Natl. Acad. Sci. USA. 79, 665. 1982.
15. Osborne, N.N., Nesselhut, T., Nicholas, D.A. and Cuello, A.C. Neurochem. Int. 3, 171. 1981.
16. Cuello, A.C., Wells, C., Chaplin, A.J. and Milstein, C. Lancet (Submitted).
17. Hornykiewicz, O. Pharmacol. Rev. 18, 925. 1966.
18. Pearson, J., Brandeis, L. and Cuello, A.C. Nature 295, 61. 1982.

HUMAN IMMUNOGLOBULIN EXPRESSION IN HYBRID CELLS

Carlo M. Croce [1]

The Wistar Institute of Anatomy and Biology
Philadelphia, Pennsylvania
U.S.A.

I. INTRODUCTION

Somatic cell hybrids between mouse myeloma cells and lymphocytes derived from either mice or rats immunized with various antigens have been extensively used to produce monoclonal antibodies specific for the antigens of interest (1-3). These hybrids, or "hybridomas", are phenotypically similar to the myeloma cell parent and produce extremely large quantities of the antibodies expressed by the lymphocytes derived from the immunized animals (1-3). Since one lymphocyte produces only one antibody, the hybrid cells derived from a single lymphocyte produce only one kind of antibody molecule, provided that the parental mouse myeloma cell line does not produce immunoglobulin chains. Therefore the antibody molecules produced by each hybrid clone are identical (monoclonal). Since successful hybridization of myeloma cells is much more efficient with lymphocytes that have been stimulated to divide than with resting lymphocytes, somatic cell hybridization occurs preferentially with those lymphocytes that have responded to antigenic stimulation (4). For this reason it is quite straightforward to generate hybridomas that secrete monoclonal antibodies of

[1] This work was supported in part by USPHS grants CA-10815, CA-23568, GM-20700, CA-16685, CA-21124, CA-20741, CA-25875, CA-27712 from the National Institutes of Health, and grant I-522 from the National Foundation — March of Dimes.

the desired specificity. The logical extension of the study of hybridomas secreting rodent monoclonal antibodies is to determine whether it is possible to produce stable hybridomas that produce and secrete human monoclonal antibodies. The availability of a cell system that allows one to immortalize human antibody-producing cells through somatic cell hybridization procedures could be of enormous value in the study of human autoimmune diseases and, possibly, in human immunotherapy.

II. MOUSE X HUMAN B CELL HYBRIDS

At first we attempted to produce hybrids between mouse myeloma cells deficient in hypoxanthine phosphoribosyltransferase (1-3) and human B cells derived from different sources. The hybrids were found to segregate into hybrid clones that produce human heavy chains and hybrids that do not (5, 6). The same was observed in the case of expression of human light chains (7). We also observed that the loss of human chromosomes in these hybrids is not a random event and that some human chromosomes are preferentially retained (in particular, human chromosome 14 is retained by the large majority of hybrid clones) and others are preferentially lost (Table I) (8). Therefore if the chromosomes carrying the human immunoglobulin genes were in the group of chromosomes that are preferentially lost, it would be difficult to produce mouse x human hybridomas that continuously secrete human monoclonal antibodies.

III. MAPPING OF THE CHROMOSOMAL LOCATION OF HUMAN IMMUNOGLOBULIN GENES

We have analyzed somatic cell hybrids between mouse myeloma cells and human B cells secreting IgM for the expression of human μ chains as detected by immunoprecipitation procedures and polyacrylamide gel electrophoresis (5, 7, 8). The hybrids were also studied for the expression of isozyme markers assigned to each of the different human chromosomes in order to assess the human chromosomal constitution of the hybrid cells. The results of this analysis are reported in Table II. Because of the phenomenon of allelic exclusion (9), we can expect to find hybrids that do not express human heavy chains but do contain the human chromosome carrying the heavy chain gene cluster. For this reason

TABLE I. Expression of Human Isozyme Markers in Mouse
x Human Hybridomas

Human chromosome no.	Isozymes tested	Total no. of clones analyzed	No. of clones containing human chromosome	% of clones containing human chromosome
1	ENO-1	56	5	8.9
2	IDH_S	42	0	0
3	β-GAL	22	5	23.0
4	PGM_2	40	16	40.0
5	HEX_B	41	29	70.7
6	GLO-1	32	11	34.3
7	β-GUS	23	4	17.4
8	GSR	44	11	25.0
9	$ACON_S$	45	5	11.1
10	GOT_S	44	8	18.1
11	LDH_A	54	26	48.0
12	LDH_B	54	12	22.2
13	Est-D	43	5	11.6
14	NP	46	46	100.0
15	MPI	47	20	42.5
16	APRT	33	11	33.3
17	GK	46	19	41.3
18	PEP A	53	14	26.4
19	GPI	57	13	22.8
20	ADA	48	9	18.7
21	SOD-1	44	14	31.8
22	ARS	21	13	61.9
X	G6PD	30	16	53.3

the critical information regarding the chromosome mapping is contained in the second column (+/-), since it allows the identification of the chromosomes that do not carry the μ chain gene. As shown in Table II, only one human chromosome cannot be excluded and that is human chromosome 14. Therefore we assigned the μ chain gene to this human chromosome. As indicated in Table II, 11 of 26 independent hybrids contain human chromosome 14 but do not express μ chains. This indicates that the chromosome 14 present in the μ-negative hybrids is the one carrying the excluded μ allele. Because of the phenomenon of allelic exclusion, the informative column is the +/-, and thus we have preselected mouse x human hybrid clones producing human γ chains and have studied them

TABLE II. Expression of Human Antibody Heavy Chains in
Mouse x Human Hybridomas

Human chromo- some	μ/chromosome				γ/chromosome	
	+/+	+/-	-/+	-/-	+/+	+/-
1	2	11	0	12	2	6
2	0	13	0	8	0	8
3	4	7	7	1	4	2
4	2	7	10	3	0	8
5	10	5	9	0	3	3
6	6	7	2	0	0	6
7	0	6	0	2	1	5
8	5	9	4	6	0	7
9	2	15	3	6	0	8
10	3	18	0	2	0	8
11	11	11	11	1	0	8
12	2	20	7	5	0	8
13	3	18	2	0	0	8
14	15	0	11	0	8	0
15	5	10	7	7	0	8
16	4	9	1	1	1	7
17	3	11	8	4	0	7
18	3	18	5	6	1	7
19	3	19	0	14	3	5
20	1	13	1	12	0	8
21	8	7	2	10	0	6
22	0	5	0	2	0	6
X	6	6	2	2	0	6

for the expression of human γ chains and isozymes to determine whether the human γ genes are on the same human chromosome. As shown in Table II, all human chromosomes with the
exception of human chromosome 14 can be excluded (8). We
then used a similar approach to map the chromosomal location
of the human α chain genes to the same chromosome (5, 6).

   We have also used a similar approach to map the genes
for human immunoglobulin λ chains to human chromosome 22 (7)
(fig. 1). As shown in Table III, only two human chromosomes,
14 and 22, are candidates to carry the λ chain genes. Since
the heavy and light immunoglobulin genes are on different

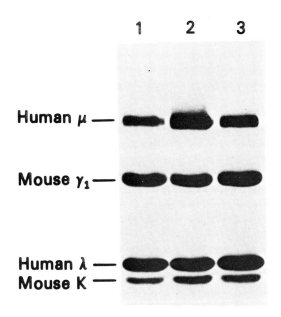

Figure 1.   SDS-polyacrylamide gel electrophoresis of
immunoglobulin chains secreted by somatic cell hybrids between P3 x 63Ag8 mouse myeloma cells and human peripheral
blood lymphocytes.   The human and mouse immunoglobulin
chains were labeled and immunoprecipitated as described in
figure 1 legend.   The independent hybrids (lanes 1-3), pre-
selected for isozyme analysis, expressed human λ chain.
Immunoprecipitation of cytoplasmic immunoglobulin chains
present in the hybrids was also carried out, with identical
results (data not shown).

chromosomes, this result indicates that the λ chain genes
are on chromosome 22.   In order to confirm this finding we
have subcloned five independent human λ-producing hybrid
clones and we have studied the hybrid subclones for the expression of human λ chains (fig. 2) and of isozyme markers
assigned to each one of the different human chromosomes.   As

TABLE III. Human Chromosomes Present in Mouse x Human
Hybridomas Producing Human Immunoglobulin λ Chains

Human chromosome	No. of hybridoma clones that are:	
	+/+	+/-
1	0	11
2	0	11
3	5	7
4	6	6
5	9	3
6	10	10
7	3	9
8	6	14
9	0	12
10	4	8
11	8	4
12	2	10
13	3	9
14	20	0
15	9	3
16	2	10
17	18	2
18	5	6
19	1	11
20	4	8
21	2	10
22	14	0
X	18	2

Hybrid cells were studied for the expression of isozyme markers assigned to each of the different human chromosomes by starch gel or cellulose acetate gel electrophoresis: chromosomes 1, enolase 1 (EC 4.2.1.11); 2, isocitrate dehydrogenase (EC 1.1.1.42); 3, β-galactosidase (EC 3.2.1.23); 4, phosphoglucomutase 2 (EC 2.7.5.1); 5, hexosaminidase B (EC 3.2.1.30); 6, glyoxalase-1 (EC 4.4.1.5) and phosphoglucomutase 3 (EC 2.7.5.1); 7, β-glucuronidase (EC 3.2.1.31); 8, glutathione reductase (EC 1.6.4.2); 9, aconitase (EC 4.2.1.3); 10, glutamate oxaloacetic transaminase (EC 2.6.1.1); 11, lactate dehydrogenase A (EC 1.1.1.27); 12, lactate dehydrogenase B (EC 1.1.1.27); 13, esterase D (EC 3.1.1.1); 14, nucleoside phosphorylase (EC 2.4.2.1); 15, mannosephosphate isomerase (EC 5.3.1.8); 16, adenine phosphoribosyltransferase (EC 2.4.2.7); 17, galactokinase (EC 2.7.1.60); 18, peptidase A (EC 3.4.11.); 19, glucose phosphate isomerase (EC

(Legend to Table III continued) 5.3.1.9); 20, adenosine deaminase (EC 3.5.4.4); 21, superoxide dismutase 1 (EC 1.15.1.1); 22, arylsulphatase (EC 3.1.6.1); X chromosome, glucose-6-phosphate dehydrogenase (EC 1.1.1.49). +/+, Clones that both produce human λ chains and carry the numbered chromosome; +/-, clones that produce human λ chains and do not carry the numbered chromosome.

TABLE IV. Expression of Human λ Chains in Mouse x Human Hybridoma Subclones

Human chromosome	No. of hybridoma subclones that are:			
	+/+	+/-	-/+	-/-
1	0	9	0	7
2	0	9	0	7
3	8	5	4	3
4	4	5	4	3
5	3	6	5	2
6	16	2	6	1
7	0	9	2	5
8	0	9	0	7
9	0	10	0	7
10	0	9	0	7
11	5	4	6	1
12	2	7	2	5
13	1	8	1	6
14	9	2	6	1
15	3	6	4	3
16	0	9	0	7
17	13	1	6	1
18	2	7	2	5
19	0	9	0	7
20	1	8	1	6
21	0	8	0	8
22	7	0	0	7
X	5	2	6	1

+/+, +/-, Same as for Table I; -/+, clones that do not express human λ chains and carry the numbered chromosome; -/-, clones that neither express human λ chains nor carry the numbered chromosome.

Figure 2. SDS-polyacrylamide gel electrophoresis of immunoglobulin chains secreted by four hybrid subclones of a hybrid clone between P3 x 63Ag8 cells and human peripheral blood lymphocytes (see fig. 1). The expression of the human µ and λ chains segregated independently in the hybrid subclones. The subclones in lane 1 have lost the ability to secrete (produce) human µ and λ chains, whereas those in lanes 2 and 3 have retained the expression of these chains. The subclones in lane 4 have lost the ability to secrete (produce) human λ chains. Immunoprecipitation of cytoplasmic immunoglobulin chains present in the hybrids was also carried out, with identical results (data not shown).

shown in Table IV, chromosome 22 is the only human chromosome present in all human λ chains producing hybrids (7). By using a different approach consisting either in Southern blot hybridization analysis (11) of hybrid cell DNAs or <u>in situ</u> hybridization with specific probes, T. Rabbitts at

Cambridge and W. McBride and P. Leder at NIH have mapped the human κ chain genes to human chromosome 2 (personal communications).

These results taken together are of particular interest because they suggest a possible relationship between specific chromosome translocations, immunoglobulin genes and the expression of malignancy in Burkitt's lymphoma (12). A translocation between chromosomes 8 and 14, described as t (8, 14) (q24; q32), has been found in numerous cases of the disease (10, 11). Recently, two different translocations which involve the same segment of human chromosome 8 and either human chromosome 22 (that carries the λ chain genes) or chromosome 2 (that carries the κ chain gene) have been described in non-African Burkitt's lymphoma (12, 13). More recently, similar translocations, t (2, 8) (p12; q24) and t (8, 22) (q24; q11), have also been found in cases of African Burkitt's lymphoma (14, 15). These findings indicate that approximately the same segment of human chromosome 8 is translocated to each of the human chromosomes carrying immunoglobulin genes in Burkitt's lymphoma. It would be interesting to determine whether the break points on human chromosomes 2, 14 and 22 affect chromosomal DNA segments that lie close to or contain the human immunoglobulin chain genes.

IV. EXPRESSION OF HUMAN IMMUNOGLOBULIN mRNAs IN MOUSE HUMAN HYBRIDOMAS

Many of the somatic cell hybrids tested (Table V) secreted significantly more human H chains than did their human parental B cells. This was observed with a number of hybridomas secreting different classes or subclasses of human H chains (μ, α_1, α_2, or γ_2). Because hybridomas produced with the mouse P3 x 63Ag8 cell line or its nonsecreting subline (NP3) still showed an increased rate of secretion of human H chain compared to their human B cell parents (Table V, compare GM607, CSK-10-2B1-C14, and CSK-NS6-201-C7; GM1500, 106.2-B4-3G6, and FSK-4-2B2-C1), the coupling of a more rapid mouse immunoglobulin L chain secretion with the human H chain cannot solely account for the increased secretion rate. These H chain- "hypersecreting" hybrids have enabled us to purify human H (and L) chain mRNAs in much greater yields than from an equivalent mass of any of the human B cell lines examined. As can be seen in Table V, the hybrid αD5-DH11-BC11 secreted 4-6 times more human μ chain than did the human B cell lines GM607 or SED. We recovered

6-7 times more μ mRNA sedimenting at 20S (per g of hybrid cells) than from GM607 or another IgM κ-producing B cell line (SED) obtained from a patient with chronic lymphocytic leukemia (data not shown).

We have shown the results of polyacrylamide gel separations of the labeled polypeptides synthesized with partially purified human μ and κ mRNAs as well as their in vivo synthesized and secreted polypeptides (16). The in vivo secreted μ chain had a mobility, on reducing gels, corresponding to 77,000 daltons, whereas its in vitro directed product migrated as 69,000 daltons. We observed this in two human cell lines (GM607, and SED) and in two somatic cell hybrids (57-77-F7 and αD5-DH11 BC11) (16). Our observations are contrary to the report by Klukas et al. (17) that, in RPMI 1788 cells, the in vitro directed immunoprecipitated μ polypeptide comigrated with the in vivo secreted form. It is possible that RPMI 1788 secretes a nonglycosylated μ chain. In vitro translations of μ mRNA with [^{14}C]mannose instead of [^{35}S]methionine showed no labeling of any polypeptides on polyacrylamide gel autoradiograms.

These observations and the lack of functional endoplasmic reticulum and Golgi apparatus in reticulocyte lysates, combined with the known 10-12% carbohydrate component of human μ chains secreted in vivo (18) all are consistent with the absence of glycosylation of the in vitro synthesized μ polypeptides and their faster mobility on reducing gels. Furthermore, mouse H chains contain an additional NH$_2$-terminal signal peptide and still show similar mobility relationships between in vitro and in vivo synthesized polypeptides (19). Human κ polypeptide (26,500 daltons) synthesized in vitro appears to be about 2500 daltons larger than its in vivo secreted form (24,000 daltons) (16). A similar observation has been made with human λ (20). This is most likely due to signal peptides that are known to be cleaved from the NH$_2$ terminus during Ig assembly in the mouse system (21) and that are not cleaved during in vitro translation unless microsomes are supplemented. The somatic cell hybrid αD5-DH11-BC11 does not secrete any detectable human L chain (Table V) nor does it contain any translatable human L chain mRNAs. Human γ$_2$, α$_2$, and λ mRNAs were isolated and characterized in a similar fashion and their properties are shown in Table V.

These mRNAs were converted into dscDNA and cloned. The μ recombinant bacteria were screened sequentially with [^{32}P]cDNA or mRNA probes of (i) homologous derived μ mRNA; (ii) heterologous μ mRNA; and (iii) sucrose gradient RNA fractions containing no detectable μ mRNA based on translation (data not shown). These procedures narrowed the potential number of μ cDNA recombinants to about 250.

TABLE V. Properties of Human Immunoglobulin mRNAs and Their In Vivo and In Vitro Synthesized Polypeptides

Cell	Source[a]	Ig secreted[b]	H chain ng/10^6 cells/hr	$M_r \times 10^{-3}$ H chain In vivo	H chain In vitro	L chain In vivo	L chain In vitro	mRNA $S_{20}R, C_S$ H chain	L chain
GM607	Lymphoblastoid	IgM κ	10	77	69	24	26.5	20	13
SED	Chronic lymphocytic leukemia	IgM κ + IgD κ	14	77	69	24	26.5	20	13
GM1500	Myeloma	IgG2 κ	27	50	49	23	25	17	12
GM1056	Lymphoblastoid	IgA2 λ	6	60	58	24	26.5	18	13
GM923	Lymphoblastoid	IgA1 λ	2	60	ND	24	ND	ND	ND
αD5-DH11-BC11	HPL-P3	μ(IgG1 κ)	58	77	69	--	--	20	--
D3 D24.3	HPL-P3	μ(IgG2 κ)	46	77	ND	--	--	ND	--
57-77-F7	HPL-P3	μ(IgG κ)	32	77	69	--	--	20	--
CSK-10-2B1-C14	GM607-P3	μ(IgG κ)	21	77	ND	--	--	ND	--
CSK-NS6-201-C7	GM607-NP3	μ	30	77	ND	--	--	ND	--
106.2-B4-3G6	GM1500-P3	γ2;(IgG1 κ)	35	50	ND	24	ND	ND	ND
FSK-4-2B2-C1	GM1500-NP3	γ2 κ	40	50	ND	24	ND	ND	13
DSK-13-2A5-C8	GM1056-P3	α2 λ(IgG1 κ)	171	60	58	24	26.5	18	13
ESK-12-1D2-C3	GM923-P3	α1; λ(IgG1 κ)	63	60	ND	ND	ND	ND	ND
P3 x 63Ag8	Plasmacytoma	(IgG1 κ)	0	57	54;50	23	25;25	17	13
NP3	Nonsecreting P3	--	0	--	--	--	--	--	--

Human Ig secreted and class identification were determined by NaDodSO$_4$/polyacrylamide gel analysis of cell medium class-specific immunoprecipitates labeled with [^{35}S]methionine (9), quantitative immunofluorescence, Ouchterlony precipitin rings with class-specific antisera, radioimmunoassays with class-specific reagents, and subclass Marchalonis assays. ND, not determined.

[a]HPL, human peripheral lymphocytes.

[b]Mouse Ig secreted is shown in parentheses.

[c]$S_{20}R$ was determined in neutral 5–25% sucrose gradient of twice-purified oligo(dT)-cellulose polyadenylated RNA relative to agarose gel purified 4S and 18S rRNA.

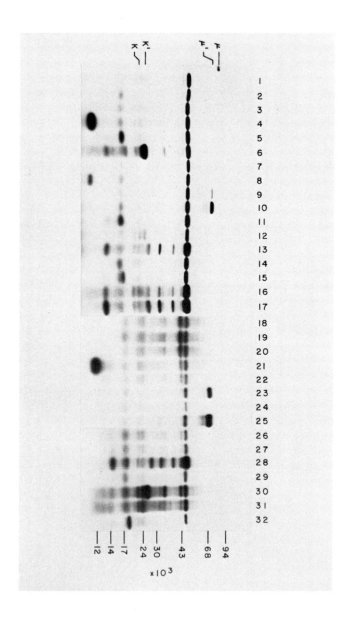

Figure 3.

Sixty of these recombinant plasmids were isolated, linked to DBM-paper, and used for hybridization-selection of their complementary mRNA. Four of these plasmids (two from GM607 and two from αD5-BC11-DH11) were capable of selectively hybridizing mRNA that coded predominantly for a polypeptide with the properties of in vitro synthesized human μ chain (fig. 3, lanes 9, 10, 23, and 25). Plasmid hybridization-selected mRNA translations from human κ (GM607) cDNA clonings and several others were included on these gels but will be thoroughly discussed elsewhere with accompanying sequence data.

Figure 4 shows part of the DNA nucleotide sequence of one of the 40 μ cDNA recombinants identified [pTD-H μ-(αD5: 11-16)]. This Pst I-rescued μ cDNA insert (655 base pairs) contains enough sequences to code for amino acids glycine (residue 421) through the terminal tyrosine (residue 576) of the human μ chain as well as the entire 3'-noncoding sequence including 12 residues of poly(A). The partial amino acid sequence as determined from the base sequence is in almost total agreement with the primary structures of both human OU and GAL μ chains isolated from patients with immune disorders (22, 23). In the human OU μ chain, glutamate residues occur at positions 487 and 493 (22) whereas our sequence data indicate glutamine codons. The human GAL μ chain has an extra glutamine residue at position 488 (20), where no glutamine codon occurs. These inconsistencies are most likely attributed to partial deamination of glutamine during sequential degradation and amino acid analysis.

Figure 3 (Opposite page). Autoradiogram of [^{35}S]methionine-labeled proteins synthesized in vitro with mRNAs hybridselected from total human B cell mRNA by recombinant plasmid cDNAs. GM607 polyadenylated RNA (1 mg) was hybridized with H and L recombinant plasmids linked to diazobenzylmethoxy-paper and the products of in vitro translation directed by eluted mRNA were resolved on composite NaDodSO$_4$/polyacrylamide gels. Lanes: 7 and 29, endogenously labeled products of the reticulocyte lysate; 1 and 24, background translation products of mRNAs eluted from vector-linked filters (pBR322). The other lanes show the translation products of mRNAs eluted from 28 independent recombinant filters: 6, pTD-H κ(607:1-31); 12 and 30, pTD-H κ-(607:8-14), plasmids that selectively hybridize human κ mRNA; 9, pTD-H μ-(αD5:11-10); 10, pTD-H μ-(αD5:11-16); 23, pTD-H μ-(607:6-9) and 25, pTD-H μ-(607:7-12) all selectively hybridize human μ mRNA. M_r x 10^{-3} of standard proteins are shown. μ, μ', κ, and κ', Positions of in vivo and in vitro synthesized polypeptides.

Comparison of the human 3' noncoding sequence with that published for mouse μ (24) shows a striking conservation in certain regions (fig. 4), amino acid residues shared in common. When the common amino,(as well as the codons of many acid codons differ) in the mouse and human mRNA, they involve third base neutral codon substitutions (i.e., 560, Tyr; 563, Asn; 564, Valp 569, Ser; and 576, Tyr) (fig. 4). Unlike several other 3' noncoding regions, the sequence A-A-U-A-A begins 12 residues from the poly(A) addition site in the human μ mRNA rather than the usual 20-30 residues observed

Residue	421	430	440
Amino Acid	GlyGluArgPheThrCysThrValThrHisThrAspLeuProSerProLeuLysGlnThr		

```
Residue              421                      430                           440
Amino Acid    GlyGluArgPheThrCysThrValThrHisThrAspLeuProSerProLeuLysGlnThr
     5'        GG(31)CCGGGGAGAGGTTCACGTGCACCGTGACCCACACAGACCTGCCCTCGCCACTGAAGCAGACC
     3'    ACGTCC(31)GGCCCCTCTCCAAGTGCACGTGGCACTCGGTGTGTCTGGACGGGACGGTGACTTCGTCTGG

                            450                      460
          IleSerArgProLysGlyValAlaLeuHisArgProAspValTyrLeuLeuProProAlaArgGluGlnLeu
     5'    ATCTCCCGCCCCAAGCGGGTGGCCCTGCACAGGCCCGATGTCTACCTGCTGCCACCAGCCCGGGAGCAGCTG
     3'    TAGAGGGCCGGGTTCCCCCACCGGGACGTGTCCGGGCTACAGATGGACGACGGTGGTCGGGCCCTCGTCGAC

                        470                      480
          AsnLeuArgGluSerAlaThrIleThrCysLeuValThrGlyPheSerProAlaAspValPheValGlnTrp
     5'    AACCTGCGGGAGTCGGCCACCATCACGTGCCTGGTGACGGGCCTTCTCTCCCGCGGACGTCTTCGTGCAGTGG
     3'    TTGGACGCCCTCAGCCGGTGGTAGTGCACGGACCACTGCCCGAAGAGAGGGCGCCTGCAGAAGCACGTCACC

                        490                      500                      509
          MetGlnArgGlyGlnProLeuSerProGluLysTyrValThrSerAlaProMetProGluPro
     5'    ATGCAGCGGGGGCAGCCCTTGTCCCCGGAGAAGTATGTGACCAGCGCCCCTATGCCGGAACCC-
     3'    TACGTCGCCCCCGTCGGGAACAGGGGCCTCTTCATACACTGGTCGCGGGGATACGGCCTTGGG-

                        560                      570                      576
          ThrLeuTyrAsnValSerLeuValMetSerAspThrAlaGlyThrCysTyr *
     5'    ACCCTGTACAACGTGTCCCTGGTCATGTCAGACACAGCTGGCACCTGCTACTGACCCTGCTGGCCTGCCCAC
     3'    TGGGACATGTTGCACAGGGACCAGTACAGTCTGTGTCGACCGTGGACGATGACTGGGACGACCGGACGGGTG

                                            ‡
     5'    AGGCTCGGGCGGCTGGCCGCTCTGTGTGTGCATGCAAACTAACCGTGTCAACGGGGTCGAGATGTTGCATCT
     3'    TCCGAGCCCGCCGACCGGCGAGACACACACGTACGTTTGATTGGCACAGTTGCCCCAGCTCTACAACGTAGA

                            †
     5'    TATAAAATTAGAAATAAAAAGATCCATTCA(12)C(26)CTGCA
     3'    ATATTTTAATCTTTATTTTTCTAGGTAAGT(12)G(26)G
```

Figure 4.   Partial nucleotide sequence of human μ[pTD-H μ(αD 5:11-16)]-cDNA insert rescued from a Pst I diges-tion of the recombinant plasmid (16).   Residue refers to amino acid residue from NH$_2$ terminus of human OU.   The nu-cleotide sequence between residues 510 and 559 is pending. *, Termination codon UGA; ‡, termination codon UAA; †, be-ginning of poly(A) tail.   Sequences underlined are homologous sequences observed in the mouse μ untranslated sequence (16).

with other mammalian 3' untranslated sequences (see fig. 4). The differences observed in the 3' noncoding sequences of mouse and human mRNA tend to be clustered, high in G+C content, and remote from the highly homologous regions near the A-A-U-A-A and the initial 3' noncoding sequences. A more thorough analysis comparing the codon preferences and coding and noncoding sequence homologies and divergences should await completion of the mouse and human µ genomic sequence data.

The production of a number of independently derived human H and L chain probes obtained from normal circulating lymphocytes via hybridomas and lymphoblastoid cell lines would be of considerable value in examining the molecular basis of several well-characterized human immunoglobulin disorders.

V. PRODUCTION OF HUMAN HYBRIDOMAS SECRETING HUMAN MONOCLONAL ANTIBODIES

Mouse x human hybridomas segregate human chromosomes and often lose the expression of one or both human immunoglobulin chains so that it could become quite laborious to obtain stable hybrid cells that produce specific human monoclonal antibodies. For this reason we decided to attempt to produce human intraspecific hybridomas.

Since we have also determined that the human myeloma-derived cell line GM1500 produces relatively large amounts of human immunoglobulins, we have attempted to select for hypoxanthine phosphoribosyltransferase-deficient mutants of this cell line in order to have a cell parent able to immortalize antigen-stimulated normal human lymphocytes. We have obtained such mutants following mutagenization of GM1500 human myeloma-derived cells with ethyl methanesulfonate and selection in high concentrations of 6-thioguanine (25). The mutant clones were then tested for the ability to form hybrids with human peripheral blood lymphocytes. Since hybrid clones producing immunoglobulin chains derived from the GM1500-6TG parent and new immunoglobulin chains were obtained, we concluded that the GM1500 mutant cells could be used in an attempt to produce human hybridomas secreting specific human monoclonal antibodies. Therefore we chose to fuse the myeloma-derived cells with peripheral lymphocytes derived from a patient with subacute sclerosing panencephalitis (SSPE), a disease due to a measles virus infection of the central nervous system. This patient had an extremely high titer of antibodies against measles virus. By fusing lymphocytes derived from 10 ml of peripheral blood we obtained more than twenty independent hybrid clones, two of which were capable of immunoprecipitating the major nucleocapsid protein (NP) of measles virus and its cleavage product. Subcloning of the positive hybrids resulted in the selection of subclones most of which (>80%) continued to produce the antiviral monoclonal antibodies, suggesting that the human x human hybridomas are quite stable.

VI. CONCLUSION

The results described in this paper indicate that it is possible to produce inter- and intra-specific hybrids secreting human immunoglobulins. Because of chromosome segregation in interspecific rodent x human hybrid cells and the stability of human x human hybridomas, it appears that human hybrid clones are much more promising if we intend to immortalize specific human antibody-producing lymphocytes in order to study the development of human autoimmune diseases and to produce specific human reagents to be used in human immunotherapy.

REFERENCES

1. Kohler, G., and Milstein, C., Nature 256, 445 (1975).
2. Koprowski, H., Gerhard, W., and Croce, C. M., Proc. Natl. Acad. Sci. USA 74, 2485 (1977).
3. Martinis, J., and Croce, C. M., Proc. Natl. Acad. Sci. USA 75, 2320 (1978).
4. Gerhard, W., Croce, C. M., Lopes, D., and Koprowski, H., Proc. Natl. Acad. Sci. USA 75, 1510 (1978).
5. Croce, C. M., Shander, M., Martinis, L., Cicurel, L., D'Ancona, G. G., Dolby, T. W., and Koprowski, H., Proc. Natl. Acad. Sci. USA 76, 3416 (1979).
6. Shander, M., Martinis, J., and Croce, C. M., Transplant. Proc. 12, 417 (1980).
7. Erikson, J., Martinis, J., and Croce, C. M., Nature 294, 173 (1981).
8. Croce, C. M., Shander, M., Martinis, M., Cicurel, L., D'Ancona, G. G., and Koprowski, H., Eur. J. Immunol. 10, 486 (1980).
9. Early, P., and Hood, L., Cell 24, 1 (1981).
10. Manolov, G., and Manolova, Y., Nature 273, 33 (1972).

11. Zech, L., Haglund, V., Nillson, K., Anotklin, G., Int. J. Cancer 17, 47 (1976).
12. Van den Berghe, H., et al., Cancer Genet. Cytogenet. 1, 9 (1979).
13. Miyoshi, I., Hiraki, S., Kimura, I., Miyamoto, K., and Sato, J., Experientia 35, 742 (1979).
14. Berger, R., et al., Human Genet. 53, 111 (1979).
15. Bornheim, A., Berger, R., and Lender, G., Cancer Genet. Cytogenet. 3, 307 (1981).
16. Dolby, T. W., DeVuono, J., and Croce, C. M., Proc. Natl. Acad. Sci. USA 77, 6027 (1980).
17. Klukas, C. K., Cramer, F., and Ganlot, H., Nature 269, 262 (1977).
18. Metzger, H., in "Advances in Immunology" (F. J. Dixon and H. G. Kunkel, eds.), p. 57. Academic Press, New York, (1970).
19. Jilka, R. L., and Pestka, S., Proc. Natl. Acad. Sci. USA 74, 5692 (1977).
20. Yaffe, L., and Pestka, S., Arch. Biochem. Biophys. 190, 495 (1978).
21. Burstein, Y., and Schechter, I., Proc. Natl. Acad. Sci. USA 74, 716 (1977).
22. Putman, F. W., Florent, G., Paul, C., Shimada, T., and Shimozu, A., Science 182, 287 (1973).
23. Watanabe, S., Barbikal, H. V., Horn, J., Bertram, J., and Hilochmann, N. N., Physiol. Chem. 354, 1505 (1973).
24. Calame, K., Rogers, J., Early, P., Davis, M., Livant, D., Wall, R., and Hood, L., Nature 284, 452 (1980).
25. Croce, C. M., Linnenbach, A., Hall, W., Steplewski, Z., and Koprowski, H., Nature 288, 488 (1980).

DISCUSSION

J.W. LARRICK: How stable have the SSPE-derived clones been over the past 15 months?

C.M. CROCE: The clones are very stable. Subcloning of the two positive clones resulted in the growth of hybridomas that produced anti-measles virus antibodies in more than 80% of the subclones.

J.W. LARRICK: Is it true that GM1500 is in fact an EBV$^+$ lymphoblastoid cell line?

C.M. CROCE: Yes, the GM1500-6TG lines are EBNA-positive. In order to avoid the possibility of obtaining EBV transformed cells it is preferable to use GM1500-6TG mutants that are resistant to ouabain. The hybrids can be selected in HAT medium containing 10^{-5}M ouabain.

J.W. LARRICK: Have you performed 2D-gels on the supposedly monoclonal antibodies to demonstrate that they are monoclonal?

C.M. CROCE: We did not do 2D-gel analysis but we have carried out SDS polyacrylamide gel electrophoresis of the secreting chains. Since, as you know, one lymphocyte produces only one antibody, one hybridoma clone produces only one specific antibody.

J.W. LARRICK: Have you done a chromosome analysis of your human-human hybrids? Over a period of time to test stability?

C.M. CROCE: The human hybridoma are nearly tetraploid. Chromosome segregation is slow, but it occurs.

S.I. CHAVIN: You showed a μ' chain, translated in vitro, which had faster electrophoretic migration than the secreted counterpart. Do you know whether this μ' chain possesses a leader sequence?

C.M. CROCE: The μ polypeptide chains produced in vitro are identical to those produced in vivo. The difference is only in glycosylation. No glycosylation of the μ chains occurs in the in vitro translation system. If we treat μ-producing cells with tunicamycin, the size of the μ chain collapses to the size of the in vitro product.

J.L. STROMINGER: With what did you fuse the B cell from the juvenile diabetes patient?

C.M. CROCE: The lymphocytes were fused with the human myeloma derived cell line.

IN VITRO SYNTHESIS OF DNA
AND THE GENERATION
OF PROTEIN ANALOGS

THE ROLE OF SYNTHETIC DNA IN THE PREPARATION OF STRUCTURAL GENES CODING FOR PROTEINS[1]

John J. Rossi
Ryszard Kierzek[2]
Ting Huang
Peter Walker
Keiichi Itakura

Department of Molecular Genetics
City of Hope Research Institute
1450 East Duarte Road
Duarte, California 91010 USA

I. INTRODUCTION

One of the key steps in the production of proteins by recombinant DNA technology is the preparation of structural genes coding for those desired proteins. There may be two approaches to this goal, artifical gene synthesis (1) and a combination of cloned cDNA sequences and synthetic DNAs (2).

A. Artificial Gene Synthesis

The present method for the preparation of artificial genes was developed for the synthesis of genes for tRNAs by Khorana and coworkers in the late 1960's (3). Oligonucleotides (10-15 bases long) are chemically synthesized and joined together with DNA ligase using the inherent nature of oligonucleotide chains to form ordered duplexes by virtue of base pairing (3). A gene for somatostatin was synthesized using this approach and successfully expressed in *E. coli* to produce the peptide hormone (Figure 1) (1). The longest gene synthesized to date is that coding for a human leukocyte interferon $\alpha1$ gene (4), no fewer than 514 base pairs (b.p.)---

---

[1]Supported by USPHS Grant GM 26408 (KI).
[2]Present address: Institute of Organic Chemistry, Polish Academy of Sciences, Poznan, Poland.

FROM GENE TO PROTEIN:
TRANSLATION INTO BIOTECHNOLOGY

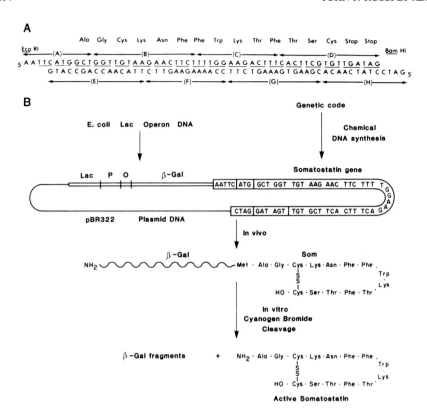

FIGURE 1. (A) Somatostatin gene. Eight oligodeoxynucleotides (fragments A to H) were synthesized by the
modified triester method (4). The codons are underlined and
their corresponding amino acids are given. The eight fragments were designed to have at least five nucleotide complementary overlaps to ensure correct joining by T4 DNA ligase
to give duplex DNA 60 nucleotides long. (B) Schematic overview of the somatostatin project. The gene for somatostatin
was fused to the *E. coli*; the plasmid directs the synthesis
of a chimeric protein that has somatostatin on a short tail
on the end of β-galactosidase. Somatostatin is clipped off
by treatment with cyanogen bromide, which cleaves specifically at methionine residues.

more then 1,000 nucleotides---have been assembled from 67
oligonucleotide building blocks which are synthesized by the
solid phase method. Although this result demonstrates that
relatively large genes can be synthesized by the combination

of short synthetic oligonucleotides and DNA ligase, the synthesis of genes coding for the average sized proteins (300
amino acids) would still be time consuming.

Recent progress in the solid phase synthesis of polydeoxyribonucleotides permits the development of another
approach for the construction of double stranded DNA of desired sequences. As described below (Figure 3), four polynucleotides (43-mer, 42-mer, and two 39-mers) are synthesized
on a solid support. The sequences of these four fragments
are designed to form ten and nine base pair duplexes through
their 3'-termini, respectively. When these fragments are
allowed to anneal with each other at their 3' termini and
used as substrates for a DNA polymerase I reaction with the
four deoxynucleoside triphosphates, two 72 base pair double
stranded DNAs are produced. The resultant DNA fragments are
joined together by DNA ligase in the presence of a large
fragment of the plasmid PXJOO1(5) cut with restriction endonucleases EcoRI and PstI. After cloning, the sequence of a
synthetic 136 b.p. DNA fragment was confirmed by the Maxam-
Gilbert method (6).

For the chemical synthesis of a 136 b.p. DNA, 272
nucleotides are required by the classical approach. However,
by the new approach, the requirement for the synthetic DNA is
reduced to only 163 nucleotides.

B. Combination of Cloned cDNA Sequences and Synthetic DNA

This approach has been developed for the preparation of a
gene coding for the human growth hormone (2) (Figure 2). The
total cDNA sequence is digested with an appropriate restriction endonuclease and ligated to the synthetic DNA fragment
encoding part of the N-terminal amino acid sequence of the
hormone. Thereby, the cDNA clone can be edited to code only
for the amino acid sequence essential for biological activity.

The most difficult task in the cloning of double stranded
cDNA sequences is likely to be the identification of the
desired colony, particularly when the desired mRNA is of low
abundance. Very recently, we have developed a colony screening method using oligonucleotides as hybridization probes (7,
8, 9). The general approach is to chemically synthesize a
mixture of oligonucleotides which represent all possible
codon combinations for a small portion of the amino acid sequence of a given protein. Under stringent hybridization
conditions only the perfectly matched duplex will form, allowing use of the mixture of oligonucleotides as a specific
hybridization probe. This method was successfully applied to
identify the cDNA clones for β2-microglobin (8) and H-2K[b]

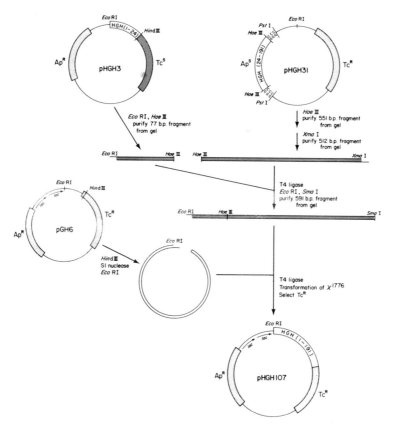

FIGURE 2. Construction of a Plasmid (pHGH107) Containing
a Gene for Human Growth Hormone. The plasmid pHGH3 was
cleaved with *EcoR1* and *Hae*III restriction endonucleases and
the 77 base pair fragment containing coding sequences for HGH
amino acids 1-23 was isolated. The plasmid pHGH31 was successively cleaved with *HaeIII and Xma*I and the resulting 512
base pair fragment containing coding sequences for amino
acids 24-191 was isolated. These two fragments, *EcoRI-Hae*III
and *Hae*III-*Xma*I, were ligated with T4 DNA ligase and then
treated with *EcoRI* and with *Sma*I to yield a 591 base pair
fragment. This fragment was inserted into the expression
plasmid pGH6 between *EcoR1* and *Hind*III restriction endonucleases and the resulting plasmid pHGH107 was used to
transform X1776.

antigen (9). Therefore, the identification of a cloned cDNA
sequence of low abundance mRNA would no longer be a time
consuming step, provided that a partial amino acid sequence
of the desired protein is available.

In this paper, we present the details of a new approach for the synthesis of artificial genes. In the Appendix, the details of a rapid and simple solid phase method for the synthesis of oligodeoxyribonucleotides are described. The details of the colony screening method using oligonucleotide probes have already been published (7, 8, 9).

II. RESULTS

A. Synthetic Design

The complete amino acid sequence of human leukocyte interferon (Le-IF) α2 has been deduced by analysis of cDNA (10) and genomic (11) clones. We have utilized the amino acid sequences and codons of the most abundant *E. coli* tRNA species (12) in order to design a gene for IF α2 (residues 126 through 165). Therefore, it differs considerably from that found in the corresponding natural IF α2 (Table I).

Four oligonucleotides, 43, 42, 39 and 39 bases long, were chemically synthesized on the solid support by the phosphotriester approach. Each pair of fragments was designed to

TABLE I.   Comparison of Codon Usage for Synthetic Versus Corresponding Natural α2 Gene Segment[a]

		U		C		A		G	
U	Phe	0-*1*	Ser	5-*2*	Tyr	0-*1*	Cys	1-*1*	U
		1-*0*		1-*0*		2-1		-	C
	Leu	0-*1*		0-*1*	Ochre	-	Opal	0-*1*	A
		0-*2*		-	Amber	1-*0*	Trp	1-*1*	G
C	Leu	1-*0*	Pro	0-*1*	His	-	Arg	3-*0*	U
		0-*1*		-		-		-	C
		-		-	Gln	0-*1*		-	A
		4-*1*		1-*0*		1-*0*		-	G
A	Ile	-	Thr	2-*1*	Asn	-	Ser	0-*2*	U
		2-2		-		1-*1*		0-*1*	C
	Met	-		0-*1*	Lys	2-2	Arg	0-*3*	A
		1-*1*		-		2-2		-	G
G	Val	1-*1*	Ala	2-*1*	Asp	-	Gly	-	U
		0-*1*		-		-		-	C
		1-*0*		0-*1*	Glu	4-*3*		-	A
		-		-		1-2		-	G

[a] Italicized numbers represent natural IFα2 codon usage for amino acids 126 to the stop codon (10).

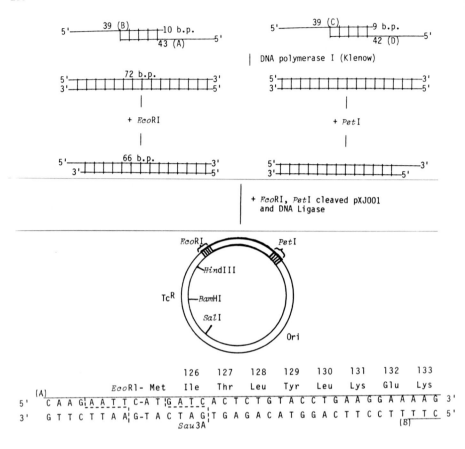

	126	127	128	129	130	131	132	133
	EcoR1- Met	Ile	Thr	Leu	Tyr	Leu	Lys	Glu	Lys

```
       (A)        EcoRl- Met Ile Thr Leu Tyr Leu Lys Glu Lys
    5'  C A A G A A T T C-A T G A T C A C T C T G T A C C T G A A G G A A A A G 3'
    3'  G T T C T T A A G-T A C T A G T G A G A C A T G G A C T T C C T T T T C 5'
                        Sau3A                                          (B)
```

	134	135	136	137	138	139	140	141	142	143	144	145	146
	Lys	Tyr	Ser	Pro	Cys	Ala	Trp	Glu	Val	Val	Arg	Ala	Glu

```
                       (A)                                            (C)
    5' A A A T A C T C T C C G T G T G C T T G G G A A G T T G T A C G T G C T G A A 3'
    3' T T T A T G A G A G G C A C A C G A A C C C T T C A A C A T G C A C G A C T T 5'
                                                                   (B)
```

	147	148	149	150	151	152	153	154	155	156	157	158	159
	Ile	Met	Arg	Ser	Phe	Ser	Leu	Ser	Thr	Asn	Leu	Gln	Glu

```
                                                                (C)
    5' A T C A T G C G T T C T T T C T C C C T G T C T A C T A A C C T T C A G G A G 3'
    3' T A G T A C G C A A G A A A G A G G G A C A G A T G A T T G G A A G T C C T C 5'
                                                       (D)
```

	160	161	162	163	164	165
	Ser	Leu	Arg	Ser	Lys	Glu Stop	PstI

```
    5' T C T C T G C G T T C T A A A G A A T A G C T G C A G T G G
    3' A G A G A C G C A A G A T T T C T T A T C G A C G T C A C C
                                                          (D)
```

have a short region of complementary sequences at their
3' termini (Figure 3). For the A and B fragments, this
stretch is 10 b.p. long, and for the C and D fragments, 9
b.p. When these pairs of fragments were annealed, they were
substrates for DNA polymerase I (Klenow) in the presence of
four deoxyribonucleoside triphosphates. In addition, restriction endonuclease recognition sites EcoRI and PstI have
been introduced into the synthetic design to facilitate cloning of the synthetic fragments (I and II) into a bacterial
plasmid vector.

B. *In vitro* Construction of Double-Stranded Gene Fragments

Prior to polymerization, the synthetic fragments were
phosphorylated at their 5' termini, using polynucleotide
kinase and $[\gamma\text{-}^{32}\text{P}]$ATP. This was done to facilitate detection
and subsequent cleavage of the double stranded products by
the appropriate restriction endonucleases. In a typical
experiment, equimolar amounts of 5' terminally phosphorylated
fragments, A and B or C and D, were mixed together, heated in

FIGURE 3. Schematic of Strategy for Assembly and Cloning
of the IFNα2 Gene Segment. The synthetic gene fragments consist of the following deoxyribooligonucleotides: A, 43 mer;
B, 39 mer; C, 42 mer; and D, 39 mer. Fragments A and B share
10 bases of a complementary sequence at their 3' termini,
while fragments C and D share 9 bases of a 3' complementary
sequence. The appropriate pairs of fragments were annealed,
and the double-stranded products polymerized by the addition
of all four deoxyribonucleoside triphosphates and *E. coli* DNA
polymerase I (Klenow fragment). The double-stranded products
formed are 72 base pairs long in each case.
    Subsequent to the polymerization reaction, the double
stranded polymerization products were cleaved with either
EcoRI for A and B, or PstI for C and D. The desired products of these cleavages were purified by gel electrophoresis and ligated with the EcoRI, PstI cleaved plasmid pXJ001
as depicted. See text for additional details concerning
screening and selection of colonies containing the cloned
IFNα2 gene segment.
    The complete nucleotide sequence of the synthetic, poly-
merized products is depicted below the schematic. The heavy
underlining depicts the original, synthetic oligonucleotides
A, B, C and D. The underlinings overlap at the regions of 3'
complementary sequences.

a boiling water bath for 3 minutes, quickly chilled, and then allowed to form the desired annealing at their 3' termini (see Figure 4). The fragments were then incubated with all four deoxyribonucleoside triphosphates and DNA polymerase I for 30 minutes at room temperature. The reaction products were electrophoresed in acrylamide to resolve the duplexes from the single-stranded oligonucleotides. From the relative intensities of the bands in the autoradiograph presented in Figure 4, it can be seen that approximately 40-50% of the starting single stranded oligonucleotides were polymerized into duplex structures. To prepare the synthetic sequences for biological cloning, 62 picomoles (pmoles) of each oligonucleotide were treated as described above. Subsequent to polymerization, the duplex formed from A plus B was cleaved with the restriction endonuclease *Eco*RI, while the C plus D duplex was cleaved with *Pst*I. After restriction endonuclease digestions were deemed complete, the appropriately cleaved fragments were electrophoresed in a non-denaturing acrylamide gel (data not shown), and electroeluted from the gel slices in preparation for biological cloning.

C.   Strategy for Cloning the Synthetic IFNα Fragments

The bacterial vector used for cloning the IFα2 fragments is a derivative of pBR327, designated pXJ001 (5). This vector has unique *Eco*RI and *Pst*I sites into which the synthetic fragments were inserted. Removal or replacement of the ca. 650 b.p. plasmid *Eco*RI-*Pst*I fragment destroys the plasmid coded ampicillinase gene. The strategy used for ligation of the synthetic oligonucleotides into the plasmid vector is depicted in Figure 3 and involves a three-part ligation of

FIGURE 4.   *In Vitro* Polymerization of Double-Stranded Fragments. Using E. coli DNA Polymerase I (Klenow). A, B, C and D were individually phosphorylated at their 5' termini with polynucleotide kinase and a mixture of $[\gamma^{32}]$ATP and non-radioactive ATP. The reactions were terminated by extractions with phenol and chloroform, and finally, ethanol precipitated after the addition of carrier RNA. The precipitates were dried *in vacuo*, resuspended in distilled water, and equal molar amounts of fragments A and B or C and D mixed together in a final volume of 41 μl at individual concentrations of 1.5 μm. The mixtures were heated in a boiling water

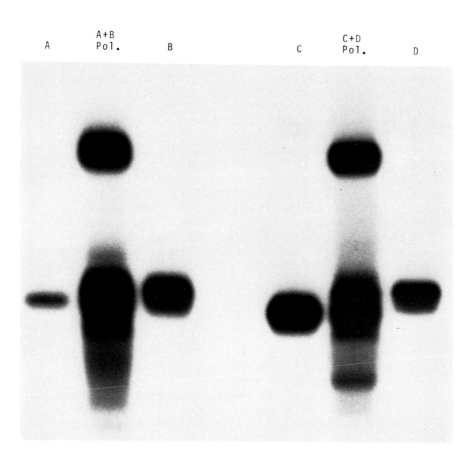

bath for 3 minutes, cooled quickly in an ice water bath, and brought to 10 mM Tris pH 7.9, 60 mM NaCl, 10 mM β-mercapto-ethanol. A mixture of all four deoxyribonucleoside triphos-phates were added to a final concentration of 60 μgm per ml for each dNTP. The polymerization reaction was initiated by the addition of 1.5 units of DNA polymerase I (Klenow) and allowed to proceed for 30 minutes at room temperature. The polymerization was stopped by 2 phenol extractions, followed by 2 chloroform extractions and finally, an ethanol precipi-tation. Aliquots of the individual (A, B, C or D) and poly-merized (A + B Pol.) or (C + D Pol.) products were electrophoresed in an 8% polyacrylamide gel. The single stranded 43 (A), 39 (B), 42 (C), and 39 (D), as well as the polymerized 72 base pair (A+B Pol.) and 72 base pair (C + D Pol.) fragments are indicated. The background products in the polymerized reactions are due to the 3-5'- exonuclease activity of DNA polymerase I.

*Eco*RI ends, *Pst*I ends, and a blunt end ligation of the two synthetic fragments. Approximately 30 picomoles (pmoles) each of the double stranded *Eco*RI or *Pst*I cleaved fragments were ligated with 2 pmoles of *Eco*RI-*Pst*I cleaved, alkaline phosphatase treated pXJ001. The ligation mixture was used to transform the *E. coli* strain MC1061, and selection was made for colonies resistant to 30 µg/ml tetracycline. From this mixture there were approximately 1,000 transformants. Twenty out of approximately 400 clones tested were ampicillin sensitive. Plasmid DNA was prepared from 10 of these ampicillin sensitive colonies using a mini-screen procedure (13). Eight of the 10 colonies tested had plasmids which, when cleaved with *Eco*RI and *Pst*I, yielded a restriction fragment of the expected length, 132 b.p. Plasmid DNA from one of these transformants was purified on large scale and the insert DNA was subjected to DNA sequence analysis (Figure 5). The results presented demonstrate that the desired gene fragment sequence had been successfully cloned by the combined chemical-biochemical approach.

III. DISCUSSION

The present paper describes an alternative approach to the synthesis of polydeoxyribonucleotides with defined sequences longer than those accessible by conventional chemical methods. The original method developed by Khorana and his coworkers (3) has a few disadvantages. In order to obtain correct ligation of synthetic oligonucleotides, the target sequence must be very carefully designed to avoid both complementary and repeated sequences. The total double stranded DNA fragment is constructed by template-dependent joining of synthetic oligonucleotides (10-15 bases long) with DNA ligase. Therefore, if a 136 base pair DNA is desired, 272 nucleotides must be chemically synthesized.

Recent advances in the chemical synthesis of oligonucleotides (by the solid phase phosphotriester approach) permit us to rapidly synthesize single stranded oligonucleotides of up to 40 bases in length (see Appendix). Accordingly, it is now possible to synthesize double-stranded DNA fragments using DNA polymerase I in the presence of four deoxynucleoside triphosphosphates. Although DNA polymerase has been utilized previously to complete the synthesis of a double-stranded DNA fragment, only one strand served as a primer (3). With the methodology presented here, the amount of chemical synthesis needed to prepare a 136 base pair DNA is reduced to 163 nucleotides as compared with 272 nucleotides with the ligation method. Therefore,

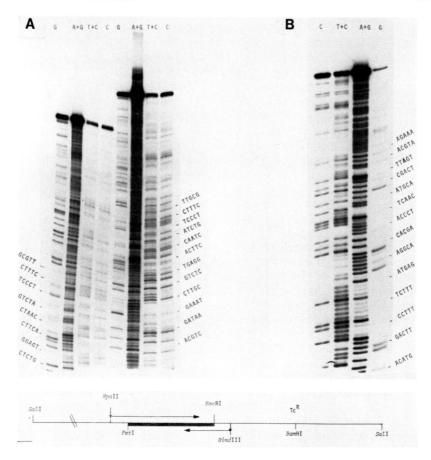

FIGURE 5. DNA Sequence of Cloned IFNα2 Gene Segment. The DNA sequences presented were determined from 3'-labelled fragments using the DNA sequencing procedure of Maxam and Gilbert (6). For the sequence presented in panel A, a *Hpa*II site corresponding to the site at position 2460 of the pBR327 sequence (19) was 3' labelled with α[³²P]dCTP and DNA polymerase I (Klenow). The plasmid was secondarily cleaved with *Hind*III and the appropriate fragment was purified from the gel and sequenced. The nucleotide sequence presented in Panel A corresponds to the upper strand of the Figure 1 sequence beginning with the *Pst*I site. For clarity, two different lengths of sequencing gel runs are presented. For the panel B sequence the *Hind*III site was 3' labelled with α[³²P]dATP and DNA polymerase I (Klenow). The plasmid was secondarily cleaved with *Pst*I. The sequence presented corresponds to the lower strand of the Figure 1 sequence beginning with amino acid 128. The correct sequence upstream of this position has also been confirmed (data not presented).

the new approach requires approximately 40% less chemical synthetic work than the original approach. In the design of our approach, examination of the self-complementary sequences at the 3'-end of oligonucleotides is an essential requirement, which can easily be done. Thus, our approach may have more flexibility in the design of target DNA sequences than the ligation approach. With further improvement of the chemical synthesis of long oligodeoxyribonucleotides by the solid phase method, our approach should prove to be a generally useful method for the synthesis of double-stranded DNA fragments encoding average sized proteins, as well as regulatory sequences.

One of the important goals in synthetic chemistry is to make molecules function better than natural products. It would be surprising if it were found that the double stranded cDNA sequence of a mRNA coding for a human protein contains the most optimal codons for expression in a different environment such as in *E. coli* or yeast.

There has been much speculation about the mechanisms by which the expression of human cDNA sequences cloned into microrganisms are controlled (14). These may include the stability of the products, copy number of plasmids, transcription and translation efficiencies, stability of mRNAs, as well as other factors. Regulation at the level of transcription has been examined and a number of promoter systems have been devised (2,15,16). These systems have allowed high levels of expression of human genes, presumably as a result of the presence of abundant copies of the mRNAs.

For the translation of mRNAs, studies on initiation have revealed that the secondary structure around the initiation site, and the distance between a ribosome binding site and the initiation codon AUG may play significant roles in gene expression (17). Comparison between the mRNA sequences of highly and weakly expressed genes in *E. coli* (12) and yeast (18) suggests that highly expressed genes have a special bias in codon usage. We have designed the gene (IFα2) using condons preferred for highly expressed genes in *E. coli* (Table I).

Obviously, we will have to await the total gene synthesis for IFα2 in order to draw any conclusion as to whether or not codon usage plays a significant role in the gene expression.

REFERENCES

1. Itakura, K.,Hirose, T., Crea, R., Riggs, A.D., Heyneker, H.L., Bolivar, F., and Boyer, H.W. *Science*, *198*, 1056 (1977).
2. Goeddel, D.V., Heyneker, H.L., Hozumi, T., Arentzen, R., Itakura, K., Yansura, D.G., Ross, M.G., Miozzari, G., Crea, R., and Seeburg, P.H., *Nature*, *281*, 544 (1979).
3. Khorana, H.G., *Science*, *203*, 614 (1979).
4. Edge, M.D., Greene, A.R., Heathcliffe, G.R., Meacok, P.A., Schuch, W., Scanlon, D.B., Atkinson, T.C., Newton, C.R., and Markham, A.F., *Nature*, *292*, 756 (1981).
5. Soberon, X., Rossi, J.J., Larson, G.P., and Itakura, K., (1982) *Nucl. Acids Res. Symp. Series*, *No.9*, in press.
6. Maxam, A., and Gilbert, W., *Proc. Natl. Acad. Sci. USA*, *74*, 560 (1977).
7. Wallace, R.B., Johnson, M.J., Hirose, T., Miyake, T., Kawashima, E.H., and Itakura, K., *Nucl. Acids Res.*, *9*, 879 (1981).
8. Suggs, S.V., Wallace, R.B., Hirose, T., Kawashima, E.H., and Itakura, K., *Proc. Natl. Acad. Sci. USA*, *78*, 6613 (1981).
9. Reyes, A.A., Johnson, M.J., Schold, M., Ito, H., Ike, Y., Morin, C., Itakura, K., and Wallace, R.B., *Immunogenetics*, *14*, 383 (1981).
10. Goeddel, D.V., Leung, D.W., Dull, T.J., Gross, M., Lawn, R.M., McCandliss, R., Seeburg, P.H., Ullrich, A., Yelverton, E., and Gray, P.W., *Nature*, *290*, 20 (1981).
11. Lawn, R.M., Adelman, J., Dull, T., Gross, M., Goeddel, D.V., and Ullrich, A., *Science*, *212*, 1159 (1981).
12. Ikemura, T., *J. Mol. Biol.*, *151*, 389 (1981).
13. Birnboim , H.C., and Doly, J., *Nucl. Acids Res.*, *7*, 1513 (1979).
14. Morgan, J., and Whelon, W.J., *Recombinant DNA and Genetic Experiment*, 141-143, (Pergamon, Oxford, 1979).
15. Gray, P.W., Leung, D.W., Pennica, D., Yelverton, E., Najarian, R., Simonsen, C.C., Derywck, R., Sherwood, R.J., Wallace, R.M., Berger, S.L., Leviwson, A.D., and Goeddel, D.V., *Nature*, *295*, 503 (1982).
16. Private communication from M. Inouye.
17. Roberts, J.M., Kacich, R., and Ptashne, M., *Proc. Natl. Acad. Sci. USA*, *76*, 760 (1979).
18. Private communication from B. Hall.
19. Soberon, X., Covarrubias, L., and Bolivar, F., *Gene*, *1*, 287 (1980).

APPENDIX

SYNTHESIS OF OLIGODEOXYRIBONUCLEOTIDES BY THE PHOSPHOTRIESTER SOLID PHASE METHOD.

In the chemical synthesis of oligodeoxyribonucleotides by the phosphotriester approach on the solid support, the 3'-terminal nucleoside of the desired sequence is initially attached to the support (2). The coupling units, mono-nucleotides (7) or dinucleotides (1), are sequentially coup-led to the growing nucleotide chain on the support until the desired sequence is built (Figure 1). The product (on the support) is composed of completely protected oligonucleotides and, therefore, after the synthesis all of the protecting groups must be removed. The final product is purified by HPLC or gel electrophoresis and can be used for biological studies. If the synthesis is started using commercially available reagents and solvents, the construction of oligo-nucleotides essentially requires three reactions (Figure 1, Table I).

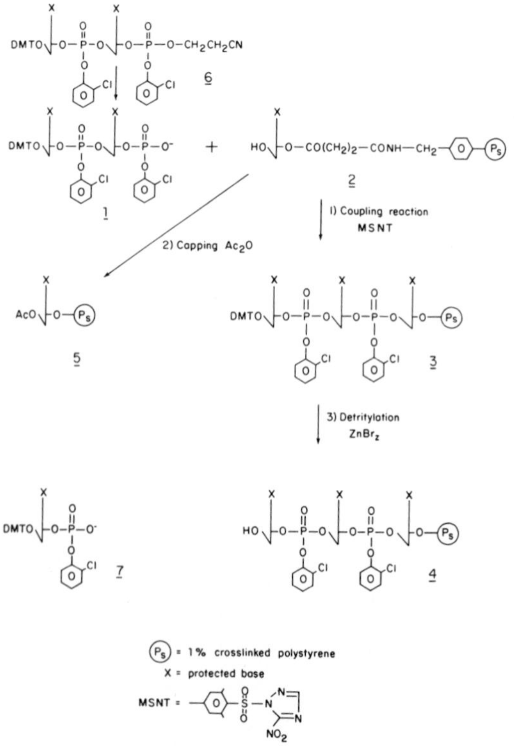

FIGURE 1.

TABLE I. Assembly of Oligonucleotides on the Support[a]

STEP		SOLVENT or REAGENT	AMOUNT	REACTION TIME	NUMBER OF OPERATIONS
1	Washing	THF	3.0 ml	10 sec.	3
2	Drying	Suction		5 min.	1
3	Coupling reaction	Nucleotide	0.3 ml	40 min.[c]	1
4	Washing	Pyridine	3.0 ml	10 sec.	1
5	Capping	THF-Pyridine-Ac$_2$O DMAP (100 mg)	1.5 ml	10 min.	1
6	Washing	Pyridine	3.0 ml	10 sec.	1
7	Washing	CH$_2$Cl$_2$-IPA (85:15 v/v)	3.0 ml	10 sec.	3
8	Detritylation[b]	1 M ZnBr$_2$	1.5 ml	5-10 min.	1
9	Washing	CH$_2$Cl$_2$-IPA (85:15 v/v)	3.0 ml	10 sec.	1
10	Washing	0.5 M Et$_3$N$^+$HAcO$^-$ in DMF	3.0 ml	10 sec.	3
11	Washing	Pyridine	3.0 ml	10 sec.	1

[a]20 mg of the support is used.
[b]Repeat two times, if necessary.
[c]If the reaction is carried out at 40°C, it can be done in 20 minutes.

1) Sequential addition of the coupling units to the support (coupling reaction);
2) Capping of any unreacted 5'-hydroxyl group (capping reaction); and
3) Removal of the 5'-hydroxyl protection DMT group to afford a new 5'-hydroxyl group for the next coupling cycle (detritylation reaction).

We strongly recommend the dinucleotide· coupling approach because the number of coupling cycles is half that of the monomer approach for synthesizing the same length of oligonucleotides, longer oligonucleotides (approximately 35 bases) can be synthesized, and the final purification is much easier. This paper will detail the experiments of the protocol using 20 mg of the support.

I. SOLVENTS AND REAGENTS

All solvents (reagent grade) are commercially available, pyridine, tetrahydrofuran (THF), dichloromethane, dimethylformamide (DMF), and isopropyl alchohol (IPA) from the J.T. Baker Co., acetic anhydride and triethylamine from Fisher Scientific, anhydrous zinc bromide (ZnBr$_2$, Alfa), tetramethylguanidine (Tridom-Fluka), *syn*-pyridine-2-carboxaldoxime and 4-dimethylaminopyridine (DMAP) from Aldrich Chemicals. All of the nucleotides (monomers and dimers), the coupling reagent 2- (mesitylbenzenesulfonyl) -3-nitro -1,2,4-triazole

(MSNT), and the solid supports which carry the first nucleo-sides are available from Bachem, Inc. or other commercial sources.

When anhydrous solvents are required (pyridine, THF, and triethylamine), the reagent grade solvents should be kept in the presence of molecular sieve 4A (J.T. Baker Co.) for one week before use. For convenience, a small amount of anhydrous solvents (pyridine and THF) can be kept in a round bottomed flask using a septum. Transfer of these solvents can be done with a hypodermic syringe through the septum. From time to time, low pressure inside the flask should be released by nitrogen gas. Disposable pipets are very con-venient for handling other solvents.

II. EQUIPMENT

Simple glass equipment is used for the coupling reaction and coevaporation of the nucleotides with pyridine, as depicted in Figure 2. Nitrogen gas, a heat gun (hair dryer), vortex, oil pump and a solvent trap are essential (Figure 2b). All reactions described are carried out at room tem-perature, unless otherwise mentioned.

III. OPERATIONS

A. Reaction 1: Coupling Reaction

In order to obtain high yields of the coupling reaction, anhydrous conditions are essential. Steps 1 and 2 should be carried out very carefully to avoid any source of moisture.

<u>1. Steps 1 and 2: Drying Solid Support (Table II).</u> The solid support (20 mg, 0.1 mmole nucleoside/g resin is put in the reaction vessel filter (Figure 2a) and a septum is applied to the top of the filter. The system is vacuumed weakly through the suction flask and then the stopcock is closed. Dry THF (3 ml) is injected slowly into the filter through the septum using a hypodermic syringe.

The mixture of the support and the solvent is allowed to stand for approximately 10 seconds and then the solvent is removed by suction. The washing step does not require a mix-ing operation nor a long reaction time (10 seconds is enough)

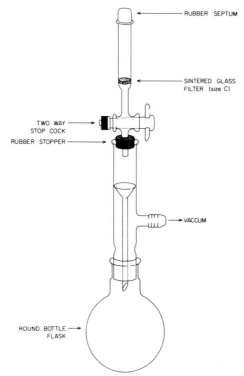

FIGURE 2a.   Reaction apparatus.

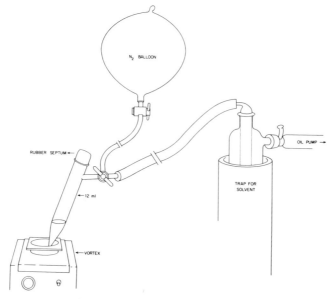

FIGURE 2b.   Evaporation apparatus.

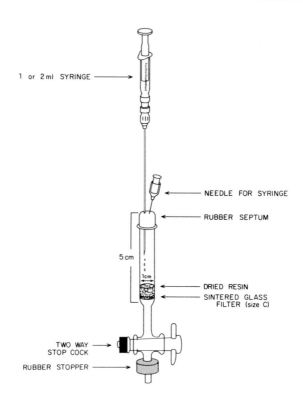

FIGURE 2c. Syringe/septum detailing.

since the polystyrene support swells very quickly in various
solvents. This operation is repeated three times. The stopcock of the filter is closed and separated from the suction
flask. Vacuum is applied directly to the filter (5 minutes
with warm air from a dryer), and the vacuum is then released
by nitrogen gas. The coupling mixture (described below) is
injected through the septum with a syringe.

    2. Step 3: Coupling Reaction, Monomer Coupling. Usual-
ly commercially available monoucleotides 1 are not completely
dry and a coevaporation step with pyridine is essential before the coupling reaction. A mixture of 25 µM of a nucleotide and 0.5 ml of pyridine (the pyridine should be injected
by a syringe from the top of the flask) in the evaporation
flask is set on a vortex and vacuum is applied (Figure 2b).
Evaporation is sped up with warm air. After the first evaporation, the stopcock is closed and the vacuum released by
nitrogen (N_2) gas (or dry air). Secondly, pyridine (0.5 ml)
is injected into the flask and evaporated as described above,

and the vacuum is released by N_2 gas. The coupling reagent, MSNT, dissolved in pyridine (0.45 M solution, 0.25 ml) is taken from the stock solution and added to the nucleotide using a syringe through the septum. The flask is shaken until the nucleotide is dissolved completely in pyridine (approximately 5 minutes). The mixture is taken up using a syringe and injected into the reaction vessel through the septum. A syringe needle must be inserted into the filter through the septum in order to release the pressure which is building up when the reaction mixture is injected. The reaction is gently shaken for 40 minutes at room temperature. If the reaction is carried out at 40°C, the coupling can be done in 20 minutes.

During this coupling reaction, the next nucleotide should be prepared. Although the stock solution of MSNT in pyridine is stable for a week at room temperature, it should not be prepared in large amounts. The stock solution is prepared simply by mixing the reagent and solvent. In order to avoid moisture, a septum is put on the top of the flask.

Coupling Reaction, Dimer Coupling. Dimers 6 are usually stored in their fully protected form and the cyanoethyl group ($-OCH_2CH_2CN$) must be removed before the coupling reaction (Figure 1), a mixture of a dinucleotide (25 μM), pyridine (0.5 ml) and triethylamine (0.5 ml) is allowed to stand for 45 minutes in the evaporation vessel and then evaporated under vacuum as described above. The evaporation is repeated twice with pyridine (0.25 ml) and used for the coupling reaction as mentioned for the monomer coupling. The reaction time is 40 minutes at room temperature when a 0.1 Molar solution of the dimer is used.

3. Step 4: Washing. After the coupling reaction, the septum is removed and the mixture is filtered and the support is washed quickly with pyridine (approximately 3 ml) once.

B. REACTION 2: Capping

1. Step 5: Acetylation of Unreacted 5'-Hydroxyl Groups. A stock solution of THF-pyridine-acetic anhydride (7:2:1 v/v, 3 ml) containing freshly dissolved DMAP (100 mg) is added to the filter (without need of the syringe). The mixture is left standing for 10 minutes, covered with the septum. After the reaction, the mixture is filtered and the support washed with pyridine (Step 6) (1 x 3 ml) and CH_2Cl_2-IPA (85:15 v/v, 3 x 3 ml) (Step 7). The stock solution is prepared using THF (70 ml), pyridine (20 ml), and acetic anhydride (10 ml), and stored in a bottle with a tight cap.

C. REACTION 3: Detritylation

1. Step 8: ZnBr$_2$ Treatment. 1 Molar solution of the ZnBr$_2$ in CH$_2$Cl$_2$-IPA (85:15 v/v) is prepared as follows: ZnBr$_2$ (22.5 g) is first dissolved in 15 ml of IPA (warm, if necessary) and cooled to room temperature using a drying tube on the top of the flask. To this mixture is added 85 ml of CH$_2$Cl$_2$ to give a 1 M solution, which is stored with a tight cap. It is recommended that the ZnBr$_2$ solution is prepared immediately before each oligonucleotide synthesis because HBr may gradually be evolved. The ZnBr$_2$ solution (1 M, 1 ml) is added to the support in the filter using a pipet and the reaction mixture is allowed to stand for 5 or 10 minutes. A red color develops immediately after the addition. The color of the reaction mixture is very informative. The mixture is filtered and the support is washed with CH$_2$Cl$_2$-IPA (85:15 v/v, 3 ml) three times, then a fresh ZnBr$_2$ solution is added to the filter to make sure that all DMT groups have been removed from the oligonucleotide chains. The ZnBr$_2$ solution and the washing solvent (CH$_2$Cl$_2$-IPA) are collected in one flask and used for the estimation of the coupling yield. The solvents are evaporated *in vacuo* and then EtOH is added to prepare exactly 100 ml of the ethanol solution. This ethanol solution (1 ml) is mixed with HClO$_4$ (70% v/v, 1.5 ml) and the mixture is then further diluted with 5 ml of an EtOH-HClO$_4$ solution (2:3 v/v). Comparisons of the optical density at 500 nm (ε = 88,700) of each step gives coupling yields of 85-95%.

The reaction rate is dependent upon the nucleoside at the 5'-position from which the DMT group is removed. The order of the reactivity is G>A>T>C. For G and A five minutes with ZnBr$_2$ are usually enough, but for T and C a ten-minute reaction time is essential. Generally, the color in the second treatment is very weak and, therefore, the reaction can be terminated by filtration in a few minutes. The washing step with CH$_2$Cl$_2$-IPA between the two treatments is essential for the detection of the color.

2. Steps 9, 10 and 11: Washing. After the second reaction, the support is washed with the CH$_2$Cl$_2$-IPA solution (2 X 3 ml), 0.5 M Et$_3$N$^+$HAcO$^-$ solution (3 x 3 ml) in dimethylformamide (DMF), and then with dry pyridine once using a pipet. The stock solution (500 ml) of Et$_3$N$^+$HAcO$^-$ is prepared simply by mixing the reagents, Et$_3$N (34.9 ml) and AcOH (14.3 ml), in DMF and stored in a bottle with a tight cap.

For the synthesis of oligonucleotides, the operation of these steps is repeated until the desired oligonucleotide sequence is constructed on the support. After the last coupling, the DMT group should not be removed from the 5'-terminal nucleotide.

IV. DEBLOCKING AND PURIFICATION

The final support (approximately 10 mg) is reacted with 0.5 M *syn*-pyridine-2-carboxaldoxime tetramethylguanidium solution (400 µl) in pyridine-water (9:1 v/v) overnight at 37°C. Fresh ammonia (28%, 2 ml) is then added to the mixture and the reaction mixture is allowed to stand at 60°C for several hours. The reaction is carried out in a centrifuge tube (12 ml) which is capped tightly with a septum and then masking tape is used to ensure that the septum can withhold the inside pressure of the ammonia gas. If the ammonia leaks, the deblocking reaction slows down significantly. After the reaction mixture is cooled, the ammonia solution is transferred to another tube and evaporated using a water aspirator (do not use an oil pump until ammonia gas is evaporated) at approximately 35°C. The residue is treated with 80% acetic acid (1 ml) for 20 minutes and evaporated. The product is dissolved in 0.05 M triethylammonium bicarbonate (TEAB) buffer and extracted with ether (2 X 0.5 ml). The aqueous layer is applied to a G-50 Sephadex column (2 x 50 cm) equibriumed with 0.05 M TEAB buffer. The first eluted peak containing the desired product is evaporated and further purified by the standard method using polyacrylamide gel (20%) electrophoresis in the presence of 7 M urea. The gel is put on silica gel TLC plates containing fluorescence (silica gel 60 F-254, E. Merck) and the oligonucleotide band can be detected under an UV lamp (the limitation is about 0.1 O.D. unit of the products). The slowest moving band is cut out and electroeluted.

DISCUSSION

A.A. ANSARI: My comment pertains to the use of synthetic oligonucleotides for screening purposes. Assuming an average degeneracy of the codons to be three possible codons per amino acid, the number of possible oligonucleotide sequences corresponding to five amino acids comes out to be $(3)^5=243$. This is a very high number of oligonucleotides to be synthesized for the screening. Your data also show that screening by this method is very specific, so specific that only one nucleotide mismatch is enough for the oligonucleotide not to bind to the RNA/DNA of interest. This means only one of the 243 oligonucleotide sequences will bind to the desired RNA or DNA. Also, what is the possibility that a "wrong" RNA or DNA is detected by one of these large number of oligonucleotides during the screening?

K. ITAKURA: Your argument is based entirely on statistics; 11 of 20 amino acids have only one or two codons. It is not difficult to find a short stretch of three amino acids in the desired protein. Therefore, there is no need to synthesize 243 kinds of oligonucleotide.

In order to avoid the detection of any undesired DNA, synthesis of two sets of the probes is recommended. One set of the probes is used for screening cloned DNA and the second set is used for confirmation by the Southern blot technique. Using this approach, you can detect any undesired cloned DNA sequence.

It should be noted that a 14-base-long nucleotide is required for the colony hybridization and a 11-base-long nucleotide for the Southern blot.

U.R. MULLER: Have you tried to use the synthetic DNA probe while it is still on the resin to purify the mRNA before preparing cDNA?

K. ITAKURA: The oligonucleotides synthesized on the resin are completely protected and, therefore do not hybridize to any DNA or RNA sequence. When the protecting groups are removed, the oligonucleotides are also released from the resin. Therefore, we cannot use the purification for the mRNA. Our probe is so specific that there is no advantage in the purification of the mRNA before cloning.

THE SYNTHESIS AND BIOCHEMICAL REACTIVITY OF BIOLOGICALLY IMPORTANT GENES AND GENE CONTROL REGIONS

Marvin H. Caruthers[1]

Department of Chemistry
University of Colorado
Boulder, Colorado
U.S.A.

I. INTRODUCTION

Recent developments in the recombinant DNA field have created a need for sequence defined deoxyoligonucleotides. These compounds are being used as probes for isolating natural genes, for experiments involving site-directed mutagenesis, and for the synthesis of genes and gene control regions. However until recently the synthesis and isolation of deoxyoligonucleotides has been a difficult and time-consuming task. Ideally chemical methods should be simple, rapid, versatile and accessible to the nonchemist. This paper outlines a synthetic methodology which satisfies all these criteria. The approach has been used successfully by nonchemists in my own laboratory and more recently by other research groups with limited background in nucleic acid chemistry.

II. OUTLINE OF THE PROCEDURE

A. Introduction

The general synthetic strategy involves adding mononucleotides sequentially to a nucleoside covalently attached to an

---

[1]Supported by NIH Grants GM21120, GM25680 and a Career Development Award from NIH, 1 KO4 GM00076.

insoluble polymer support. Reagents, starting materials and
side products are removed simply by filtration. After various
additional chemical steps, the next mononucleotide is joined
to the growing, polymer supported deoxyoligonucleotide. At
the conclusion of the synthesis, the deoxyoligonucleotide is
chemically freed of blocking groups, hydrolyzed from the support, and purified to homogeneity by polyacrylamide gel electrophoresis.

B. The Support

We chose   to use   high performance liquid chromatography (hplc) grade silica gel as an insoluble polymer support.
Silica gel is a rigid, nonswellable matrix in common organic
solvents. Additionally hplc grade silica gel is stable to all
our reagents and is designed for efficient mass transfer.
These features make hplc grade silica gel an attractive starting material for deoxyoligonucleotide synthesis.
The support is prepared using the steps outlined schematically in Figure 1. The initial step involves forming

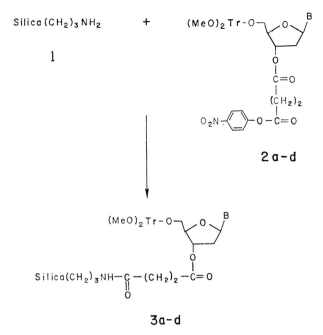

Figure 1. Synthesis of the Polymer Support. B refers to thymine in 2a and 3a; to N-benzoylcytosine in 2b and 3b; to N-
benzoyladenine in 2c and 3c; and to N-isobutyrylguanine in 2d
and 3d. (MeO)$_2$Tr designates the di-p-anisylphenylmethyl
protecting group.

compound 1 by refluxing 3-aminopropyl triethoxysilane with
silica gel in dry toluene for 3 h. The next step is synthesis
of compounds 2a-d. This can most easily be accomplished using
a procedure modified from published protocols (1-3). The
complete details for synthesizing these compounds are outlined
in the experimental section. When any one of the compounds
2a-d is added to compound 1 in a mixture of dimethylformamide, dioxane and triethylamine, an intense yellow color
rapidly forms, indicating the elimination of p-nitrophenol and
the formation of compounds 3a, 3b, 3c, or 3d. Usually the
ratio of reagents is adjusted so that approximately 50 μmole
nucleoside per gram silica gel is obtained. Thus 1-2 μmole
of a decanucleotide can be produced from 100-200 mg of silica
gel. Unreactive amino groups are blocked against further
reaction by acylation with acetic anhydride. This overall
procedure is quite satisfactory for attaching any nucleoside
to the polymeric support. Additionally we observe more consistent, reproducible loading than when our previously published method is followed (4).

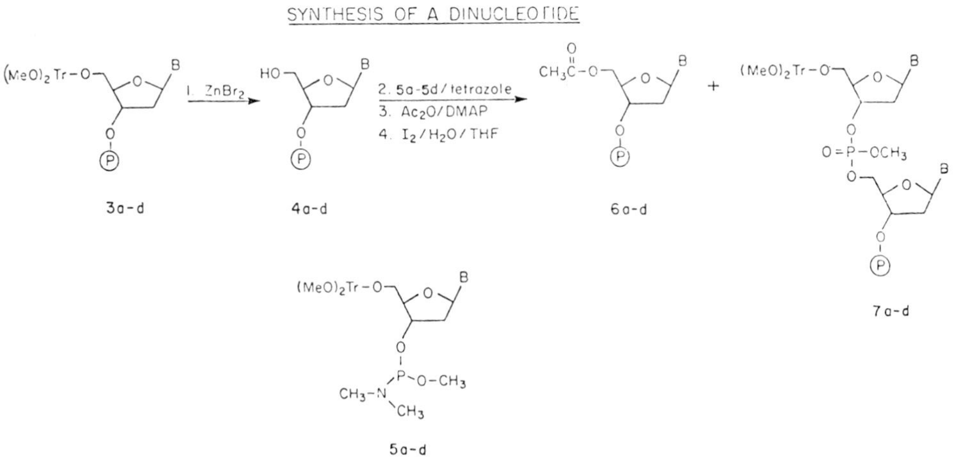

SYNTHESIS OF A DINUCLEOTIDE

Figure 2.   Steps in the Synthesis of a Dinucleotide. B refers
to thymine in 3a, 4a, 5a, 6a and 7a; to N-benzoylcytosine in
3b, 4b, 5b, 6b and 7b; to N-benzoyladenine in 3c, 4c, 5c, 6c
and 7c; to N-isobutyrylguanine in 3d, 4d, 5d, 6d and 7d.
(MeO)$_2$Tr refers to the di-p-anisylphenylmethyl group when 5d
is being used. For 5a, 5b and 5c, trityl groups as defined
in Table 2 are used.

C. The Synthesis Cycle

The addition of one mononucleotide to compound 3a, 3b, 3c or 3d requires the following four steps: (1) removal of the trityl protecting group with ZnBr$_2$; (2) condensation with the appropriate 5'-O'di-p-anisylphenylmethyl-3'-methoxy-N,N-dimethylaminophosphine nucleoside; (3) acylation of unreactive 5'-hydroxyl groups; (4) oxidation of the phosphite triester to the phosphate triester. These steps are summarized in Figure 2. Thus the synthesis proceeds in a 3'- to 5'- direction by adding one nucleotide per cycle. The individual steps for one synthesis cycle are listed in Table I and outlined in the following paragraphs.

Zinc bromide in nitromethane completely removes trityl ethers from support bound deoxyoligonucleotides without any associated depurination (4,5). In order to reduce the time allotted to this step to 5 minutes, 5% methanol has been added to the solution. A recent innovation in this step has been to use it to monitor the progress of the synthesis. As can be seen by examining Figure 2, a di-p-anisylphenylmethyl group is removed during each synthetic cycle. More importantly, this di-p-anisylphenylmethyl group is part of the mononucleotide (5a, 5b, 5c or 5d) most recently added to the growing oligonucleotide attached to silica gel. The di-p-anisylphenylmethyl group forms an orange color when removed from the nucleotide during the ZnBr$_2$ catalyzed detritylation step. If four triarylmethyl protecting groups that form distinct colors in ZnBr$_2$/nitromethane could be attached to the different nucleotides, then each condensation step could be individually monitored. For example, if the di-p-anisylphenylmethyl group were assigned only to 5d, then an orange color during the subsequent ZnBr$_2$/nitromethane cleavage step would indicate that N-isobutyryldeoxyguanosine had been added to the growing oligonucleotide during the preceding condensation step. By measuring the color intensity, an approximate quantitation of the condensation yield is also possible. A different color resulting from another triaryl protecting group would indicate that one of the other mononucleotides (5a, 5b or 5c) had been added during the condensation. By attaching different triarylmethyl protecting groups to the four mononucleotides, condensation reactions can be monitored as to yield and nucleotide sequence. The triarylmethyl protecting groups that we presently use and their associated colors in ZnBr$_2$/nitromethane solutions are listed in Table II. Thus, the additions of thymidine, cytidine, adenosine and guanosine are monitored by the formation of red, blue, yellow and orange colors respectively, during the detritylation step.

Table I. A Summary of the Chemical Steps for One
          Synthetic Cycle

Reagent or solvent	Purpose	Time (min)
Satd. $ZnBr_2$ in 5% methanol/CH_3NO_2	Detritylation	4
CH_3NO_2	Wash	1
Methanol	Wash	1
Acetonitrile	Wash	1
Activated Nucleotide in Acetonitrile	Add one Nucleotide	5
Tetrahydrofuran:Lutidine:H_2O (2:2:1)	Wash	3
I_2 Solution	Oxidation	5
Methanol	Wash	1
Tetrahydrofuran	Wash	1
Capping Solution	Acylation	5
Methanol	Wash	2
CH_3NO_2	Wash	1

Table II. Assignment of Colored 5'-Triarylmethyl Groups
           to the Protected Deoxynucleotides

Nucleotide	Color
5'-0-*p*-anisylphenylmethyl-N-isobutyryl-deoxyguanosine-3'-N,N-dimethylaminomethoxy-phosphine	Orange
5'-0-*p*-anisyl-1-napthylphenylmethyldeoxy-thymidine-3'-N,N-dimethylaminomethoxy-phosphine	Red
5'-0-*p*-anisyl-1-napthylmethyl-N-benzoyl-deoxycytidine-3'-N,N-dimethylaminomethoxy-phosphine	Blue
5'-0-*p*-tolyldiphenylmethyl-N-benzoyldeoxy-adenosine-3'-N,N-dimethylaminomethoxy-phosphine	Yellow

Appropriately protected nucleoside phosphoramidites (5a-d) are used for the sequential addition of mononucleotides. These compounds have several attractive features. They can be synthesized using standard organo-chemical procedures (6) and can be stored as white powders. They are also resistant to aqueous hydrolysis and to oxidation. We therefore store, measure and transfer these reagents in a manner analogous to any other solid compound. Activation in dry acetonitrile is achieved by the addition of a mild acid. We have selected tetrazole as the activating agent since it is a nonhygroscopic commercially available material. Other mild acids such as amine hydrochlorides can also be used, but are not recommended. Amine hydrochlorides are hygroscopic and would introduce water into the condensation step. Our synthesis procedure therefore involves dissolving the nucleoside phosphoramidite and tetrazole in acetonitrile, mixing, and adding the solution to a silica gel support. The condensation reaction is complete within a minute, but we usually allow the reaction to proceed for five minutes.

Based on detailed and careful analysis of various condensation reactions, approximately 1-5% of the deoxynucleoside or deoxyoligonucleotide bound to the support (compound 4a-d) does not react with the activated nucleotide to form 7a-d. These unreactive compounds must be blocked or capped in order to prevent the formation of several deoxyoligonucleotides with heterogeneous sequences. Moreover, in the absence of a capping step, most of these failure sequences will contain one or two less nucleotides than the expected product. Purification will therefore be more difficult. If a capping step is included, the failure sequences will be quite disperse relative to size with only 1-5% of the reaction mixture having one nucleotide shorter than the product. Currently this capping step can best be accomplished using a tetrahydrofuran solution of acetic anhydride and dimethylaminopyridine. The reaction is complete within 1 to 2 minutes and does not lead to any detectable side products.

The internucleotide phosphite triester is oxidized to the phosphate triester using I_2 in water, 2,6-lutidine and tetrahydrofuran. Oxidation is extremely rapid (1-2 min) and side products are not generated. We have attempted to postpone the oxidation until after all condensation steps, but the results have not been encouraging. Several uncharacterized side products are observed.

We have used several devices in order to aid in the synthesis of deoxyoligonucleotides. Initially our syntheses were completed in manually operated machines (4). The principles outlined in these earlier devices are now being engineered into a completely automatic and programable machine. Other devices are also being used. For example we have successfully synthesized deoxyoligonucleotides in test tubes and on sintered glass funnels. In the former case, each synthesis step is followed by a low-speed centrifugation and decantation in order to remove solvents and reagents. For the latter case, reagents and solvents are removed simply by filtration. When test tubes or sintered glass funnels are used, we routinely synthesize 6 to 8 deoxyoligonucleotides simultaneously. The appropriate mononucleotide is added to each test tube during a synthesis cycle. These additions are monitored by observing the appropriate color during the subsequent detritylation step.

D. Isolation of Deoxyoligonucleotides

Once a synthesis has been completed, the deoxyoligonucleotide is freed of protecting groups and isolated by polyacrylamide gel electrophoresis. Silica gel containing the reaction product is first treated with triethylammonium thiophenoxide in dioxane (4) in order to remove the methyl groups from internucleotide phosphotriesters. This step is followed by treatment with concentrated ammonium hydroxide at 20°C for 3 h in order to hydrolyze the ester joining the deoxyoligonucleotide to the support. After centrifugation and recovery of the supernatant containing the deoxyoligonucleotide, the N-benzoyl groups from deoxycytosine and deoxyadenosine, and the N-isobutyryl group from deoxyguanosine are removed by warming the concentrated ammonium hydroxide solution at 50°C for 12 h. Finally the trityl ether is hydrolyzed with 80% acetic acid. This is the preferred detritylating agent after the amino protecting groups have been removed. Depurination is not observed with completely deprotected deoxyoligonucleotides and unlike ZnBr$_2$, 80% acetic acid is volatile and easily removed. The reaction mixture containing deprotected deoxyoligonucleotides is then fractionated by polyacrylamide gel electrophoresis. When a standard slab gel device is used, as many as eight compounds can be purified simultaneously. Usually these reaction mixtures are fractionated on 20% acrylamide gels with the standard tris-borate buffer system. The appropriate deoxyoligonucleotide is visualized with an ultraviolet light. Elution from the gel is completed using standard procedures (8).

III. Experimental Section

A. General Methods

Thiophenol, 4-dimethylaminopyridine, anhydrous $ZnBr_2$ and iodine were purchased from Aldrich and used without further purification. Reagent grade acetic anhydride, triethylamine and t-butylamine were used as received. Common solvents such as tetrahydrofuran, dioxan, acetonitrile, nitromethane and methanol were stored over activated 4A molecular sieves and used without further purification. Dry acetonitrile was obtained by refluxing reagent grade solvent over CaH_2 for several hours and distilling just prior to usage. 2,6-Lutidine was obtained by refluxing reagent grade solvent over CaH_2 for one hour followed by distillation from 4A molecular sieves. 2,6-Lutidine was stored in the dark. Tetrahydrofuran was dried over sodium-benzophenone and used freshly distilled. 1-H-Tetrazole (Aldrich) was sublimed at 110-115°C at 0.05 mm Hg prior to use.

All solution transfers involving dry reagents were completed with clean syringes dried in ovens at 50°C.

B. Synthesis of the Support

The initial step was synthesis of the succinilated deoxy-nucleosides (compounds 2a, 2b, 2c and 2d). All four were prepared using the same general procedure. To a solution of 5'-dimethoxytritylnucleoside (5 mmol) in anhydrous pyridine was added 4-dimethylaminopyridine (0.61 g; 5 mmol) and succinic anhydride (6.6 g; 6 mmol). The reaction was monitored by tlc (acetonitrile:water; 9:1) and was usually complete after 12 h at 20°C. Occasionally a second portion of succinic anhydride (0.1 g; 1 mmol) was added. The reaction was next quenched with water (0.1 ml) for 10 min at 20°C. The reaction mixture was evaporated *in vacuo*, and then co-evaporated twice with dry toluene (2 x 20 ml). The residue was redissolved in dichloromethane (40 ml) and the solution was washed successively, once with 10% citric acid (10 ml) and twice with water (2 x 10 ml). The organic solution was dried over anhydrous sodium sulfate, and evaporated *in vacuo*. The residue was redissolved in 10 ml dichloromethane (containing 5% pyridine) and the product precipitated into pentane:ether (200 ml; v/v). The precipitate was dried *in vacuo* (yield: 70-85%). In order to derivatize silica gel containing an amino function (2), the succinylated nucleoside (1 mmole) was dissolved in dioxane (4 ml) containing dry pyridine (0.2 ml) and then *p*-nitrophenol (140 mg;

1 mmol) was added. A solution of dicyclohexylcarbodiimide
(220 mg) in anhydrous dioxane (1 ml) was added and the reaction was monitored by tlc (silica gel plate; benzene:dioxane,
3:1). The reaction was virtually complete after 2 h at room
temperature. Dicyclohexylurea was removed by centrifugation,
and the supernatant containing the desired product was used
directly in the condensation reaction.

   Deoxynucleosides were attached to the support using the
following general procedure. HPLC grade silica gel (12 g,
Fractosil 200, Merck) was exposed to a 15% relative humidity
(saturated LiCl) for at least 24 h. The silica was then
treated with 3-triethoxysilylpropylamine (13.8 g, 0.01 M in
dry toluene) for 12 h at 20°C and 18 h at reflux. It was
isolated by centrifugation, washed successively (3 times each)
with toluene, methanol and 50% aqueous methanol. The silica
was then shaken with 50% aqueous methanol (200 ml) at 20°C for
18 h. After isolation by centrifugation, the silica was
washed with methanol and ether and dried *in vacuo*. The dried
silica was suspended in anhydrous pyridine and treated with
trimethylsilyl chloride (15 ml) for 12 h at 20°C. After isolation by centrifugation, the silica was washed four times
with methanol, twice with ether, and then dried *in vacuo*. The
dry silica (3 g) was suspended in DMF (5 ml) and a solution of
the 5'-dimethoxytritylnucleoside-3'-*p*-nitrophenylsuccinate in
dioxane and 1 ml of triethylamine was added. The suspension
was shaken at 20°C for 4 h. Ninhydrin test at this stage
indicated the existance of free amino groups on the resin. To
cap these groups, acetic anhydride (0.6 ml) was added and the
mixture was shaken for another 30 min, after which time a negative ninhydrin test was obtained. The silica was isolated by
centrifugation, washed successively (3 times each) with DMF,
95% ethanol, dioxane and ethyl ether, and then dried *in vacuo*.
Analysis for the extent of dimethoxytritylnucleoside attached
to the support was done spectrophotometrically. An accurately
weighed sample of silica (10-15 mg) was treated with 0.1 M
p-toluenesulfonic acid in acetonitrile and the optical density
of the supernatant obtained after centrifugation was measured
at 498 nm. (The extinction coefficient of dimethoxytritanol
is 7.0×10^4.) The extent of derivation was found to be the
following: (MeO)$_2$TrdT, 62 μmol/g; (MeO)$_2$TrdibG, 56 μmol/g;
(MeO)$_2$TrdbzA, 65 μmol/g; (MeO)$_2$TrdbzC, 68 μmol/g.

   C. Synthesis of Deoxynucleosidephosphoramidites

   The careful preparation of compounds 5a-d is of critical
importance. These compounds are prepared essentially as
described previously (6). The synthesis begins with the

preparation of chloro-N,N-dimethylaminomethoxyphosphine
[CH$_3$OP(Cl)N(CH$_3$)$_2$] which is used as a monofunctional phosphit-
ylating agent. A 250 ml addition funnel was charged with 100
ml of precooled anhydrous ether (-78°C) and precooled (-78°C)
anhydrous dimethylamine (45.9 g, 1.02 mol). The addition funnel was wrapped with aluminum foil containing dry ice in order
to avoid evaporation of dimethylamine. This solution was
added dropwise at -15°C (ice-acetone bath) over 2 h to a
mechanically stirred solution of methoxydichlorophosphine
(47.7 ml, 67.32 g., 0.51 mol) in 300 mol of anhydrous ether.
The addition funnel was removed and the 1 ℓ, three-necked
round bottom flask was stoppered with serum caps tightened
with copper wire. The suspension was mechanically stirred for
2 h at room temperature. The suspension was filtered and the
amine hydrochloride salt was washed with 500 ml anhydrous
ether. The filtrate and washings were combined and ether was
distilled at atmospheric pressure. The residue was distilled
under reduced pressure. The product was collected at 40-42°C
at 13 mm Hg and was isolated in 71% yield (51.1 g, 0.36 mol).
d^{25} = 1.115 g/ml. ^{31}P-N.M.R., δ = -179.5 ppm (CDCl$_3$) with
respect to internal 5% v/v aqueous H$_3$PO$_4$ standard. ^{1}H-N.M.R.
doublet at 3.8 and 3.6 ppm J$_{P-H}$ = 14 Hz (3H, OCH$_3$) and two
singlets at 2.8 and 2.6 ppm (6H, N(CH$_3$)$_2$). The mass spectrum
showed a parent peak at m/e = 141.

　　Compounds 5a-d were prepared by the following procedure.
5'-O-Di-_p_-anisylphenylmethyl nucleoside (1 mmol) was dissolved
in 3 ml of dry, acid free chloroform and diisopropylethylamine
(4 mmol) in a 10 ml reaction vessel preflushed with dry nitrogen. [CH$_3$OP(Cl)N(CH$_3$)$_2$] (2 mmol) was added dropwise (30-6- sec)
by syringe to the solution under nitrogen at room temperature.
After 15 min the solution was transferred with 35 ml of ethyl
acetate into a 125 ml separatory funnel. The solution was
extracted four times with an aqueous, saturated solution of
NaCl (80 ml). The organic phase was dried over anhydrous
Na$_4$SO and evaporated to a foam under reduced pressure. The
foam was dissolved with toluene (10 ml) (5d was dissolved with
10 ml of ethyl acetate) and the solution was added dropwise to
50 ml of cold hexanes (-78°C) with vigorous stirring. The
cold suspension was filtered and the white powder was washed
with 75 ml of cold hexanes (-78°C). The white powder was
dried under reduced pressure and stored under nitrogen. Isolated yields of compounds 5a-d were 90-94%. The purity of the
products was checked by ^{31}P-N.M.R. Compounds 5a-d are characterized as two peaks between -146 and -145 ppm. Various
impurities are sometimes observed with peaks between 0 and +10
relative to phosphoric acid. These impurities do not appear
to inhibit the condensation reactions.

D. Outline of the Synthesis Cycle

The appropriately derivatized deoxynucleoside attached covalently to silica gel (compound 3a, 3b, 3c or 3d) was treated with a saturated solution of anhydrous $ZnBr_2$ in nitromethane:methanol (95:5) for 4 min. The support was then washed with nitromethane followed by methanol. Before the condensation step the silica gel was carefully washed several times with dry acetonitrile under an inert atmosphere (N_2). Stock solutions of sublimed tetrazole and appropriately protected 2'-deoxynucleoside-3'-N,N-dimethylaminomethoxyphosphines were prepared. Usually these stock solutions were sufficient for at least three condensations. For each μmole of deoxynucleoside attached covalently to silica gel, tetrazole (60 μmole) and the deoxynucleoside-3'-N,N-dimethylaminomethoxyphosphine in the condensation mixture was about 0.1 M. The condensation reaction was stopped after 5 min. Immediately following the condensation reaction, the silica gel was washed with a tetrahydrofuran:water:2,6-lutidine solution (2:2:1) for 3 min. Oxidation of trivalent phosphorus to pentavalent phosphate was completed using a 0.2 M solution of iodine in tetrahydrofuran:water:2,6-lutidine (2:2:1) for five minutes. The silica gel was washed with methanol followed by tetrahydrofuran. Acylation of unreactive hydroxyl groups was completed by adding first a solution of 4-dimethylaminopyridine in dry tetrahydrofuran (2 ml; 6.5% w/v) and then a solution of acetic anhydride in 2,6-lutidine (0.4 ml; 1:1) to the support. After 5 min, this acylation solution was removed and the silica gel was washed with methanol and nitromethane. This step completes one synthesis cycle.

E. Isolation of Synthetic Deoxyoligonucleotides

After completion of the appropriate synthesis cycles, deoxyoligonucleotides free of protecting groups and side products were isolated using the following procedure. Silica gel containing the deoxyoligonucleotide was first treated with a solution containing thiophenol:dioxane:triethylamine (1:2:2) for 90 min at room temperature. This deprotection step removes the methyl phosphotriester protecting group. The support was next treated with concentrated ammonium hydroxide for 24 h at 60°C. The liquid phase was evaporated to dryness *in vacuo* and the residue was treated with a solution of t-butylamine in methanol (1:1) for 24 h at 60°C in a screw cap vial. The reaction mixture was evaporated to dryness *in vacuo* and then loaded on a Sephadex G-50 column. Fractions containing deoxyoligonucleotides were pooled and the DNA isolated by ethanol precipitation. The crude DNA pellet was dissolved

in formamide and loaded on a 20% denaturing polyacrylamide
gel. After electrophoresis, the band containing deoxyoligonucleotide product was visualized using an ultraviolet lamp.
The gel slice containing the product was eluted and desalted
using standard procedures (8). Usually the product is the
major u.v. light absorbing band on the gel.

ACKNOWLEDGMENTS

    This is paper 9 in a series on Nucleotide Chemistry.
This research was completed by several excellent scientists
in my laboratory. Those most directly involved in developing
the chemistry include S. L. Beaucage, C. Becker, W. Efcavitch,
E. F. Fisher, R. Goldman, P. de Haseth, F. Martin, M.
Matteucci and Y. Stabinsky.

REFERENCES

1.   Gait, M. J., Singh, M. and Sheppard, R. C., *Nucleic
     Acids Res. 8*, 1081-1096 (1980).
2.   Chow, F., Kempe, T. and Palm, G., *Nucleic Acids Res. 9*,
     2807-2817 (1981).
3.   Caruthers, M. H., Stabinsky, Y., Stabinsky, Z and Peters,
     M. in "Proceedings Symp. Promoters: Structure and Func-
     tion" (R. Rodriguez and M. Chamberlin, eds.), Praeger
     Scientific, New York, in press.
4.   Matteucci, M. D. and Caruthers, M. H., *J. Am. Chem. Soc.
     103*, 3185-3191 (1981).
5.   Matteucci, M. D. and Caruthers, M. H., *Tetrahedron Lett.
     21*, 3243-3246 (1980).
6.   Beaucage, S. L. and Caruthers, M. H., *Tetrahedron Lett.
     22*, 1859-1862 (1981).
7.   Maniatis, T., Jeffrey, A. and van de Sande, J., *Biochem-
     istry 14*, 3787-3794.
8.   Maxam, A. M. and Gilbert, W., in "Methods in Enzymology,
     Vol. 65" (L. Grossman and K. Moldave, eds.), pp 499-560,
     Academic Press, New York (1979).

DISCUSSION

*UNKNOWN SPEAKER: You mentioned that your reactions are essentially over in 30 seconds but, although you ran them for 5 minutes, the yield was only 95%. What do you think is happening to the other 5%? Also, do you have evidence that you really need the capping step?

*M.H. CARUTHERS: We are investigating whether we can come up with a better matrix. I think that as the synthesis proceeds a proportion of the oligomers become embedded in sites on the matrix that just won't react. The advantage of the capping step is that it changes the distribution of the failure sequences. Without the capping step most of the failure sequences will be minus only one or two nucleotides because, on the average they miss one or two steps along the way. However, after a capping step the failure sequences do not react and so they terminate as monomers, dimers, trimers, tetramers, etc. to the extent of a few percent for each step. This makes the purification simpler, especially on polyacrylamide gels when you want to cut out a specific band.

H. HEYNEKER: Have you analysed the different promoter sequences in an "in vivo" system like the Rosenberg galactokinase system and correlated promoter strength with the "in vitro" abortive initiation assay?

M.H. CARUTHERS: We have attempted to clone the lambda P_R promoter into a vector which will allow us to measure expression as a function of beta-galactosidase synthesis. However, these experiments have not been successful simply because this is an extremely strong promoter. We have however, cloned several lac promoter variants into this vector and have measured production of beta-galactosidase. Because of time restrictions, I was unable to discuss this system.

SYNTHESIS OF HUMAN INTERFERONS
AND ANALOGS IN HETEROLOGOUS CELLS

R. Derynck, P.W. Gray, E. Yelverton, D.W. Leung,
H.M. Shepard, R.M. Lawn, A. Ullrich, R. Najarian,
D. Pennica, F.E. Hagie, R.A. Hitzeman,
P.J. Sherwood, A.D. Levinson and D.V. Goeddel

Department of Molecular Biology
Genentech, Inc.
South San Francisco, California

Human interferons (IFN's) are generally classified in three groups on the basis of antigenicity: the acid-stable α- (leukocyte) and β- (fibroblast) interferons and the acid-labile γ- (immune) interferon. Many studies have been performed using either impure interferon or insufficient amounts of it, often leading to ambiguous results (for review, see ref. 1). The application of the recombinant DNA technology has caused a major breakthrough in the study of these interferons and has led to the establishment of their structure. We have been able in our laboratory to isolate and characterize cDNA clones for the different human interferons and have obtained high level bacterial expression of these, thus allowing structural and biological studies with the purified proteins.

A cDNA clone, coding for human β-interferon has been isolated from a cDNA library, derived from enriched human fibroblast interferon DNA. The nucleotide sequence, when compared with the directly determined amino-terminal sequence of the mature protein[2], predicts a 21 amino acid long signal peptide, involved in the secretion, and a mature protein of 166 amino acids.[3] Similar results, in complete agreement with ours, have been independently obtained by Taniguchi et al.[4] and Derynck et al[5]. The mature polypeptide is hydrophobic, a property previously observed by its chromatographical behavior[6], and contains three cysteines. The necessity of a disulphide bond for its biological activity is not only suggested by the conservation of two cysteine residues in the human β- and α-interferons but also by experimental evidence. Indeed, we have isolated a β-IFN cDNA clone with a G to A transition in the nucleotide sequence, resulting in a cysteine to tyrosine change at amino acid position 141 in the mature protein. The resulting gene product synthesized in E. coli was completely devoid of biological activity.[7] The deduced amino acid sequence also shows a possible N-glycosylation signal, which is in agreement with the observed glycosylation of human fibroblast interferon.[8,9]

Hybridization studies with total genomic DNA and with a cloned human chromosomal DNA library have revealed the existence of a single gene corresponding to the human β-IFN cDNA. The sequence of the gene shows a complete absence of intervening sequences[10-13], an exceptional feature for genes corresponding to cellular cytoplasmic mRNAs. Some suggestive evidence has been reported for the existence of other β-IFN mRNAs[14,15], a finding which awaits confirmation from molecular cloning of the cDNAs. However, all reported sequences for human β-IFN cDNAs or genes, are in complete agreement with each other.[3,4,5,10-13,16]

Bacterial synthesis of mature human fibroblast IFN has been achieved in E. coli using either the lac- or the trp-promoter. The bacterial β-IFN is biologically indistinguishable from authentic β-IFN, indicating that the glycosylation is not needed for biological activity.[3] A yield of 6 x 10[9] units/liter is now routinely observed in large-scale cultures of E. coli.

We have been able to isolate cDNA clones for human α-IFN, using the same approach as for β-IFN.[17,18] Unlike the case of β-IFN, for which a unique sequence has been determined, we found a divergence of sequences of the

Figure 1. Comparison of the α-IFN protein sequences (A-K) predicted from the nucleotide sequences from cloned cDNAs or genes. The one-letter abbreviations recommended by the IUPAC-IUB Commission[47] are used. The numbers refer to amino acid position (S refers to signal peptide). The dash in the 165 amino acid LeIFN A sequence at position 44 is introduced to align the LeIFN A sequence with the 166 amino acid sequences of the other LeIFNs. The LeIFN E sequence was determined by ignoring the extra nucleotide at position 187 (ref. 18) in its coding region. Amino acids common to all αIFNs (excluding the pseudogene LeIFN E) are also shown. The underlined residues are amino acids which are also present in human β-IFN.[3-5]

```
         S1                 S10                 S20   S23                 10                  20                  30               40
         .                  .                   .     .                   .                   .                   .                .
LeIF A   M A L T F A L L V A L L V L S C K S S C S V G C D L P Q T H S L G S R R T L M L L A Q M R K I S L F S C L K D R H D F G F P Q
LeIF B   M A L T F Y L M V A L V V L S Y K S F S S L G C D L P Q T H S L G N R R A L I L L A Q M R R I S P F S C L K D R H D F E F P Q
LeIF C   M A L S F S L L M A V L V L S Y K S I C S L G C D L P Q T H S L G N R R A L I L L G Q M G R I S P F S C L K D R H D F R I P Q
LeIF D   M A S P F A L L M V L V V L S C K S S C S L G C D L P E T H S L D N R R T L M L L A Q M S R I S P S S C L M D R H D F G F P Q
LeIF E                                       L P L G C D L P Q A H S V G N R R A F I L L T Q M R R I S P F S Y L K D R H D F D F P H
LeIF F   M A L S F S L L M A V L V L S Y K S I C S L G C D L P Q T H S L G N R R A L I L L A Q M G R I S P F S C L K D R H D F G F P Q
LeIF G                                                                                                                     H D F G F P Q
LeIF H   M A L P F S L M M A L V V L S C K S S C S L G C N L S Q T H S L N N R R T L M L M A Q M R R I S P F S C L K D R H D F E F P Q
LeIF I   M A L S F S L L M A V L V L S Y K S I C S L G C D L P Q T H S L G N R R A L I L L A Q M G R I S P F S C L K D R P D F G L P Q
LeIF J   M A R S F S L L M V V L V L S Y K S I C S L G C D L P Q T H S L R N R R A L I L L A Q M G R I S P F S C L K D R H E F R F P E
LeIF K   M A L P F A L L M A L V V L S C K S S C S L D C D L P Q T H S L G H R R T M M L L A Q M R R I S L F S C L K D R H D F R F P Q

All      M A     F   L           V L S   K S     S     C   L     T H S L   R R       L     Q M       I S     S C L   D R     F     P

                           50                  60                  70                  80                  90                  100
                           .                   .                   .                   .                   .                   .
LeIF A   E E F - G N Q F Q K A E T I P V L H E M I Q Q I F N L F S T K D S S A A W D E T L L D K F Y T E L Y Q Q L N D L E A C V I Q G
LeIF B   E E F D D K Q F Q K A Q A I S V L H E M I Q Q T F N L F S T K D S S A A L D E T L L D E F Y I E L D Q Q L N D L E V L C D Q E
LeIF C   E E F D G N Q F Q K A Q A I S V L H E M I Q Q T F N L F S T E D S S A A W E Q S L L E K F S T E L Y Q Q L N D L E A C V I Q E
LeIF D   E E F D G N Q F Q K A P A I S V L H E L I Q Q I F N L F T T K D S S A A W D E D L L D K F C T E L Y Q Q L N D L E A C V M Q E
LeIF E   Q V F H G N H F Q K V Q A I F L F H E M M Q Q T F N L F S T K D S S D T W D E T L L D K S Y T E L Y Q Q L N D L E A C V M * K
LeIF F   E E F D G N Q F Q K A Q A I S V L H E M I Q Q T F N L F S T K D S S A T W E Q S L L E K F S T E L N Q Q L N D M E A C V I Q E
LeIF G   E E F D G N Q F Q K A Q A I S V L H E M I Q Q T F N L F S T K D S S A T W D E T L L D K F Y T E L Y Q Q L N D L E A C M M Q E
LeIF H   E E F D G N Q F Q K A Q A I S V L H E M M Q Q T F N L F S T K N S S A A W D E T L L E K F Y I E L F Q Q M N D L E A C V I Q E
LeIF I   E E F D G N Q F Q K T Q A I S V L H E M I Q Q T F N L F S T E D S S A A W E Q S L L E K F S T E L Y Q Q L N N L E A C V I Q E
LeIF J   E E F D G H Q F Q K T Q A I S V L H E M I Q Q T F N L F S T E D S S A A W E Q S L L E K F S T E L Y Q Q L N D L E A C V I Q E
LeIF K   E E F D G N Q F Q K A E A I S V L H E V I Q Q T F N L F S T K D S S V A W D E R L L D K L Y T E L Y Q Q L N D L E A C V M Q E

All      E E F       Q F Q K         I   V L H E   Q Q   F N L F   T     S S             L L           E L   Q Q   N     E       Q

                        110                 120                 130                 140                 150                 160     166
                        .                   .                   .                                       .                   .       .
LeIF A   V G V T E T P L M K E D S I L A V R K Y F Q R I T L Y L K E K K Y S P C A W E V V R A E I M R S F S L S T N L Q E S L R S K E
LeIF B   V G V I E S P L M Y E D S I L A V R K Y F Q R I T L Y L T E K K Y S S C A W E V V R A E I M R S F S L S I N L Q K R L K S K E
LeIF C   V G V E E T P L M N E D S I L A V R K Y F Q R I T L Y L I E R K Y S P C A W E V V R A E I M R S L S F S T N L Q K R L R R K D
LeIF D   E R V G E T P L M N V D S I L A V K K Y F R R I T L Y L T E K K Y S P C A W E V V R A E I M R S L S L S T N L Q E R L R R K E
LeIF E   V G V E E T P L R N V D S I L A V R K Y F Q R I T L Y L T K K K Y S P C S W E A V R A E I M R S F S L * T N L Q E R L R R K E
LeIF F   V G V E E T P L M N V D S I L A V K K Y F Q R I T L Y L T E K K Y S P C A W E V V R A E I M R S F S L S K I F Q E R L R R K E
LeIF G   V G V E D T P L M N V D S I L T V R K Y F Q R I T L Y L T E K K Y S P C A W E V V R A E I M R S F S L S A N L Q E R L R R K E
LeIF H   V G V E E T P L M N E D S I L A V R K Y F Q R I T L Y L M E K K Y S P C A W E V V R A E I M R S F S F S T N L Q K R L R R K D
LeIF I   V G M E E T P L M N E D S I L A V R K Y F Q R I T L Y L T E K K Y S P C A W E V V R A E I M R S L S F S T N L Q K I L R R K D
LeIF J   V G V K E T P L M N E D F I L A V R K Y F Q R I T L Y L M E K K Y S P C A W E V V R A E I M R S F S F S T N L K K G L R R K D
LeIF K   V W V G G T P L M N E D S I L A V R K Y F Q R I T L Y L T E K K Y S P C A W E V V R A E I M R S F S S S R N L Q E R L R R K E

All                  P L M     D   I L   V   K Y F   R I T L Y L   E   K Y S   C A W E V V R A E I M R S   S   S             L     K
```

DG:1/0006M
1/20/82

cDNAs. These cDNAs are clearly related to each other by at least 70 percent of the deduced amino acid sequences.[18] Similar results have been reported by Charles Weissmann and coworkers.[19,20,21] Some of the minor differences in the sequences might be attributed to allelism. However, in most cases the divergence is significant enough to be attributed to a multigene family for human leukocyte IFNs. We have isolated and characterized 8 distinct α-IFN cDNA clones, designated LeIF-A to -H and 7 genomic clones, designated A,D,H,I,J,K,L (ref 22, 23, 24, unpublished results). Comparison with the results from Weissmann's group points to the presence of probably some 20 human α-IFN genes. The frequency of incidence in the cDNA library suggest that LeIF-A and -D (ref. 18) might be the major species in the natural leukocyte interferon mixture. The sequence of the LeIF-E cDNA insert shows that, due to the premature occurrence of termination codons, translation into a biologically functional polypeptide cannot take place.[18] The isolation of this clone from a cDNA library is the first evidence for trancription of a pseudogene. All sequenced α-IFN cDNAs, except the pseudogene LeIF-E, code for a signal peptide of 23 amino acids and a mature polypeptide of 166 amino acids, except for LeIF-A, which contains 165 amino acids (fig. 1). The junction between the signal peptide and the mature polypeptide could be assigned by comparison with the directly determined amino-terminal sequence of an α-IFN.[25] · No potential N-glycosylation sites are present in the sequence which is in agreement with the reported lack of glycosylation of the proteins.[26] Comparison of the amino acid sequences of the α-IFNs and β-IFN reveals a homology in the amino acid sequences for the mature IFN's of 23 percent.[27,18,20] This similarity is mainly concentrated in two regions of the molecule implying their functional importance. The conservation of two Cys-residues in the α- and β-IFNs strongly suggests the importance of a disulfide bridge. The presence of two disulfide bonds, one of which cor-responds to the two conserved cysteine-residues, has been experimentally determined in LeIF-A, purified from E. coli.[28,29]

The presence of a multigene family for the human α-interferons is also evident from hybridization studies with total genomic DNA and from the isolation and analysis of the chromosomal genes.[18,22-24] The α-IFN gene sequences show that a very significant homology exists in the 5' flanking regions.[23] Comparison with the β-IFN gene sequence reveals some striking similarities in the upstream sequences which suggests a functional involvement

in the transcription and regulation of these genes.[12,13]
Also like the β-IFN gene, all α-IFN genes are devoid of
intervening sequences.[22-24,30] The study of these genes
has in addition shown that at least several human α-IFN
genes are closely linked on the chromosome.[22,24,30]
Finally, we have been able to assign the α-IFN genes as
well as the β-IFN gene to chromosome 9, by hybridization
studies with genomic DNA from a variety of mouse-human
hybrid cell lines.[31]

Synthesis of human leukocyte interferons in E. coli has
been achieved by inserting the α-IFN cDNAs under the
control of the trp-promoter.[17,32] The biological interferon activity in the bacterial extracts was consistently
lower when the signal sequence was included in the coding
sequence, resulting in the formation of pre-interferons,
than when the mature IFN-sequence was expressed.[17,32]
High level direct synthesis of the mature α-interferons
reaches 10×10^9 units per liter in large-scale cultures.
Several α-interferons have been purified to homogeneity
from the bacterial extracts.[29] This allows detailed
characterization of the proteins, which was before
impossible with the authentic leukocyte interferons. The
production of sufficient amounts of interferons does not
only allow protein biochemical studies, but makes also a
study of the biological effects of the individual purified
species possible, as well as their use as therapeutic
agents. Both the antiviral properties of these interferons
in different virus-cell systems and the cross-reactivity in
cells of different species show marked differences,
depending on the interferon species.[33-35] This prompted
us to construct hybrid genes by combining parts of the cDNA
sequences for the α-interferons, as shown in Fig 2. The
resulting hybrid interferons, synthesized in E. coli,
showed clear differences in their antiviral activities from
the parent α-IFNs.[34] This approach does not only allow
an assignment of functionally important regions in the
molecules but also leads to the design of new interferons
with clearly different properties. A higher potency is
even observed in some cases, e.g. with a LeIF A/D hybrid.

Eukaryotic expression of leukocyte interferon is
achieved in the yeast Saccharomyces cerevisiae. The
sequence coding for mature LeIF D has been incorporated in
plasmids under the control of endogenous promoters either
for the alcohol dehydrogenase I gene[36] or the 3-phospho-
glycerokinase gene. A high expression level of 6×10^8
units per liter has been obtained, thus constituting a
potential alternative method for the large-scale production
of interferons.

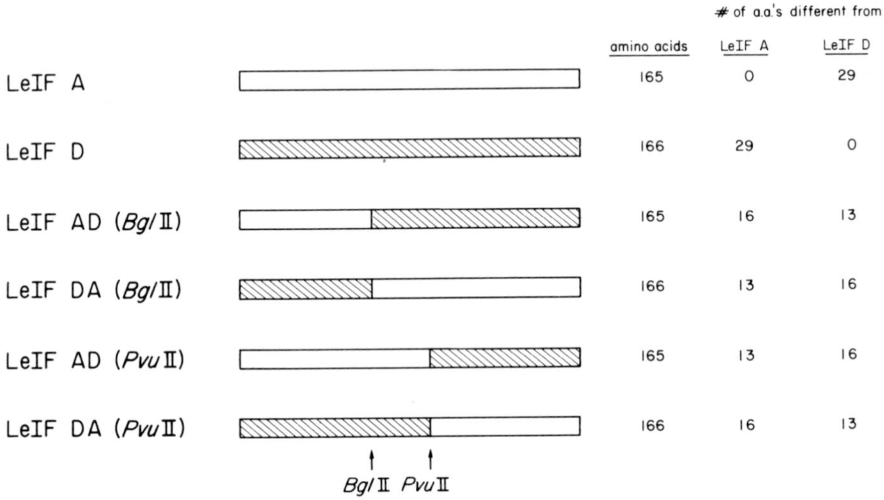

	amino acids	# of a.a.'s different from	
		LeIF A	LeIF D
LeIF A	165	0	29
LeIF D	166	29	0
LeIF AD (*Bgl* II)	165	16	13
LeIF DA (*Bgl* II)	166	13	16
LeIF AD (*Pvu* II)	165	13	16
LeIF DA (*Pvu* II)	166	16	13

Bgl II *Pvu* II

Figure 2. Representative diagram of the parental and hybrid leukocyte interferons produced and their differences from the parental molecules.

Much attention has recently been focused on the study and production of γ-interferon. Some evidence has been reported for a strong immunoregulatory or antiproliferative activity, suggesting a potential antitumoral effect.[37-40] However, its very low availability from mitogen-induced human lymphocytes has hampered research considerably. We have recently prepared a cDNA library from mRNA from mitogen induced human peripheral blood lymphocytes and have isolated a clone, coding for human immune (γ) interferon.[41] The deduced amino acid sequence shows a putative signal peptide of 20 amino acids and a mature polypeptide of 146 amino acids with two potential N-glycosylation sites (Fig. 3). The molecular weight of the unmodified mature γ-IFN is about 17,000 dalton, which is in contrast with the high molecular weights of 35,000 to 70,000 dalton[42-44], but corresponds with the lower values obtained by Yip et al.[45] This discrepancy can be due to intramolecular association or aggregation. Comparison with the human α- and β-interferons shows a lack of structural homology, except for a very short region at the junction of the putative signal peptide and the mature interferon. The amino acid sequence also shows that γ-interferon should be considered as a basic and rather hydrophilic protein, a finding that has been experimentally confirmed (ref 46, E. Rinderknecht, unpublished results).

AMINO ACID SEQUENCE OF IFN-γ

S1   S10   S20 1   10
MET LYS TYR THR SER TYR ILE LEU ALA PHE GLN LEU CYS ILE VAL LEU GLY SER LEU GLY CYS TRY CYS GLN ASP PRO TYR VAL LYS GLU

                          20   30   40
ALA GLU ASN LEU LYS LYS TYR PHE ASN ALA GLY HIS SER ASP VAL ALA ASP ASN GLY THR LEU PHE LEU GLY ILE LEU LYS ASN TRP LYS

                          50   60   70
GLU GLU SER ASP ARG LYS ILE MET GLN SER GLN ILE VAL SER PHE TYR PHE LYS LEU PHE LYS ASN PHE LYS ASP ASP GLN SER ILE GLN

                          80   90   100
LYS SER VAL GLU THR ILE LYS GLU ASP MET ASN VAL LYS PHE PHE ASN SER ASN LYS LYS LYS ARG ASP ASP PHE GLU LYS LEU THR ASN

                      110   120   130
TYR SER VAL THR ASP LEU ASN VAL GLN ARG LYS ALA ILE HIS GLU LEU ILE GLN VAL MET ALA GLU LEU SER PRO ALA ALA LYS THR GLY

                      140   146 STOP
LYS ARG LYS ARG SER GLN MET LEU PHE ARG GLY ARG ARG ALA SER GLN

Figure 3. Amino acid sequence of γ-IFN as deduced from the nucleotide sequence of the cloned cDNA. S refers to the signal peptide.

The availability of a cloned cDNA copy makes a study of the structural organization of the γ-IFN gene possible. A single gene has been detected by hybridization with total human genomic DNA, which is unlike the case of the α-IFN multigene family. This gene is, in contrast with both α- and β-interferon genes, not devoid of intervening sequences.[41]

Expression of the structural gene for human γ-interferon has been achieved in several heterologous systems. Bacterial expression in E. coli took place after insertion of the sequence coding for the mature protein under control of the trp-promoter[41]. Typical large scale cultures give a yield of 10^8 units per liter. Alternatively, expression of the gene and secretion of the protein was obtained in monkey cells. Therefore, the coding sequence of γ-interferon was introduced behind the SV40 late promoter, which itself is present on a vector capable of replication both in E. coli and monkey cells. After transfection of this plasmid into the monkey cells, a γ-interferon activity of 2,000 units per ml is observed in the medium. Eukaryotic expression was also achieved in the yeast Saccharomyces cerevisiae. The coding sequence for mature interferon has been inserted in plasmids behind the endogenous promoter for the 3-phosphoglycerokinase gene (fig. 4). Synthesis of 10^6 units per liter has been detected in extracts of yeast transformed with these recombinant plasmids. The γ-interferon which is produced

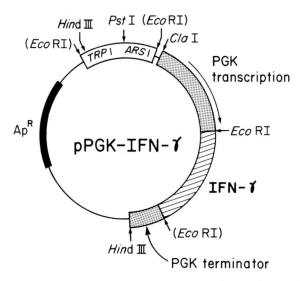

Figure 4.   Schematic representation of a plasmid for expression of γ-IFN in yeast.   The boxed region Ap^R indicates the β-lactamase gene of pBR322. TRP-1 and ARS-1 are the yeast selection marker and the yeast origin respectively.   The transcription of the IFN-γ gene starts from the phosphoglycerokinase (PGK) promoter and terminates at the 3' untranslated region of the PGK gene.

in these three heterologous host-vector systems has similar properties as authentic human γ-interferon.

    In conclusion, using recombinant DNA technology, we have isolated cDNA clones for human α, β and γ-interferons and have established the amino acid sequence of the corresponding proteins. Studies using these clones and the expression of the cDNAs in heterologous systems undoubtedly contributes to a better understanding of the biological properties and the structural organization of the inter- ferons and their genes. The availability of sufficient amounts of purified interferons allows an evaluation of these potential therapeutic agents.

REFERENCES

1. Stewart, W. E., II, The Interferon System. (Springer, Berlin, 1979).
2. Knight, E.Jr., Hunkapiller, M.W., Korant, B.D., Hardy, R.W.F. and Hood, L.E., Science 207, 525-526 (1980).

3. Goeddel, D.V., Shepard, H.M., Yelverton, E., Leung, D.V., Crea, R., Sloma, A. and Pestka, S., Nucl. Acids. Res. 8, 4057–4074 (1980).
4. Taniguchi, T.; Ohno, S.; Fuji–Kuriyama, Y. and Muramatsu, M., Gene 10, 11–15 (1980).
5. Derynck, R., Content, J., De Clercq, E., Volckaert, G., Tavernier, J., Devos, R. and Fiers, W., Nature 285, 542–547 (1980).
6. Davey, M.W., Sulkowski, E. and Carter, W. A., J. Biol. Chem 251, 7620–7625 (1976).
7. Shepard, H.M., Leung, D. Stebbing, N. and Goeddel, D.V., Nature 294, 563–565 (1981).
8. Havell, E.A., Yamazaki, S. and Vilcek, J., J. Biol. Chem. 252, 4425–4427 (1977).
9. Tan, Y.H., Barakat, F., Berthold, W., Smith–Johannsen, H. and Tan, C., J. Biol. Chem 254, 8067–8073 (1979).
10. Lawn, R.M., Adelman, J., Franke, A.E., Houck, C.M., Gross, M., Najarian, R. and Goeddel, D.V., Nucl. Acids. Res. 9, 1045–1052 (1981).
11. Houghton, M., Jackson, I.J., Porter, A.G., Doel, S.M., Catlin, G.H., Barber, C. and Carey, N.H., Nucl. Acids. Res. 9, 247–266 (1981).
12. De Grave, W., Derynck, R., Tavernier, J., Haegeman, G., and Fiers, W., Gene 14, 137–143 (1981).
13. Ohno, S. and Taniguchi, T., Proc. Natl. Acad. Sci. USA 78, 5305–5309 (1981).
14. Sehgal, P.B. and Sagar, A.D., Nature 288, 95–97 (1980).
15. Sagar, A.D., Pickering, L.A.; Sussman–Berger, P., Stewart, W.E. II and Sehgal, P.B., Nucl. Acids. Res. 9, 169–182 (1981).
16. Houghton, M., Eaton, M.A.W., Stewart, A.G.; Smith, J.C., Doel, S.M., Catlin, G.H., Lewis, H.M.; Patel, T.P., Emtage, J.S., Carey, N.H. and Porter, A.G., Nucl. Acids. Res. 8, 2885–2894 (1980).
17. Goeddel, D.V., Yelverton, E.; Ullrich, A., Heyneker, H.L., Miozzari, G., Holmes, W., Seeburg, P.H., Dull, T., May, L., Stebbing, N., Crea, R., Maeda, S., McCandliss, R., Sloma, A., Tabor, J. M., Gross, M., Familetti, P.C. and Pestka, S., Nature 287, 411–416 (1980).
18. Goeddel, D.V., Leung, D.W., Dull, T.J., Gross, M., Lawn, R.M., McCandliss, R., Seeburg, P.H., Ullrich, A., Yelverton, E. and Gray, P.W., Nature 290, 20–26 (1981).
19. Mantei, N., Schwarzstein, M., Streuli, M., Panem, S., Nagata, S. and Weissmann, C., Gene 10, 1–10 (1980).
20. Strueli, M., Nagata, S. and Weissmann, C., Science 209, 1343–1347 (1980).
21. Weissman, C., Interferon 1981, in press.

22. Lawn, R.M., Adelman, J., Dull, T.J., Gross, M., Goeddel, D. and Ullrich, A., Science 212, 1159–1162 (1981).
23. Lawn, R.M., Gross, M., Houck, C.M., Franke, A.E., Gray, P.W. and Goeddel, D.V., Proc. Natl. Acad. Sci. USA 78, 5435–5439 (1981).
24. Ullrich, A., Gray, A., Goeddel, D.V. and Dull, T.J., J. Mol. Biol., in press.
25. Zoon, K.C., Smith, M.E., Bridgen, P.J., Anfinsen, C.B., Hunkapiller, M.W. and Hood, L.E., Science 207, 527–528 (1980).
26. Allen, G. and Fantes, K.H., Nature 287, 408–411 (1980).
27. Taniguchi, T., Mantei, N., Schwarzstein, M., Nagata, S., Muramatsu, M. and Weissman, C., Nature, 285, 547–549 (1980).
28. Wetzel, R., Nature 289, 606–607 (1981).
29. Wetzel, R., Perry, L.J., Estell, D.A., Lin, N., Levine, H.L., Slinker, B., Fields, F., Ross, M.J. and Shively, J., J. Interferon Res. 1, 381–390 (1981).
30. Nagata, S., Mantei, N. and Weissmann, C., Nature 287, 401–407 (1980).
31. Owerbach, D.; Rutter, W.J.; Shows, T.B.; Gray, P., Goeddel, D.V. and Lawn, R.M., Proc. Natl. Acad. Sci. USA 78, 3123–3127 (1981).
32. Yelverton, E., Leung, D., Weck, P.W. and Goeddel, D.V., Nucl. Acid. Res. 9, 731–741 (1981).
33. Weck, P.K., Apperson, S., May, L., and Stebbing, N., J. Gen. Virol. 57, 233–237 (1981).
34. Weck, P.K., Apperson, S.; Stebbing, N., Gray, P.W., Leung, D., Shepard, H.M. and Goeddel, D.V., Nucl. Acids. Res. 9, 6153–6166 (1981).
35. Stewart, W.E. II, Sarkar, F.H., Taira, H., Hall, A., Nagata, S. and Weissman, C., Gene 11, 181–186 (1981).
36. Hitzeman, R.A., Hagie, F.E., Levine, H.L., Goeddel, D.V., Ammerer, G. and Hall, B.D., Nature 293, 717–722 (1981).
37. Rubin, B.Y. and Gupta, S.L., Proc. Natl. Acad. Sci. USA 77, 5928–5932 (1980).
38. Sonnenfeld, G., Mandel, A.D. and Merigan, T.C., Cell. Immunol. 40, 285–293 (1978).
39. Blalock, J.E., Georgiades, J.A., Langford, M.P. and Johnson, H.M., Cell. Immunol., 390–394 (1980).
40. Crane, J.L. Jr., Glasgow, L.A., Kern, E.R. and Youngner, J.S., J. Natl. Cancer Inst. 61, 871–874 (1978).
41. Gray, P.W., Leung, D.W., Pennica, D., Yelverton, E., Najarian, R., Simonsen, C.C., Derynck, R., Sherwood, P.J., Wallace, D.M., Berger, S.L., Levinson, A.D. and Goeddel, D.V., Nature 295, 503–509 (1982).

42. Langford, M.P., Georgiades, J.A., Stanton, G.J., Dianzani, F. and Johnson, H.M., Infect. Immun. 26, 36–41 (1979).
43. DeLey, M., Van Damme, J., Claeys, H. Weening, H., Heine, J.W., Billiau, A., Vermylen, C. and De Somer, P., Eur. J. Immunol. 10, 877–883 (1980).
44. Yip, Y.K., Pang, R.H.L., Urban C., and Vilcek, J., Proc. Natl. Acad. Sci. USA 78, 1601–1605 (1981).
45. Yip, Y.K., Barrowclough, B.S., Urban, C. and Vilcek, J., Science 215, 411–412 (1982).
46. Yip, Y.K., Barrowclough, B.S., Urban, C. and Vilcek, J., Proc. Natl. Acad. Sci. USA, in press.
47. Dayhoff, M.O., Atlas of Protein Sequence and Structure 5, suppl. 3 (1978).

DISCUSSION

P.A. LIBERTI: You stated that gamma interferon contains a high number of charged residues, particularly basic amino acids. In the sequence you showed, the stretch of residues around position 90 (about four basic residues in sequence) is similar to the putative complement (Clq) binding region on IgG. Is there any connection between the complement system and gamma interferon?

Also what is the significance of the 'hypervariable'-like regions in the sequences of leukocyte interferons? From your slide it looked like two regions are really hot spots for sequence variation while the remainder of the sequence seems conserved?

D.V. GOEDDEL: I am not aware of any connection between the complement system and lFN-gamma. As for the variable regions in the leukocyte interferons, these might give the leukocyte interferon family its wide range of target cell and virus specificities. The conserved regions are probably required for IFN activity, perhaps due to structural requirements.

D.C. BURKE: Interferon has two routes for its action, one on membrane mediated processes and the other via 25A and the protein kinase system. The different effects of the hybrids against the two challenge viruses may indicate that the structure/function relationship for these two mechanisms may be different. Are the different alpha genes transcribed in different proportions in different cells or induced with different viruses?

D.V. GOEDDEL: We do not have any direct evidence that the different IFN-alpha genes are transcribed in different relative proportions in different cell types and/or in response to different viruses. It does seem that types A (IFN-alpha 2) and D(IFN-alpha 1) appear to be induced under most, or all known conditions of IFN-alpha induction. Menachin Rubenstein has shown that different relative amounts of the different IFN-alpha proteins are produced when one cell type is induced to produce IFN-alpha with different viruses.

*S.A. SHERWIN: Do you have any information suggesting that the hybrid molecules differ from each other and from the parent molecule with regard to anti-proliferative activity in vitro, or with regard to effect on selected immunologic assays.

*D.V. GOEDDEL: I think I would rather leave the answer to that question to Nowell Stebbing since that is one of the things he is planning to talk about on Friday.

S.L.C. WOO: If gamma-interferon gene contains two introns, why is its cDNA clone expressable in the pSV/cos[1] cell system?

D.V. GOEDDEL: We have expressed all the IFNs: α, β, and γ in COS cells. There does not appear to be any requirement for intron splicing in this system.

S.L.C. WOO: Why did you switch from the ADH promoter to another yeast promoter?

D.V. GOEDDEL: The ADH promoter used in our leukocyte interferon work was provided by a collaborator for use in that project only. We subsequently developed the PGK system for our own use.

M. CHRETIEN: Pairs of basic amino acid residues have been recognised since 1967 as sites of cleavage to produce mature active hormones from their precursors, as for for example beta-lipotropin (Chretien, M. and C.H. Li, Can. J. Biochem, 45:1163-1174, 1967). Such pairs of basic amino acid residues are frequently present in interferons. Can interferons be cleaved at these sites with an increase in their biological activities? For example, by partial and limited tryptic digestion?

D.V. GOEDDEL: Ron Wetzel (Genentech) has attempted partial tryptic digestion of IFN-alpha A and found it very difficult to get partial digestion products. I do not know of any cases where a more active IFN molecule has been obtained through removal of any portion of the molecule.

U.R. MULLER: When you synthesize new interferon proteins by combining fragments of various alpha-interferon genes you also obtain a protein with new antigenic sites. Isn't this going to limit its use in humans?

D.V. GOEDDEL: It is possible that hybrid IFNs will contain new antigenic sites which could limit their use in humans. However, the answer to this question will come only after the hybrid IFNs have been tested clinically.

J.B. ZELDIS: Were the gamma-interferons expressed in the COS cells glycosylated? If so, were they glycosylated at both of the potential glycosylation sites?

D.V. GOEDDEL: The IFN-gamma synthesized by COS cells is glycosylated. At the present time we do not know if the glycosylation occurs at both potential glycosylation sites.

D.V. HENDRICK: In general, do you find that the level of interferon produced in E. coli and yeast is primarily a function of plasmid copy number, or are you able to effect a significant change in the level by manipulating physiological and fermentation parameters?

D.V. GOEDDEL: We find that the levels of IFN produced in E. coli are influenced by several factors. These include plasmid copy number, promoter strength, ribosome binding site structure, messenger RNA stability, IFN stability, and effects of IFN on E. coli growth. Some of these variables can be affected by fermentation conditions.

*M. BINA: When you follow the expression in the COS cell do you see excretion or not?

*D.V. GOEDDEL: Yes, all the activity we found is secreted into the medium.

C. COLBY: You demonstrated similar activities and properties of IFN-gamma from E. coli, yeast and monkey cells. Do you have information about the specific activities of IFN-gamma expressed in these three systems?

D.V. GOEDDEL: At this time only the IFN-gamma produced in monkey cells has been purified significantly. Its specific activity appears to be greater than 10^8 units per milligram. We do not yet know the specific activity of the IFN-gamma produced by E. coli or yeast.

*D. BOTSTEIN: You answered two questions which indicate to me that you made a construction without a signal sequence (SV40 hasn't got a signal sequence) and yet everything is excreted and glycosylated, is that right?

*D. GOEDDEL: No, the COS cell construction used the entire cDNA clone with 5 prime untranslated regions, signal sequence and mature coding sequence.

*D. BOTSTEIN: So that construction had all the information and therefore, presumably, was glycosylated and secreted.

*D. GOEDDEL: Yes.

V. CHOWDHRY: In your comments about gamma-interferon you mentioned that the sequence had no homology to the alpha or beta interferons. Are there cysteines in gamma IFN and does it have the disulfide found near the C-terminal in alpha and beta INF's? Is the tryptophan near the C-terminal missing?

D.V. GOEDDEL: There are cysteines at amino acid positions 1 and 3 only. The cysteine and tryptophan found near the C-termini of IFN-alpha and IFN-beta are missing.

H. YOUNG: Do you plan on making any fibroblast-leukocyte interferon hybrids?

D.V. GOEDDEL: We have considered constructing hybrids between leukocyte and fibroblast interferon genes. However, the lack of common restriction sites will make this more difficult to do than to construct hybrids between leukocyte interferons.

CLONING INTO EUKARYOTIC CELLS

MAKING MUTATIONS IN VITRO AND
PUTTING THEM BACK INTO YEAST

David Botstein[1]
David Shortle[2]
Mark Rose

Department of Biology
Massachusetts Institute of Technology
Cambridge, Massachusetts
U.S.A.

In this short review, we present the principles of three
new experimental approaches which together should greatly aid
the analysis of yeast gene function and regulation in vivo by
making it feasible to construct mutations in molecularly
cloned genes and returning them to the yeast by DNA transformation. In the first section of this paper, we review methods
developed in our laboratory which allow highly localized in
vitro mutagenesis of plasmid DNAs. In the second, we present
a new method (called "gene disruption") which allows us to
construct null mutations by insertion of a plasmid directly
into the gene at its normal location in the yeast genome;
these mutations can make it possible to assess the phenotype
of such mutations and thereby learn about gene function. The
third section briefly describes methods for the production of
gene fusions between yeast genes and the E. coli β-galactosidase gene (lacZ) which allow us to assess regulatory phenomena
in yeast by assaying β-galactosidase; fusion also help to define the regions of yeast genes which are required for normal
regulation.

---

[1]This work supported by grants from NIH (GM21253, GM18973) and
ACS (MV-90)

[2]Present address: Department of Microbiology, SUNY, Stony
Brook, New York

1. Localized Mutagenesis in Plasmid DNA

In a recent publication (1), we described a way to make nicks in plasmid DNA within predetermined regions. This method (diagrammed in Figure 1) takes advantage of the fact that the recA protein of E. coli catalyzes, in the presence of ATP, the assimilation into closed circular DNA molecules of homologous single-stranded DNA fragments. The resulting structure is called a "D-loop": the assimilated single-stranded fragment is paired with the homologous region of the plasmid DNA, displacing the other plasmid strand and producing a structure which resembles the letter "D". When D-loops are exposed to the single-stranded endonuclease S1 under conditions of very limited S1 activity, the displaced strand of the D-loop is nicked specifically, which results in the destabilization of the D-loop structure and the production of singly nicked circular plasmid DNA. The nick occurs somewhere within the displaced region, and can be used as a specific target for several kinds of in vitro mutagenesis reactions.

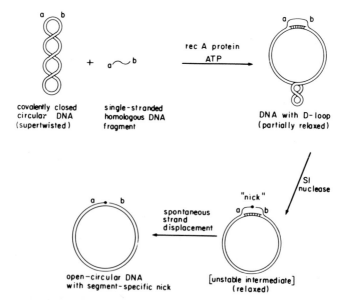

FIGURE 1. Segment-Directed Mutagenesis. Diagram shows steps in producing specifically in a closed circular DNA molecule. The nick is directed to the region between points a and b by a single-stranded fragment. Details can be found in reference 1, from which the diagram is reproduced.

The mutagenesis methods which we use all begin with the production of a short (usually about 6-8 nucleotide pairs in length) single-stranded gap which results from the action of M. luteus DNA polymerase I in the absence of nucleoside triphosphates (2). Such gaps can be mutagenized directly with sodium bisulfite (3) which deaminates cytosine residues producing uracil. After filling in of the gap using DNA polymerase, a U:A basepair is produced where a C:G basepair has been. The combination of the D-loop and bisulfite procedures was used to produce mutations in predetermined regions of the β-lactamase gene on plasmid pBR322 (see reference 1) with high specificity (more than 95% of mutations in the expected region) and efficiency (more than 30% of recovered plasmids mutant).

More recently, we have devised new ways of making mutations beginning with short gaps (4). As shown in the example given in Figure 2, a short single-stranded gap is generated from a nick and filled in under conditions which promote misincorporation. Since DNA polymerases possess 3'-5' "editing" exonuclease, misincorporations are efficiently excised. Two methods were found which avoid this undesirable competing reaction. One method (gap misrepair via nucleotide omission) promotes misincorporation by omitting one of the four nucleoside triphosphates from the reaction mixture and by adding manganese ions as well as T4 DNA ligase and ATP. Under these conditions, only non-complementary nucleotides are available for incorporation at some positions in the gap. Infrequently (the frequency is raised by the Mn^{++} ions) DNA synthesis proceeds past such a position through misincorporation; if the remainder of the single-stranded gap is filled in and the strand immediately closed by the DNA ligase present, the misincorporation nucleotide is trapped in a non-excisable state as part of a fully closed circular DNA molecule. The second method (gap misrepair via non-excisable nucleotides) uses misincorporation at the 3' terminus of the gap of "excision-resistant" α-thiophosphate analogues as a first step. Since these analogues are efficient substrates for DNA polymerase (5,6) but, once incorporated, are resistant to the 3'-5' editing exonuclease activity (7,8), their misincorporation is effectively irreversible, and misincorporations accumulate with time. Gaps containing misincorporations are then closed in the presence of natural triphosphates and sealed with DNA ligase.

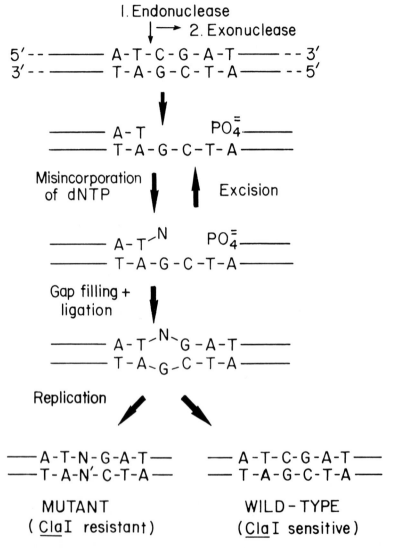

FIGURE 2. Mutagenesis by Gap Misrepair. Diagram shows an
example in which a ClaI site is mutagenized following ClaI-
catalyzed nicking in the presence of ethidium bromide. Gaps
are made with DNA polymerase I from M. luteus as described in
the text. Misrepair can be accomplished by either of the
methods explained in the text. Replication of misrepaired
molecules occurs after transformation into E. coli.

We have used these methods to construct mutations starting with nicks and found efficient (10% or more) recovery of mutations in the immediate vicinity of the nicks (4). Further, the gap misrepair methods have allowed the recovery of all types of mutations (transversions and frameshifts as well as transitions) in good yield, while the bisulfite methods produce primarily C:G-T:A transitions.

With these methods, it is now possible to mutagenize very specifically and efficiently very small regions of a plasmid DNA containing a cloned gene. The D-loop procedure allows the targeting of nicking to be within about 60 basepairs (the smallest single-stranded fragments used were about this length) most mutations made by our methods lie very close (within 20 bases) to the position of the nick. Several points about these procedures deserve emphasis. First, multiple mutations are sometimes generated, but all lie near the nick. Thus, changes which might require double events should be recoverable. Second, since all types of base changes can be produced, one need not know or guess what changes to make, as is the problem with mutagenesis via chemical synthesis. Third, the efficiency of mutagenesis is high enough to encourage individual screening of plasmids for mutations of interest even by relatively laborious biochemical methods.

2. Generation of Null Mutations in Yeast by Gene Disruption

Before one can return mutations made in vitro back into yeast, one must have some idea of the phenotype of the mutation. In the case of loss of function (the usual case for mutants), the mutation will generally be recessive, and the return of the mutation must involve a procedure by which the normal allele in the genome is replaced by the mutant one (9) or otherwise inactivated. In cases of cloned genes whose failure to function has unknown consequences, the problem is acute: even efficient methods of mutagenesis and replacement are useless unless the mutations can somehow be recognized.

We have devised a general procedure for constructing mutations which should result in total loss of gene function starting with a cloned gene. We undertook this as part of our study of the genetics of actin in yeast, where we faced the problem of a gene not knowing the mutant phenotype.

A null mutation in the chromosomal locus corresponding to the cloned yeast actin gene was constructed by the method diagrammed in Figure 3. A 1.3 kb AvaII restriction fragment which is internal to the protein coding region of the gene was inserted into the yeast plasmid vector YIp5. This transformation

vector, a derivative of the E. coli plasmid pBR322, carries
the yeast URA3 gene and can transform URA3-yeast to proto-
trophy only by integration, via homologous recombination, into
the yeast genome. Likewise, the hybrid plasmid (designated
pRB111) carrying the AvaII actin fragment also can only transform URA3-yeast to uracil-independence by integration. Since
pRB111 carries sequences homologous to both the actin and URA3
genes, integration can occur into either of these two chromosomal loci; and as shown in Figure 3, integration at the actin
locus disrupts the protein-coding portion of the actin gene.
In effect, integration by a single homologous recombinational
event between the chromosomal actin gene and the internal fragment in pRB111(which can be viewed as a double deletion mutant
of the gene) results in a direct repeat of actin gene sequences, in which only that portion of the gene carried by the
plasmid is duplicated. Consequently, each of the repeated
copies of the actin gene is now incomplete.

Disruption of an essential gene in this way should be a
lethal event in a haploid strain. However, if the loss of
gene function is, as expected, recessive, disruption of the
gene in a diploid strain would result in a recessive-lethal
mutation, a genetic lesion easily detected by standard tetrad
analysis.

When a diploid strain was transformed with pRB111, the integration at the actin locus indeed resulted in a recessive
lethal mutation. From this result, we could determine that
point mutations in actin might have a conditional lethal phenotype (e.g. temperature-sensitive), and on this basis we were
able to isolate such mutations by applying the methods described above and putting them into yeast by a modification of the
gene disruption technique.

The gene disruption technique is general in the sense that
the null phenotype of any gene can be determined with it. In
the case of essential genes, recessive lethality is expected.
In the case of genes whose loss leads to metabolic requirements, the requirements can easily by determined. In the case
of non-essential genes, gene disruption allows the discovery
of whatever accessory function is missing. In the case of duplicate genes, gene disruption allows the easy removal of one
of the duplicates so that consequent loss of the other gene
(by further mutagenesis, for example) can more easily be investigated.

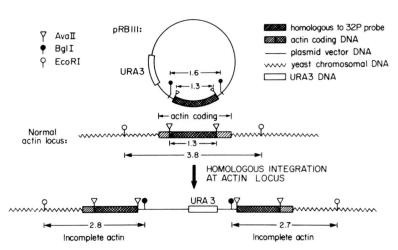

FIGURE 3. Gene Disruption. The diagram shows consequen-- ces of integration of an AvaII fragment of the yeast actin gene carried on a plasmid vector into the intact chromosomal actin gene. The diagram also shows the expected sizes (in kilobases) of fragments after integration; these expectations were confirmed by gel-transfer hybridization experiments.

3. Construction and Use of Gene Fusions in Yeast

Gene fusions have proved themselves very useful in the analysis of regulatory phenomena in prokaryotic systems. We have used fusions of yeast genes to the E. coli lacZ gene (which specifies ß-galactosidase) to investigate regulatory phenomena in yeast. In this case, we have developed the system using the yeast URA3 gene; a plasmid, which is capable of replication either in E. coli or Saccharomyces cerevisiae and which has markers selectable in either organism, is construc- ted such that the beginning of a lacZ gene which is missing its translation initiation condon immediately follows the end of the intact URA3 gene. Fusions can be made either by dele- tions in E. coli (11), or in vitro by the scheme shown in Figure 4. When introduced into yeast (as autonomous plasmids or integrated into the genome), successful fusions to yeast genes resulted in the expression of ß-galactosidase enzyme activity which can be detected by including a chromogenic sub- strate into agar plates (11,12) or by enzyme assay of extracts. Further, fusions allow the sensitive assay of regulatory res- ponse under circumstances in which the response itself has no effect. Such "gratuity" is essential for study of complicated systems like self-regulated genes.

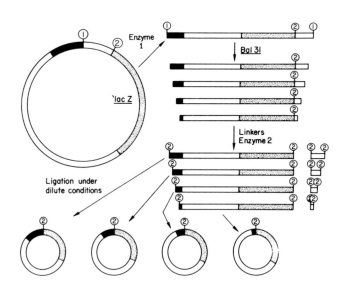

FIGURE 4. Making Fusions In Vitro. Diagram shows a general method for making fusions to a fragment of the lacZ gene (shown stippled) which is missing only the first few codons. In order to be expressed, lacZ must be ligated to another expressed gene (shown filled in black) in the correct reading frame. Enzymes 1 and 2 are restriction endonucleases; linking are synthetic oligodeoxynucleotides bearing the recognition site for one of the enzymes.

In conclusion, the technology for making random mutations more or less at will in cloned genes and the technology for determining the function of genes through construction of such mutations is becoming quite well developed. The kinds of genetic tools which led to understanding at the molecular level of some mechanisms of gene regulation and function in bacteria are now readily available to yeast molecular biologists as well.

ACKNOWLEDGMENTS

Research was supported by grants from the American Cancer Society (MV-90) and NIH (GM18973 and GM21253). D.S. was a fellow of the Helen Hay Whitney Foundation; M.R. was supported by a National Science Foundation Graduate Fellowship.

REFERENCES

1. Shortle, D., Koshland, D., Weinstock, G.M. and Botstein, D., Proc. Natl. Acad. Sci. USA 77, 5375 (1980).
2. Parker, R.C., Watson, R.M. and Vinograd, J., Proc. Natl. Acad. Sci. USA 74, 851 (1977).
3. Shortle, D. and Nathans, D., Proc. Natl. Acad. Sci. USA 75, 2170 (1978).
4. Shortle, D., Grisafi, P., Benkovic, S.J. and Botstein, D., Proc. Natl. Acad. Sci. USA in press (1982).
5. Burgers, P.M.J. and Eckstein, F., J. Biol. Chem. 254, 6889 (1980).
6. Vosberg, H.P. and Eckstein, F., Biochemistry 16, 3633 (1977).
7. Benkovic, S.J., Gupta, A. and Henrie, R., unpublished results.
8. Kunkel, T.A., Eckstein, F., Mildvan, A.S., Koplitz, R.M. and Loeb, L.A., Proc. Natl. Acad. Sci. USA in press (1981).
9. Scherer, S. and Davis, R.W., Proc. Natl. Acad. Sci. USA 76, 4951 (1979).
10. Franklin, N. Annual Rev. Genet. 12, 193 (1978).
11. Rose, M., Casadaban, M.J. and Botstein, D., Proc. Natl. Acad. Sci. USA 78, 2460 (1981).
12. Guarente, L. and Ptashne, M., Proc. Natl. Acad. Sci. USA 78, 2199 (1981).

DISCUSSION

D. SCHLESSINGER: When looking at mutant alleles in genes like those for tubulin or actin, whose products function in long polymers, shouldn't a fair number, perhaps many, t.s. alleles be dominant in diploids?

D. BOTSTEIN: Yes, one would expect that. However, in the presence of benomyl a large number of mutants are formed; eighty percent of which are recessive while twenty percent are, in some degree, dominant. Thus there is an unusually large number of dominant mutations in this system, but they are by no means the overwhelming class.

D. GELFAND: Do you have any mutants in the putative 28 amino acid polypeptide from the weak constitutive promoter (that initiates from the weak constitutive URA promoter) that have altered function?

D. BOTSTEIN: Since fusion is made, it is likely that the peptide is synthesized.   However, we don't have a single mutation of the point mutation type which would suggest involvement of the peptide in any function related to uracil.

A. SKALKA: With regard to the "gap missrepair" methods, is the lack of purity of deoxynucleotides a serious problem?   Is it necessary to purify commercial products before use?

D. BOTSTEIN: No.   We tried several suppliers and in the case of half the suppliers the background is so low that purification is not necessary.   In fact, we found unlabeled nucleotides to be more pure than the radiolabeled ones.

K. SIROTKIN: Would I be wrong in saying that some of the regulatory sequences seem to be internal as is the case with 5S xenopus genes?

D. BOTSTEIN: No, you wouldn't be wrong.   The point I did not make is that in the uracil gene what is important to regulation, initiation, transcription, or translation is 5' to the 11th nucleotide.   This is the longest deletion we've looked at.   We are now introducing point mutations in this region to determine what is important.

EXPRESSION OF THE HUMAN GROWTH HORMONE GENE IS REGULATED IN MOUSE CELLS

Diane M. Robins[1]
Inbok Paek
Richard Axel[2]

Institute of Cancer Research
College of Physicians and Surgeons
of Columbia University
New York, N.Y.
U.S.A.

Peter H. Seeburg

Genentech, Inc.
South San Francisco, Calif.
U.S.A.

I. INTRODUCTION

Steroid hormones regulate the expression of restricted sets of gene products in a tissue-specific manner. One simple model of hormone action compatible with available data assumes that the interaction of steroid-receptor complex with appropriate DNA sequences enhances transcription. The introduction of hormonally responsive genes into cells provides an experimental system to determine whether inducibility is a property inherent in discrete nucleotide sequences. Thus, rat α-2u globulin genes (1), as well as mouse mammary tumor virus (MMTV) genes (2,3), remain responsive to glucocorticoids following transfer into heterologous recipients. Further, the fusion of the promoter element of MMTV to the structural gene encoding dihydrofolate reductase renders this gene inducible with glucocorticoids (4).

In vertebrates, growth hormone synthesis is restricted to the pituitary gland. In cultures of pituitary cells, either glucocorticoid or thyroid hormone generate a 3-fold increase in the level of GH mRNA; when added together, a 10-fold induction is observed (5,6). In this study, we

[1] Supported by a Jane Coffin Childs Postdoctoral Fellowship.
[2] Supported by grants from the NIH.

have used cotransformation to introduce from 1 to 20 copies of the human growth hormone (hGH) gene into thymidine kinase deficient (tk⁻) murine fibroblast cells which express functional glucocorticoid receptors (7). The administration of hormone to these cotransformed cells results in a 3- to 5-fold induction of hGH mRNA and a similar induction in secreted growth hormone protein. The DNA sequences responsive to induction reside within 500 nucleotides of DNA flanking the putative site of transcription initiation. Fusion of this segment of DNA to the tk structural gene now renders the tk gene reponsive to hormone action.

II. RESULTS

A. The Regulated Expression of Human Growth Hormone Genes in Mouse Cells

The human growth hormone gene is a member of a small multi-gene family composed of at least two additional hormones, chorionic somatomammotropin and prolactin (8,9). Several different recombinant phage containing hGH sequences have been isolated from a library of human DNA (P. Seeburg, manuscript in preparation) constructed in the Charon phage λ4A (10). Two different phage, designated λ20A and λ2C, contain the entire growth hormone gene within a 2.6 kb Eco RI fragment (Fig. 1). The complete nucleotide sequence of this Eco RI fragment derived from λ20A indicates that this fragment contains the entire hGH gene along with 500 5' and 525 3' flanking nucleotides (P. Seeburg, manuscript in preparation). The restriction map of the 2.6 kb Eco RI fragment of λ2C is identical to that of λ20A. Partial sequence analysis of λ2C from the 5' Eco RI site to the Bam HI site adjacent to the first exon (Fig. 1) is identical to that of λ20A. Both λ20A and λ2C, which presumably encode the major form of growth hormone, were utilized in our studies.

Mouse Ltk⁻ cells express functional glucocorticoid receptor (7). We have utilized a viral tk gene as a selectable marker to introduce the 2.6 kb Eco RI fragment into this cell line by DNA-mediated gene transfer (11,12). Cotransformants were identified by blot hybridization (13) and tested for the capacity to regulate the expression of exogenous human growth hormone sequences. DNA was isolated from eleven transformants obtained following cotransfer with tk and hGH DNA, restricted with the enzyme Eco RI, and analyzed by blot hybridization utilizing highly radioactive hGH probes (Fig. 2). Nine of the eleven transformants integrate at least one intact RI fragment containing the hGH

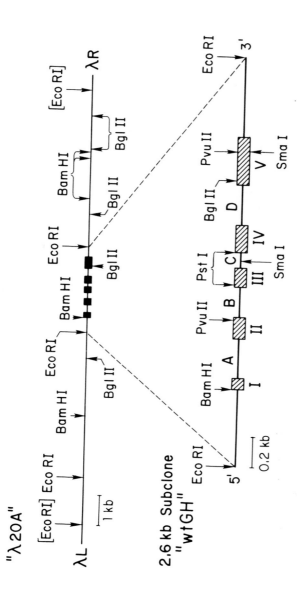

FIGURE 1. Recombinant hGH clones. λ20A is a Charon 4A clone with a 14 kb insert of human DNA. The hGH gene is contained in a 2.6 kb Eco RI fragment. Exons are shown by solid bars. A more detailed map of the sequenced 2.6 kb Eco RI fragment, "wtGH", shows the five exons (hatched bars) interrupted by introns A–D.

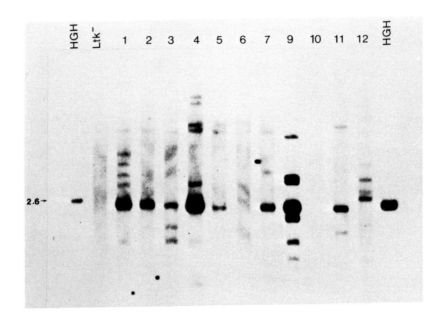

FIGURE 2. wtGH DNA in cotransformed L. cells. Ltk⁻ cells were cotransformed with 1 μg ptk, 1 μg wtGH and 20 μg Ltk⁻ DNA. Colonies surviving in HAT medium were picked and grown into mass culture. High molecular weight DNA was digested with Eco RI and blotted and hybridized as before. The hGH lane on the left contains 20 pg of wtGH DNA and the hGH lane on the right contains 100 pg of wtGH DNA, along with 20 μg of chicken DNA, digested with Eco RI; these amounts are equivalent to approximately 1 and 5 hGH genes per genome, respectively. 20 μg each of DNA digested with Eco RI from Ltk⁻ cells and 11 independently isolated tk⁺ transformants (1-12) are shown. An arrow indicates the faint endogenous mouse GH gene present as a high molecular weight Eco RI fragment in all the L cell lanes.

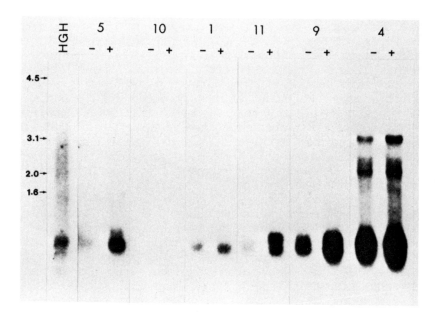

FIGURE 3. hGH RNA induction in L cells containing the 2.6 kb Eco RI hGH fragment. Cotransformant colonies (numbered as from Fig. 2) were grown for 72 hours in the absence (-) or presence (+) of 10^{-6} M dexamethasone. Total cellular poly A^+ RNA (12 g/lane) was fractionated on an 0.8% formaldehyde gel and blotted onto nitrocellulose. 50 ng of human pituitary RNA was run in the first lane, along with OD size markers visualized by ethidium bromide staining. The nitrocellulose filter was hybridized to ^{32}P-labeled hGH DNA (the 2.6 kb Eco RI fragment). As the 5 cotransformant cell lines contain different amounts of hGH RNA, three autoradiographic exposures were examined. hGH, clone 5 and clone 10 (a tk^+ non-cotransformant) lanes were examined 2 days, clones 1 and 11 were exposed 18 hours, and clones 9 and 4, 12 hours.

gene. The hGH copy number in these lines is variable. Some lines contain 10-20 copies of hGH DNA.

These cotransformants were then tested for their ability to regulate the expression of hGH RNA. Poly A^+ RNA was then prepared from five clones and examined by Northern blot analysis (Fig. 3). All five lines synthesize an 825 bp hGH mRNA which is inducible by the addition of glucocorticoid. The levels of hGH mRNA, as well as the extent of induction, differ among the different cell lines. Some lines show a 3-fold induction (lines 9 and 11), other lines show intermediate levels of induction. A rough correlation exists between the amount of mRNA produced and the number of hGH copies integrated into transformed cell DNA. Thus, line 4 synthesizes the largest amount of mRNA and has integrated the largest number of intact GH genes. Line 5 synthesizes perhaps the least amount of mRNA and has probably integrated only one or a few genes. Control transformants lacking an hGH gene fail to express detectable human or mouse GH mRNA. Similar data has emerged from an analysis of several additional cotransformants obtained following exposure to hGH λ phage (Table I). The frequency with which cotransformants express elevated levels of GH mRNA upon exposure to glucocorticoid, even if only a single intact hGH gene is integrated into the chromosome, strongly suggests that the information required for induction resides within the 2.6 kb Eco RI fragment. Therefore, hormonal control is not likely to result solely from the integration of hGH DNA into pre-existing hormonally responsive sites in the recipient cell. Further, hormonal control is not observed for all cotransformed genes, but is specific for the hGH gene. Analysis of the level of tk mRNA in these cell lines by blot hybridization indicates that tk RNA levels remain constant in the presence or absence of steroid hormone.

B. Induction of Growth Hormone Protein

We have asked whether the growth hormone mRNA sequences present within our transformants direct the synthesis of a secreted growth hormone polypeptide. In this manner, we could determine whether the levels of induction of mRNA are reflected by a proportionate induction in the level of secreted protein. Control cells and transformants containing hGH genes were grown either in the presence or absence of hormone for five days without media change. The media was then tested for hGH in radioimmune assays utilizing antibody-coated filter disks. As shown in Table I, cell lines which demonstrate significant levels of hGH mRNA also secrete significant amounts of hGH into the tissue culture medium. The maximum secretion of hGH is observed

TABLE I.   Quantitation of hGH mRNA and Protein
           in Mouse Cells.

Cell line		cpm/µg A$^+$ [a]	mRNA copies/[b] cell	Fold induction	Secreted GH (ng/ml) [c]
λGH-D	−	392	35	1.2	1.0 (below control)
	+	473	42		3.8
λGH-E	−	93	8	2.9	1.8 (below control)
	+	267	24		4.9
λGH-I	−	291	26	4.8	5.0
	+	1401	125		19.0
λGH-A	−	30	>3	1.0	N.D.
	+	38	3		
wtGH-1	−	58	5	2.5	N.D.
	+	146	13		
wtGH-4	−	1147	102	1.7	19.0
	+	2054	183		66.0
wtGH-5	−	65	5	1.1	N.D.
	+	71	6		
wtGH-9	−	487	43	3.0	18.4
	+	1455	129		66.0
wtGH-11	−	71	6	2.6	N.D.
	+	185	17		

[a] Determined from dot blots containing 3-5 concentrations
of Poly A$^+$ RNA of each sample.

[b] There are 2×10^5 mRNA molecules per L cell, assuming
0.5 pg poly A$^+$ RNA/cell and an average mRNA size of 1 kb.
From a standard curve derived by dot blotting human
pituitary RNA in which approximately 20% of the mRNA is
hGH, there are 450 cpm/100 ng of pituitary RNA, or 2.25
cpm/pg hGH mRNA. All cpm's have been normalized to the
same standard curve.

[c] From radioimmune assay, in which the negative control
is < 2.5 ng/ml.

N.D. = Not Determined.

with cell line 9, which also exhibits a high level of mRNA
in our series. This cell line synthesizes about 15 µg per
liter in the absence of hormone and 60 µg of hGH per liter
in the fully induced state. Thus, induction in the levels
of GH mRNA result in a concomitant induction in the levels
of secreted protein.

C. Regulation of an hGH-tk Fusion Gene

We next asked whether the hGH sequence element
responsive to hormonal induction can impart hormone
sensitivity when fused to other structural genes.
Expression of the 2.6 kb Eco RI fragment is hormonally
regulated in mouse cells. In contrast, cotransformed tk
gene expression is not regulated by glucocorticoids. We
therefore constructed a fusion gene consisting of the 5'
flanking sequences of hGH and the structural sequences
encoding tk to discern whether the tk gene in this
configuration is responsive to glucocorticoid induction.
The 5' flanking region of hGH DNA extending from the Eco RI
site to the Bam HI site, 3 nucleotides into the 5'
untranslated region, was ligated to a fragment of tk DNA
beginning in the 5' untranslated region of the tk message 60
nucleotides upstream from the translation start site (Fig.
4). This tk fragment contains the entire structural gene
but lacks all essential promoter elements (14). Thus, a
fusion gene was introduced into Ltk$^-$ cells, and poly A$^+$ was
then isolated from resulting tk$^+$ transformants. The
patterns of the tk RNAs on Northern blot analysis are
complex.

In control cells containing a wild-type tk gene, two
transcripts, 1.3 and 0.9 kb, are observed. The 1.3 kb RNA
is always expressed in transformants containing an intact tk
gene and encodes the wild-type enzyme. The 0.9 kb RNA is
infrequently observed in tk$^+$ transformants. This transcript
initiates internal to the tk gene, consists solely of tk
structural gene sequences, and must encode a truncated
protein (15). RNA from one transformant containing the
wild-type tk gene (line A), demonstrating these two
transcripts, is shown in the Northern blot in Figure 5. RNA
from three tk$^+$ transformants containing the hGH-tk fusion
gene reveals two species of tk RNA at 1.25 and 0.9 kb. The
1.25 kb species is precisely the size expected if the growth
hormone promoter initiates appropriately in hGH sequences to
generate fusion mRNA. The 0.9 kb RNA presumably represents
the aberrant tk transcript. In two of three transformants
containing the fusion gene, we observe that the fusion
transcript is inducible upon glucocorticoid administration.
The 0.9 kb transcript generated from an internal tk promoter

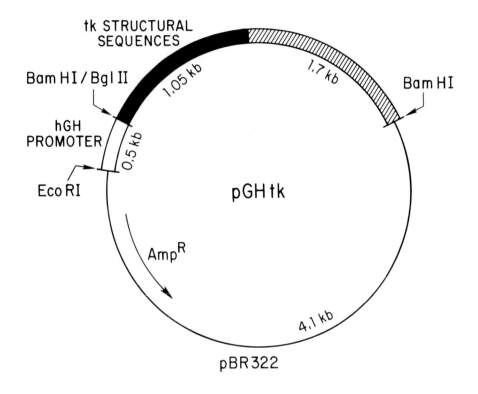

FIGURE 4.   Recombinant plasmid pGHtk contains an hGH-tk fusion gene.   Plasmid pGHtk contains an 0.5 kb Eco RI/Bam HI fragment of hGH (▭) (from the 5' end of the 2.6 kb Eco RI fragment) inserted at the Bgl II site of ptk, replacing the tk promoter.   Tk information includes the entire coding sequence of the tk gene (▬▬) as well as 1.7 kb of 3' flanking DNA sequences (▨).   The initiator AUG is located 50 nucleotides 3' to the Bgl II site.   The Bam HI site of the hGH fragment, including the putative promoter, is 3 nucleotides beyond the transcription initiation site of hGH (9).

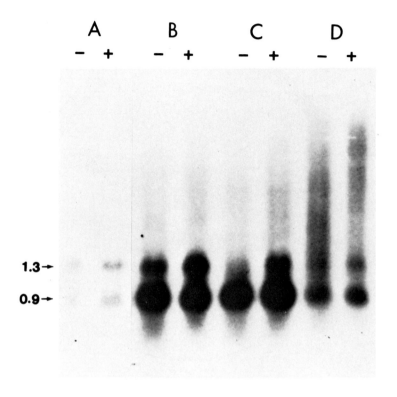

FIGURE 5.   Induction by glucocorticoid of a tk gene fused
to hGH DNA. pGHtk was donated to Ltk⁻ cells and tk⁺ colonies
selected in HAT medium. Several colonies were grown in the
absence (−) or presence (+) of glucocorticoid for 48 hours,
poly A⁺ RNA was isolated, and subjected to Northern blotting.
10 µg of poly A⁺ RNA was loaded in each lane. Cell line A
is a wild-type tk⁺ transformant; lines B-D contain the pGHtk
fusion gene. The filter was hybridized with ³²P-labeled tk
DNA (the 3.6 kb Bam HI fragment of ptk) and exposed to X-ray
film for 1 day.

provides a fortuitous internal control. Induction of RNA levels is observed only for the fusion transcript; the 0.9 kb RNA remains unresponsive to glucocorticoid administration. It is apparent from these studies that sequences responsive to induction reside within the 500 nucleotides flanking the 5' terminus of the hGH gene. Fusion of this element to other structural genes now renders these sequences hormonally responsive.

III. DISCUSSION

The introduction of recombinant clones containing the human growth hormone gene into mouse fibroblasts results in the regulated expression of hGH mRNA. Our results suggest that the induction we observe results from transcriptional activation rather than RNA stabilization. First, hGH-tk fusion presumably generates an inducible mRNA with very few 5' hGH nucleotides; the remainder of the RNA encodes tk enzyme. In control cells, wild-type tk mRNA is not inducible by hormone. We consider it highly unlikely that the short segment of hGH can confer stability on the remaining length of mRNA independent of sequence or source. Second, we have identified hGH transformants synthesizing mature hGH mRNA constitutively. If induction results solely from stabilization, we would expect all transformants which synthesize hGH RNA to express enhanced levels in the presence of hormone. These results do not exclude mRNA stability as a contributing factor in the inductive process, nor do they argue that transcription level control is the sole determinant of induction. Our data do indicate that an element present in 500 bp of 5' flanking DNA is sufficient to render a gene responsive to hormone and that this element most likely operates to enhance transcription.

Evidence is accumulating for several genes that regulatory elements controlling the rate of transcription reside in 5' DNA quite close to the structural gene. Thus, MMTV (4), α-2u globulin (D. Kurtz, personal communication), hGH, tk (14), mouse globin (16), metallothionein (17) and Drosophila heat shock (18) genes remain responsive to widely differing inducing agents with 1 kb or less of 5' flanking DNA. It is possible that such elements exert their effects locally and do not "transduce" information over long distances. The hGH-tk fusion gene generates two mRNAs: a 1.25 kb RNA presumably initiated in GH sequences and a second 0.9 kb species initiating in the tk structural gene about 300 nucleotides downstream. Only the 1.25 kb RNA is hormonally inducible, suggesting that the hGH regulatory element acts locally to activate transcription and has little or no effect upon the frequency of close, but downstream initiations. This argument must be tempered by the fact that cell lines expressing the fusion gene have integrated multiple copies of this gene. It is therefore possible that the smaller transcript derives solely from genes which remain unresponsive to hormone action.

One striking observation is that newly introduced hGH genes express significant levels of mRNA and protein while the endogenous murine GH gene in fibroblasts remains inactive either in the absence or presence of glucocorticoid. Since GH synthesis is restricted to the pituitary and is never expressed physiologically by fibroblasts, we are obliged to consider why newly introduced GH genes function in the recipient cell. It should be noted that although our control fibroblasts do not synthesize significant levels of endogenous GH mRNA, we do not know whether the murine gene is active in transformants expressing exogenous GH genes. However, in an analogous system, hormonal induction of exogenous rat α-2u globulin genes in a mouse fibroblast is not associated with activation of the endogenous mouse genes (1).

A pattern is emerging from gene transfer experiments which suggests that the mere introduction of exogenous genes into cells is sufficient to assure their expression. Thus, several genes, including globin (16,19); hGH and MMTV (2) when introduced into cells, synthesize significant quantities of RNA whereas their endogenous counterpart genes remain transcriptionally silent. Further, if appropriate control signals exist in the recipient cell, expression of the newly introduced genes may be properly regulated.

Studies of gene transfer together with studies of gene expression during normal development therefore define at least three states of genetic activity: "off", "on" and "regulated". The mere presence of a glucocorticoid receptor complex is inadequate to activate genes in the "off" state as is apparent for the endogenous GH gene in non-pituitary cells. Maintenance of this state perhaps reflects the chromosomal location of the endogenous gene or alternatively may result from prior developmental events about this gene which are self-perpetuating through cell division. Whatever mechanism is responsible for maintenance of the off state, it appears to be "cis" acting; a single exogenous gene introduced into a cell in which the endogenous gene is off can function in a regulated manner. Transformed genes may therefore escape the developmental history of the cell and immediately conform to the "on" state, a state accessible to appropriate regulators. In this state, genes may be regulated if appropriate controlling elements exist within the cell. The regulated state may therefore involve "trans" acting factors such as steroid hormone receptor complexes.

ACKNOWLEDGMENTS

We wish to thank Tom Livelli and John Fleming for technical assistance and Sandra Hayenga and Pam Ross for preparing the manuscript.

REFERENCES

1. Kurtz, D. T., Nature 291, 629 (1981).
2. Hynes, N. E., Kennedy, N., Rahmsdorf, U., and Groner, B., Proc. Natl. Acad. Sci. USA 78, 2038 (1981).
3. Buetti, E., and Diggelmann, H., Cell 23, 335 (1981).
4. Lee, F., Mulligan, R., Berg, P., and Ringold, G., Nature 294, 228 (1981).
5. Martial, J., Baxter, J., Goodman, H., and Seeburg, P. H., Proc. Natl. Acad. Sci. USA 74, 1816 (1977).
6. Tushinski, R., Sussman, P., Yu, L., and Bancroft, F. C., Proc. Natl. Acad. Sci. USA 74, 2357 (1977).
7. Lippman, M. E., and Thompson, E. B., J. Biol. Chem. 249, 2483 (1974).
8. Niall, H. D., Hogan, M. L., Sayer, R., Rosenblum, I. Y., and Creenwood, R. C., Proc. Natl. Acad. Sci. USA 68, 866 (1971).
9. DeNoto, F. M., Moore, D. D., and Goodman, H. M., Nucl. Acids Res. 9, 3719 (1981).
10. Maniatis, T., Hardison, R. C., Lacy, E., Lauer, J., O'Connell, C., Quon, D., Sim, G. K., and Efstratiadis, A., Cell 15, 687 (1978).
11. Graham, F. L., and van der Eb, A. J., Virology 52, 456 (1973).
12. Wigler, M., Sweet, R., Sim, G. K., Wold, B., Pellicer, A., Lacy, E., Maniatis, T., Silverstein, S., and Axel, R., Cell 16, 777 (1979).
13. Southern, E. M., J. Mol. Biol. 98, 503 (1975).
14. McKnight, S. L. Gavis, E. R., Kingsbury, R., and Axel, R., Cell 25, 385 (1981).
15. Roberts, J. M., and Axel, R., Cell, in press.
16. Chao, M., Mellon, P., Wold, B., Maniatis, T., and Axel, R., In: "Proc. of Symp. on Hemoglobin Synthesis" (E. Goldwasser, ed.) Elsevier, New York, in press.
17. Brinster, R. L., Chen, H. Y., Trumbauer, M., Senear, A. W., Warren, R., and Palmiter, R., Cell 27, 223 (1981).
18. Corces, V., Pellicer, A., Axel, R., and Meselson, M., Proc. Natl. Acad. Sci. USA 78, 7038 (1981).
19. Mantei, N., Boll, W., and Weissmann, C., Nature 281, 40 (1979).

DISCUSSION

M. NILSEN-HAMILTON, :   Have you determined the number of endogenous copies of either of the regulatory sequences you have identified?

R. AXEL: It has not been well characterized in a mammalian cell.   In Drosophila we know that there are about four or five genes coding for HSP70.   So there would be four or five 5' specific elements.   Investigators are just beginning to isolate appropriate probes to allow careful determinations in the mouse.

M. PATER: (1) Are high levels of tk toxic to cells?
(2) We get high frequency of expression with promoterless tk plasmid using the proplast fusion method but no expression with the calcium phosphate technique.   Can you offer an explanation for this observation?

R. AXEL: Too high a level of tk may indeed be toxic since the cells are untimely forced into thymidine block.   Although our evidence is weak, it is suggestive of this mechanism. For instance, the cell in which 50 thymidine kinase genes were inserted died when the concentration of thymidine of the medium supporting its growth was increased.

   With regard to your second question, we don't use protoplast fusion technique.   I would guess that the promoterless gene may be integrating next to the elements that serve as promoter, or alternatively, protoplast fusion may introduce more copies into the cell than DNA mediated gene transfer, and that itself would give you a tk^+ phenotype.

R. MEAGHER: Is it possible that the primate globins present problems because their unique regions are conserved?   If you had selected globin gene from another organism you might have had a better chance of obtaining expression in a heterologous system, e.g. chick globin?

R. AXEL: It is possible, but I wouldn't place too much emphasis on our negative data.   These data can be explained in a number of ways.   For example, one possibility is that induction of the human globin gene requires yet another human protein that is not present in the murine cell lines.   This protein could be encoded, perhaps near to the globin cluster, on the same chromosome as suggested by transfer experiments. I can't make any serious conclusions at present since the regulatory elements have not been identified.

TRANSFER OF THE MOUSE METALLOTHIONEIN-I GENE INTO CULTURED CELLS AND INTO ANIMALS

Richard D. Palmiter[*] and Ralph L. Brinster[+]

[*]Howard Hughes Medical Institute Laboratory
Department of Biochemistry
University of Washington
Seattle, Washington
U.S.A.

[+]Laboratory of Reproductive Physiology
School of Veterinary Medicine
University of Pennsylvania
Philadelphia, Pennsylvania
U.S.A.

Metallothioneins (MTs) are small, cysteine-rich, metal-binding proteins found in most eukaryotes. We have isolated and characterized the mouse MT-I gene and have shown that this gene is transcriptionally regulated by heavy metals in vivo and in most cultured cells. To define the mechanisms and sequences responsible for transcriptional regulation, we have constructed vectors that include the MT-I gene along with a selectable marker gene. When these vectors are transfected into mouse or human cells the MT-I gene is transcriptionally regulated by cadmium. S_1 mapping reveals that the 5' end of the mRNA is the same as that originating from endogenous MT-I genes. The promoter and regulatory region of the MT-I gene has also been fused to the structural gene of herpes virus thymidine kinase resulting in a vector (pMK) in which thymidine kinase expression is controlled by cadmium when it is injected into mouse eggs. A minimum of 90 bp of mouse sequence 5' of the mRNA cap site is required for cadmium regulation of thymidine kinase expression. Mouse eggs injected with pMK have been reimplanted into the oviducts of pseudopregnant mice. Several of the resulting mice express a high level of herpes thymidine kinase in the liver in response to cadmium. DNA analysis reveals multiple copies of the plasmid which are inherited as though they are integrated into a single chromosome. Some offspring of these mice also express viral thymidine kinase indicating that functional fusion plasmids can be stably incorporated into the germ line.

Research supported by NIH grants HD-09172, HD-15477 and NSF grant PCM 81-07172.

FROM GENE TO PROTEIN:
TRANSLATION INTO BIOTECHNOLOGY

INTRODUCTION

Metallothioneins are proteins thought to be involved in heavy metal detoxification and in zinc and copper homeostasis (1,2). Most eukaryotes synthesize MT; vertebrates possess at least two genes, designated MT-I and MT-II (2). A mouse MT-I genomic clone has been characterized (3,4) and used to show that synthesis of MT-I mRNA is transcriptionally regulated by heavy metals and by glucocorticoid hormones (5-7).

To explore the mechanisms and sequences involved in the cellular regulation of this gene, we have prepared vectors suitable for transforming tissue culture cells. We have also fused the MT-I promoter/regulatory sequences to a herpes virus thymidine kinase gene and microinjected this plasmid into mouse eggs to study acute gene expression and developmental regulation. Several groups have shown recently that DNA molecules injected into eggs can be stably integrated into the genome (8-13). Here we describe our observations on the regulation of MT-I genes in transformed tissue culture cells and animals.

Figure 1. Plasmids used for cell and animal transformation

Plasmid pMT-TK was generated by inserting a 4 kb Eco R1 fragment that carries the MT-I gene and a 3.5 kb Bam H1 fragment that carries the TK gene into the Eco R1 and Bam H1 sites of pBR322. Plasmid pMK was prepared from pMT-TK by restriction with Bgl II followed by religation of the large fragment. pMKΔ was created from pMK by deletion of a 3 kb Pvu II fragment as indicated. pMK-MT was generated by substituting a Kpn I-Pvu II fragment from pMT-TK for a Pvu II-Bam H1 fragment from pMK with the aid of linkers and adapters as indicated. pMT-GPT was generated by insertion of the 4 kb Eco R1 fragment carrying the MT-I gene into pSV2-gpt (15). pBR322 sequences are shown as a single line; other sequences are shown as double lines; coding regions are solid; SV40 sequences are cross-hatched; the direction of transcription is indicated by arrows.

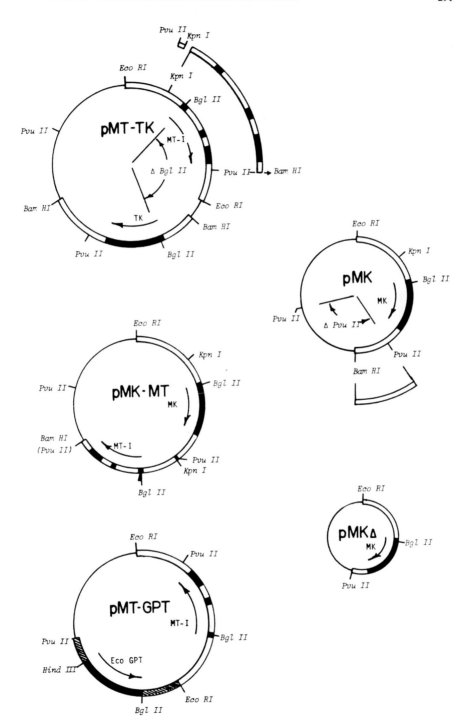

RESULTS AND DISCUSSION

Expression of the Metallothionein-I Gene
in Transformed Cells

A 4.0 kb *Eco* R1 fragment that includes the MT-I gene was subcloned into pBR322 from the original λ phage clone which contains an additional 11.5 kb of 3' mouse sequences (3). This 4.0 kb fragment was then introduced into eukaryotic selection vectors (14,15) to generate plasmids pMT-TK for selection of transformants in thymidine kinase-deficient L^{tk-} mouse cells or pMT-GPT for selection of transformants in hypoxanthine-guanine phosphoribosyl-transferase-deficient (HeLa^{hgprt-}) human cells (see Figure 1). A large number of transformants were selected and then individual clones were tested for expression and regulaton of the mouse MT-I gene.

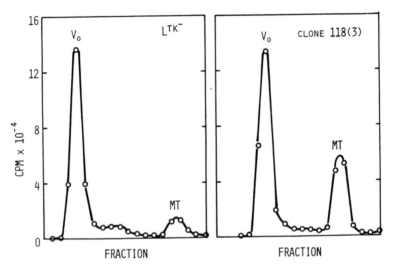

Figure 2. Increased synthesis of metallothionein in transformed cells

L^{tk-}cells and a clone resulting from transfection with pMK-MT were labeled for 14 hr with [^{35}S]cysteine in the presence of 40 μM Cd. The cells were lysed with Triton X-100, the extract was heated at 68° for 10 min, centrifuged, and an aliquot of the supernatant was chromatographed on Sephadex G-75. The position of the void volume (Vo) and metallothionein (MT) peak is shown. In the absence of Cd treatment, the MT peak is barely detectable.

Table 1. Cadmium regulation of the metallothionein-I gene
in transformed human and mouse cells

Species	Clone	MT-I mRNA[a]		MT-I Gene Transcription[b]	
		Control	+Cd[c]	Control	+Cd[c]
		molecules/cell		ppm	
Human	HeLa hgprt-(control)	0	0	6	6
"	pMT-GPT-5	840	4645	100	400
Mouse	L tk-(control)	300	1100	60	290
"	pMT-TK-206	300	3800	70	660
"	pMK-MT-118(3)	ND	5000	ND	ND

[a] MT-I mRNA was measured by solution hybridization as
described by Beach and Palmiter (20).

[b] MT-I gene transcription was measured by labeling isolated
nuclei with a[^{32}P]UTP and hybridizing RNA to genomic
plasmids immobilized on nitrocellulose. Data are
presented as parts per million, i.e. cpm hybridized per
10^6 cpm input and then corrected for efficiency of   50%
(21).

[c] CdSO$_4$ (40 μM) was added for 8 hr to monitor mRNA
accumulation or for 1 hr to measure transcription.

Table 1 shows representative results with a single
clone of each type. The human transformants are easiest to
analyse because the human MT genes are sufficiently
different so that human mRNA does not cross hybridize with
the mouse MT-I probes. Human transformant #5, which carries
several copies of the plasmid, expresses the mouse MT-I gene
constitutively, but the rate of transcription and the

accumulation of MT-I mRNA can be enhanced 4-5 fold by addition of Cd to the medium. This result is similar to the regulaton of the endogenous mouse MT-I genes in L[tk-]cells (Table 1). When mouse cells are transformed with extra copies of the MT-I gene (e.g. clone 206), then an elevated rate of transcription and accumulation of MT-I mRNA are observed (Table 1). S_1 nuclease mapping reveals that the mRNAs synthesized in both human and mouse transformants start at the same site (64 nucleotides 5' of the *Bgl* II site) as transcripts from the endogenous MT-I gene (16). We conclude from these experiments that the DNA sequences responsible for Cd regulation are included within the 4 kb *Eco* R1 fragment and that these sequences respond to human as well as mouse regulatory molecules.

To ascertain whether the MT-I mRNA synthesized in transformed cells is functional, cells were labeled with [^{35}S]cysteine and the heat-stable proteins were then chromatographed on a Sephadex G-75 column. Figure 2 shows

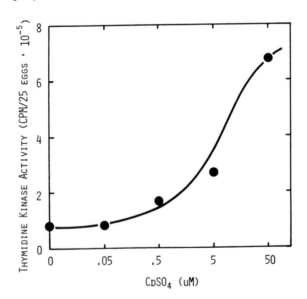

Figure 3.   Cadmium induction of MK gene expression
in mouse eggs

About 2000 copies of pMK were injected into mouse eggs. Batches of 25 eggs were incubated for 22 hr with the indicated concentration of CdSO₄; then TK activity was measured by monitoring the conversion of [³H]thymidine to TMP (12,17).

that a clone #118(3), resulting from transformation with plasmid pMK-MT (see Figure 1), incorporates about 28% of the cysteine into proteins that migrate in the position of MT whereas the L^{tk-} control cells incorporate about 8% of the cysteine into MT. The difference in MT synthesis is proportional to the difference in MT-I mRNA levels (Table 1); thus, we conclude that the mRNAs derived from pMK-MT are functional.

Localization of the Cadmium Regulatory Sequences

To determine which sequences are essential for heavy metal regulation, we created a fusion gene and adopted a more rapid assay system. Plasmid, pMK, was created by deletion of a *Bgl* II fragment from pMT-TK (Figure 1). This plasmid fuses 1.8 kb of 5' MT-I sequences to the thymidine kinase (TK) structural gene; the abbreviation for this fusion gene is MK. Figure 3 shows that this gene is functional when injected into mouse eggs and, more importantly, synthesis of thymidine kinase is enhanced in

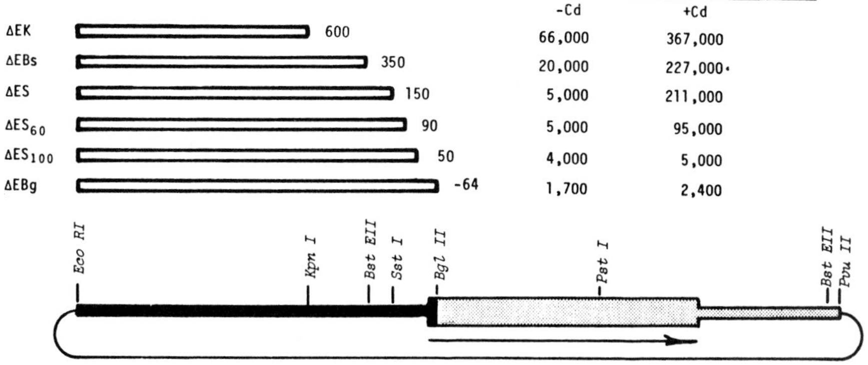

	Thymidine Kinase Activity TMP (cpm/25 eggs)	
	-Cd	+Cd
ΔEK	66,000	367,000
ΔEBs	20,000	227,000
ΔES	5,000	211,000
ΔES_{60}	5,000	95,000
ΔES_{100}	4,000	5,000
ΔEBg	1,700	2,400

Figure 4. Effect of 5' deletions on cadmium regulation of MK gene expression

A set of deletions extending from the Eco R1 site towards the MK gene in plasmid pMK Δ was created. The number of nucleotides of mouse DNA sequence remaining 5' of the MT-I mRNA cap site is indicated to the right of each deletion. About 2500 molecules of each deletion were injected into mouse eggs and the eggs were incubated for 22 hr plus or minus 50 μM Cd prior to TK assay.

response to Cd. The assay requires only a few thousand
molecules of plasmid and TK acitivity is measured 22 hr
after injection. Thus, it is ideal for analysis of mutants.
For technical reasons, plasmid pMK was simplified further by
deletion of a 3 kb Pvu II fragment to generate pMK   (Figure
1). Deletions were then made which extend from the Eco R1
site towards the 5' end of the MK gene; the number of
nucleotides remaining 5' of the transcription start site is
indicated to the right of each deletion in Figure 4. Assay
of these deletions for TK expression reveals that plasmids
with 90 nucleotides 5' of the mRNA cap site or more respond
to Cd whereas mutants with 50 nucleotides or less do not.
Although 90 nucleotides is sufficient for Cd regulation,
there appear to be potentiating effects of sequences further
upstream. The deletion with only 50 nucleotides remaining
5' of the cap site has lost a large palindrome that may be
involved with binding regulatory molecules; however, it
retains the TATAA box $\sim$30 nucleotides upstream of the cap
site and it can be transcribed in a cell-free system,
whereas the next deletion (-64) is not transcribed in the
cell-free system (L. Beach, personal communication).

Insertion of the Metallothionein-Thymidine Kinase Fusion
Gene into Mice

      The pMK fusion gene was inserted into mice following
the generalized protocol shown in Figure 5. Fertilized
mouse eggs are injected with a few hundred copies of the
plasmid and then inserted into the oviducts of foster
mothers. When the resulting pups are weaned, nucleic acids
are isolated from a piece of tail and hybridized to a
plasmid probe in a simple dot hybridization procedure to
ascertain which mice carry plasmid sequences. Figure 6
illustrates the dot hybridization method. Mice designated
MaK-67 and MyK-84 carry about 100 copies of the MK gene;
hence their DNA gives intense hybridization. MyK 116
carries 2 copies and MaK-122 carries 10 copies of MK.
Offspring #10 of MyK-84 and siblings of MyK-116 and MaK-122
carry no MK sequences.

      Thus far, about 10-20% of the mice have been positive
for MK sequences. Those that are positive are then injected
with $CdSO_4$ to induce MK gene expression and a partial
hepatectomy is performed 18 hr later. The liver is used to
measure TK activity and to isolate nucleic acids. Tissues
can also be prepared for cell culture. The mouse is then
mated to see if the MK gene is in the germline.

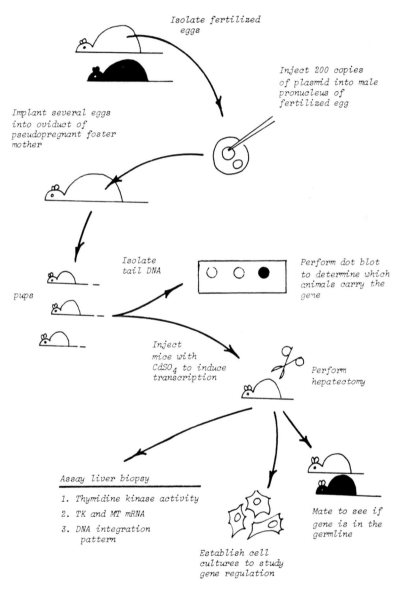

Figure 5. *Scheme illustrating the introduction of foreign DNA into mice and detection of its expression.*

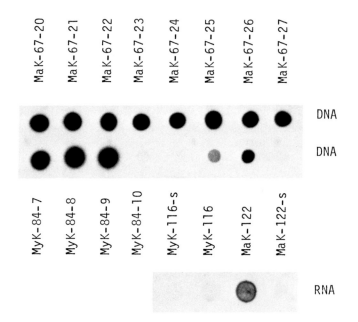

Figure 6.   Dot   hybridization   procedure   used   to   detect DNA or RNA

Aliquots of DNA (5 µg) or RNA (20 µg) were spotted onto nitrocellulose, baked, and hybridized with a TK-specific probe. The mice are named as described in Figure 7; hyphenated numbers represent   offspring; siblings that do not carry MK genes are designated by s.

Figure 7 *(stippled bars)* shows the level of TK activity in the liver of 7 mice that express the MK gene.   The TK activity in these mice is 3 to 100 times greater than the TK activity of control mice.   Furthermore, this activity is almost completely suppressed by an antibody specific for herpes TK (12).   MK mRNA has also been detected by solution hybridization (12) and by dot blotting (Figure 6).   To ascertain whether the TK activity is indeed induced by heavy metals, a second hepatectomy was performed a month later without   prior   treatment   with   Cd.   Figure 7 *(solid bars)* shows that with two mice the TK activity declined more than 20-fold when Cd treatment was omitted.

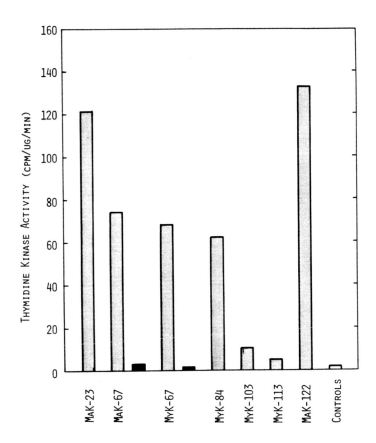

Figure 7. Expression of thymidine kinase in the liver of mice resulting from injection of pMK into fertilized eggs

Mice were injected with $CdSO_4$ (1 mg/kg; stippled histograms); 18 hr later a partial hepatectomy was performed and TK activity was measured. For two of the mice a second hepatectomy was performed a month later without prior $CdSO_4$ treatment (solid histograms). The average TK activity of control mice of the same age is shown by an open histogram. Male mice that express viral TK are called MaK followed by the number of their foster mother; female mice that express viral TK are called MyK.

Inheritance of the Metallothionein-Thymidine Kinase Fusion Gene

(a) MaK-67

Inheritance of the MK gene has been studied in two cases. Figure 8 shows the lineage of the MK gene through two generations of MaK-67. Restriction enzyme analysis shows that the offspring of this mouse either receive the entire set of MK genes or none at all (18). This observation, coupled with the segregation shown in Figure 8, suggest that the MK genes (~100 copies) are all integrated into a single chromosome of the father and that they are stably transmitted to the offspring. Surprisingly, of the 13 MK-positive first-generation offspring analyzed so far, only two (#6 and #8) express viral TK and their level of expression is less than the father. We have given higher doses of $CdSO_4$ and monitored the induction of MT-I mRNA to insure delivery of the inducer, but still obtained no induction of herpes TK from offspring #7,14,16 or 19.

Because DNA methylation controls the inducibility of the MT-I gene in some cell types (19), we investigated the DNA methylation pattern of MaK-67 and offspring that do or do not express viral TK to see if differential methylation might explain the loss of Cd response of the MK genes. DNA

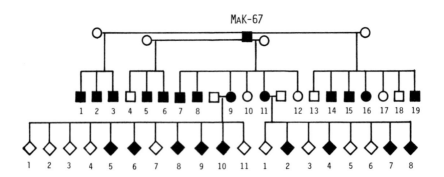

Figure 8. Inheritance of the MK genes in two generations of offspring of MaK-67

The presence of the MK genes was detected by dot-hybridization as shown in Figure 6. The sex of the offspring in the second generation was not determined.

was digested with several enzymes that are sensitive to cytidine methylation, including *Hpa* II (CCGG), *Hha* I (GCGC), *Sma* I (CCCGGG) and *Sst* II (CGGCCG), electrophoresed on agarose gels, transferred to nitrocellulose and hybridized with MK gene probes. Essentially all of the restriction sites were methylated, i.e. uncut, in both father and offspring, and there was no significant difference that correlated with expression. These results indicate that there is a *de novo* methylation system that methylates foreign DNA, but they do not provide insight into the loss of MK expression. However, because there are many MK gene copies in MaK-67 it is possible that only one or a few copies are expressed; a change in methylation of one copy would be difficult to detect.

(b) MyK-84

This female resulted from injection of a blunt-ended 2.1 kb *Bst* EII fragment containing the MK gene (see Figure 4). Restriction analysis reveals that this fragment is tandemly duplicated in a head-to-tail orientation which must have occurred by recombination within the mouse cells.

MyK-84 has had 10 offspring, five of which received MK genes. Inheritance of MK sequences is more complicated than in the case of MaK-67, in that some of the offspring inherit only about half of the MK sequences as summarized in Table 2. Table 2 also shows that viral TK expression was retained by four of the offspring (the fifth one died prior to TK assay), but the enzyme activity varied widely. Two of the offspring have 2-3 fold more activity than the mother while two have 3-5 fold less activity than MyK-84.

In this case there is a correlation between DNA methylation and expression. The two offspring that express high levels of viral TK have essentially no methylation of *Hpa* II, *Hha* I, or *Sma* I sites whereas in MyK-84 and the other two offspring there is extensive methylation of these sites. The *Sst* II sites are fully methylated in all offspring (Table 2).

These data indicate that microinjected genes can be transmitted through the germline; however, developmental phenomena that are not yet understood appear to influence the expression or commitment of the genes in subsequent generations. It will be important to analyze the DNA and expression of mice with single copies of foreign genes to try to discern the molecular basis of this phenomenon.

Table 2.   Analysis of TK expression and DNA methylation in
MyK-84 and her offspring

Mouse	MK DNA[a]	TK activity[b]	DNA methylation[c]	
			Hpa II, Hha I Ava I, Sma I	Sst II
		cpm/μg/min		
MyK-84	++++	58	++	++++
MyK-84-1	0	0	-	
MyK-84-2	0	0.1	-	
MyK-84-3	++	153	0	++++
MyK-84-4	0	1.1	-	
MyK-84-5	0	0	-	
MyK-84-6	++++	ND[d]	ND	ND
MyK-84-7	++	152	0	++++
MyK-84-8	++++	13	++	++++
MyK-84-9	++++	27	++	++++
MyK-84-10	0	0	-	

[a]DNA content determined by dot blot; ++++ equals about 100 copies; ++ equals about 50 copies.

[b]TK activity was determined as described (15).

[c]DNA methylation was determined as described in text; ++++ means that these sites were totally methylated and hence uncut by the restriction enzymes listed; ++ means that some of the sites were methylated; 0 means that the enzymes cut completely.

[d]ND = not determined; animal died

REFERENCES

1. Cherian, M.G. and Goyer, R.A. (1978) *Life Sci. 23,* 1-10.

2. Kagi, J.H.R. and Nordberg, M.,eds. (1979) *Metallothionein,* Birkhauser Verlag, Basel.

3. Durnam, D.M., Perrin, F., Gannon, F. and Palmiter, R.D. (1980) *Proc. Natl. Acad. Sci. USA 77,* 6511-6515.

4. Glanville, N., Durnam, D.M. and Palmiter, R.D. (1981) *Nature 292,* 267-269.

5. Durnam, D.M. and Palmiter, R.D. (1981) J. Biol. Chem. 256, 5712-5716.

6. Mayo, K.E. and Palmiter, R.D. (1981) *J. Biol. Chem 256,* 2621-2624.

7. Hager, L.J. and Palmiter, R.D. (1981) Nature 201, 340-342.

8. Wagner, E.F., Steward, T.A. and Mintz, B. (1981) *Proc. Natl. Acad. Sci. USA 78,* 5016-5020.

9. Wagner, T.E., Hoppe, P.C., Jollick, J.D., Scholl, D.R., Hodinka, R.L., and Gault, J.B. (1981) *Proc. Natl. Acad. Sci. USA 78,* 6376-6380.

10. Constantini, F. and Lacy, E. (1981) *Nature 294,* 92-94.

11. Rusconi, S. and Schaffner, W. (1981) *Proc. Natl. Acad. Sci. USA 78,* 5051-5055.

12. Brinster, R.L., Chen, H.Y., Trumbauer, M., Senear, A.W., Warren, R. and Palmiter, R.D. (1981) *Cell 27,* 223-231.

13. Gordon, J.W. and Ruddle, F.H. (1981) *Science 214,* 1244-1246.

14. Wigler, M., Silverstein, S., Lee, L.-S., Pellicer, A., Cheng, T. and Axel, R. (1977) *Cell 11,* 223-232.

15. Mulligan, R.C. and Berg, P. (1980) *Science 209,* 1422-1427.

16. Mayo, K.E., Warren, R. and Palmiter, R.D. (1982) manuscript submitted.

17. Brinster, R.L., Chen, H.Y., Warren, R., Sarthy, A. and
    Palmiter, R.D. (1982) *Nature,* in press.

18. Palmiter, R.D., Chen, H.Y., and Brinster, R.L. (1982)
    manuscript in preparation.

19. Compere, S. and Palmiter, R.D. (1981) *Cell 25,* 233-240.

20. Beach, L.R. and Palmiter, R.D. (1981) *Proc. Natl. Acad.
    Sci. USA 78,* 2110-2114.

21. McKnight, G.S. and Palmiter, R.D. (1979) *J. Biol. Chem.
    254,* 9050-9058.

DISCUSSION

D. BOTSTEIN: Were the crosses of tk+ animals to congenic
mice? If not, then possibly the variability in methylation
and/or expression might be due to modifiers segregating in
the crosses.

R. PALMITER: In our experiments the plasmids were injected
into hybrid (SJL x C57) eggs and animals that develop from
the procedures were mated to one of the parental strains.
Thus, the variability in thymidine kinase expression could
conceivably reflect the segregation of modifier genes. We
have used hybrid eggs in our initial experiments because of
their increased viability during the injection procedures
compared to the parental eggs.

H. FABER: I would like to refer to the pedigree containing
19 offspring from one male whose genome contains copies of
the transferred DNA. The 13:6 segregation ratio of positive
to negative offspring does not look like a 1:1 ratio. A
rough chi square test supports this. The 13:6 ratio seems to
be close to a 3:1 ratio. In view of this, have you investigated the number of sites the transferred DNA has been integrated? Have in situ hybridization experiments shown that
the DNA is integrated into two different chromosomes?

R. PALMITER: We have not performed in situ hybridization
studies. Although such studies would be informative, they
are no longer possible because this mouse is dead. However,
I would point out that both F_1 and F_2 offspring inherit
pMK DNA sequences in an all-or-none fashion with a restriction pattern and hybridization intensity indistinguishable
from that of the father. This result is most easily
explained by integration of the vast majority of pMK

sequences into a single chromosome. My view is that the 13:6 ratio is a statistical anomaly. Note that the sex ratio in the F_1 generation is also skewed 13:6. We now have 54 offspring of MaK-67; analysis of the segregation pattern shows that 34 of them carry a full complement of pMK sequences and 20 carry none.

M. BINA: Do you know whether a transition of DNA structure, from B form to Z form, may regulate the activation of metallothionein gene by metal ions?

R. PALMITER: This is an interesting possibility but we have not tested it.

J. MILLS: The procedure for injecting the promoter and regulatory region of the metallothionein-I gene into mouse eggs sounds technically very difficult. Could you please furnish me with the following information:
1) What were the concentration and volume of the metallothionein-I gene injected?
2) Was this injected iontophoretically or under pressure?
3) Were the genes injected into the nucleus or just into the cytoplasm?
4: What were the success rates?

R. PALMITER: The detailed procedures are described in Cell 27: 223-231. Typically 200 to 2000 molecules of plasmid pMK were injected in a volume of 2 picoliters into the male pronucleus of fertilized eggs. The injection is performed under pressure from a Hamilton syringe. Injection is continued until the volume of the nucleus doubles. About 10-15% of the animals that develop from injected eggs carry the plasmid sequences and about 70% of these animals express the foreign genes.

INCREASING LEVELS OF GENE EXPRESSION

CONSTRUCTION OF THREE HYBRID PROMOTERS AND THEIR PROPERTIES IN ESCHERICHIA COLI

Herman de Boer[1], Herbert Heyneker[1], Lisa Comstock[1],
Alice Wieland[1], Mark Vasser[2] and Thomas Horn[2]

[1]Molecular Biology Department
and [2] Organic Chemistry Department
Genentech, Inc.
South San Francisco, California
U.S.A.

SUMMARY

This paper describes three hybrid promoters which are functional in Escherichia coli. In the case of the first hybrid promoter (tacI) sequences upstream of position -20 were derived from the trp promoter and sequences downstream of position -20 were derived from the lac-UV5 promoter. This hybrid promoter is seven times stronger than the lac-UV5 promoter. It can be repressed by the lac-repressor and induced by isopropyl-β-D-thiogalactoside (IPTG).

In the case of second hybrid promoter (tacII), we used the DNA sequences upstream of the HpaI site (which is located in the Pribnow box of the trp-promoter) and fused those sequences to a synthetic DNA fragment of 46 bp. The sequence of the synthetic fragment creates a new Pribnow-box which is followed by the lac-operator. Downstream from the lac-operator are nucleotides that code for a Shine-Dalgarno (SD) sequence. The Shine-Dalgarno sequence is flanked by two restriction sites which allows us to exchange different Shine-Dalgarno sequences. Thus, we constructed an inducible promoter with a portable Shine Dalgarno sequence which forms an active ribosome binding site when fused to the start codon of a foreign gene. The tacII promoter is as efficient as the tacI promoter.

The third hybrid promoter (rac 5-16) is a hybrid between the rrnB promoter and the lacUV5 promoter. Its structure resembles that of the tacI promoter. At the junction, in

FROM GENE TO PROTEIN:
TRANSLATION INTO BIOTECHNOLOGY

the area of −20, three unique restriction sites were introduced. This makes it possible to change the distance and the nucleotide sequence between the −35 area and the −10 area (the Pribnow-box).

I. INTRODUCTION

Recently many eukaryotic genes have been isolated and cloned into plasmids of Escherichia coli. Expression of such foreign genes has been achieved after fusion to appropriate prokaryotic transcription and translation signals. The level of expression thus obtained varies considerably and in some favorable cases exceeds 500,000 protein copies per cell (1).

The level of expression of a gene is determined by the efficiency of transcription of the gene and the efficiency of translation of its messenger RNA. Crucial steps in these processes are the initiation frequency of transcription of the gene and the initiation frequency of translation of the messenger RNA. The efficiencies of these processes are determined by the DNA sequences of the prokaryotic promoter and the ribosome binding site on the encoded mRNA.

Although the DNA sequence of many promoters (reviewed in Refs. 2 and 3) and many ribosome binding sites (reviewed in Ref. 4) is known very little is known about the features in those sequences that determine the efficiency of both initiation processes. In promoters two domains upstream of the start site of transcription have been identified for which a consensus sequence has been formulated. These domains are the Pribnow box (5,6) and the −35 area (6,7). Both domains are in close contact (3,8) with the RNA polymerase during initiation of RNA synthesis and almost all promoter mutations map in or near these domains (reviewed in Ref. 2).

The relative efficiency of only a few promoters has been measured (9). Comparisons of promoters with widely varying strengths do not give conclusive clues as to which features in the DNA sequence determine promoter strength. In this paper we show that the efficiency of a relatively weak promoter can be increased by replacing its −35 region by the −35 region of a stronger promoter. We describe the construction of two different hybrids between the rather strong trp-promoter and the weaker lac-UV5 promoter. We also describe the construction of a hybrid between the very strong ribosomal RNA promoter (rrnB) and the lac-UV5 promoter. All three hybrid promoters contain the lac-operator sequence and therefore can be repressed by the

lac-repressor and can be induced by isopropyl-β-D-thiogalactoside (IPTG).

The first hybrid promoter (called tacI) is made of
natural sequences derived from the trp and the lac-UV5
promoter joined at position -20 with respect to the
transcription initiation site.

The second hybrid promoter (called tacII) is made of a
natural trp-promoter fragment ending within the
trp-Pribnow box which is joined to a synthetic DNA
fragment. Thus a hybrid Pribnow box is formed. It is
followed by the lac-operator and an area that codes for a
Shine-Dalgarno sequence which is surrounded by two unique
restriction sites.

The third promoter (called rac5-16) contains the natural
sequences of the -35 area derived from a ribosomal RNA
promoter (rrnB) and a synthetic DNA fragment that joins the
-35 area of the Prrn part to the lac-UV5 Pribnow box and
operator. This synthetic DNA fragment contains three unique
restriction sites which makes it possible to generate
various mutants in the area between the Pribnow box and the
-35 sequence. The relative efficiencies of the natural and
the hybrid promoters were determined using Rosenberg's (9)
promoter probe plasmids.

II. RESULTS

A. Construction of Two Hybrid trp-lac Promoters

This paper describes the construction of the promoters
tacI and tacII which are two different hybrids of the trp
promoter and the lac-UV5 promoter. Both hybrid promoters
contain the -35 area of the trp-promoter and both contain
the lac-operator. In tacI sequences downstream of the TaqI
site at position -20 with respect to the start-site of
transcription of the trp-promoter are replaced by sequences
of the lac-UV5 promoter. In tacII sequences downstream of
the HpaI site in the Pribnow box of the trp-promoter are
replaced by a synthetic DNA fragment that includes
lac-operator sequences. In both cases the lac-operator is
located at its natural position with respect to the
RNA-polymerase binding site.

The construction of tacI is shown in figure 1. The
trp-promoter was derived from pHGH207-1 whose construction
has been described elsewhere (10). This plasmid contains
the entire trp-promoter on a 310 bp EcoRI fragment. This
fragment also contains the Shine-Dalgarno (11) area of the

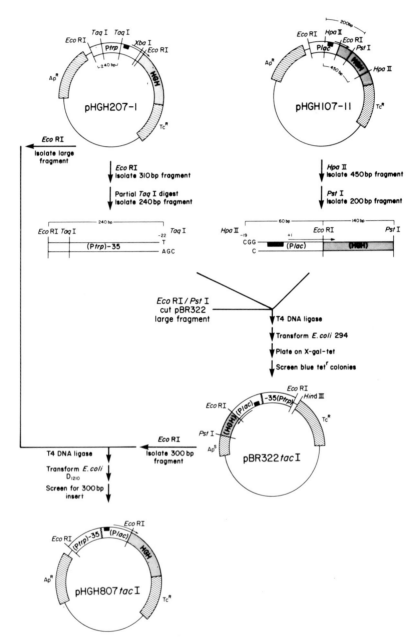

Figure 1. Construction scheme of pHGH807tacI containing the trp-lac hybrid promoter tacI.

trp-leader mRNA. The start codon for initiation of translation is provided by the adjacent human growth hormone gene.
The DNA sequence of the sense strand at this junction is:

XbaI   EcoRI

S.D.   start

5' - G G T A T C T A G A A T T C T A T G ... (HGH)

The end of the HGH gene is joined to the HindIII site of the
tetracycline resistance gene (tet-gene) of pBR322. The
HindIII site and the tet-promoter are destroyed in this
process (1,10) but the protein initiation signals on the
mRNA of the tet-gene are still intact. Hence the expression
of the tetracycline-resistance gene is under direction of
the trp-promoter. The 310 bp EcoRI fragment of pHGH207-1
was partially digested with TaqI and the indicated 240 bp
fragment was isolated. This fragment contains the -35
sequence of the trp-promoter and it ends at the TaqI site 21
bp upstream of the transcription start-site.
    The Pribnow box of the lac-promoter and the lac-operator
sequences were derived from pHGH107-11 (Fig. 1). The
structure of this plasmid, except the 90 bp EcoRI fragment
that contains the lac-UV5 promoter and the Shine-Dalgarno
area of the lac-mRNA, is identical to that of pHGH207-1.
Plasmid pHGH107-11 has a single lac-promoter fused to the
HGH-gene. It is derived as described previously (10) from
pHGH107 (1) which has two lac-promoters in a row. The
sequence at the junction of the lac-promoter/operator and
the HGH gene is:

EcoRI

S.D.   start

A G G A A A C A G A A T T C T A T G ... (HGH)

From pHGH107-11 the indicated 200 bp HpaII PstI fragment
was isolated. This HpaII site is at position -19 with
respect to the transcriptional start of the lac-UV5
promoter. The 240 bp trp-fragment, the 200 bp lac-fragment
and the large EcoRI-PstI fragment of pBR322 were covalently
joined with T4 ligase. After transformation of E. coli 294
and plating on X-gal (5-bromo-4-chloro-3-indolyl-β-D-
galactoside) tetracycline plates, plasmid DNA from blue,
tetracycline-resistant colonies was analyzed. Thus pBR322
tacI was obtained. Its structure is shown in figure 1.
From pBR322 tacI the 300 bp EcoRI fragment that contains the
newly constructed tacI promoter was purified and inserted
into the large EcoRI fragment of pHGH207-1. After
transformation of E. coliD$_{1210}$laci^q the plasmid

pHGH807tacI was obtained. In this expression plasmid the tacI containing fragment provides for the Shine–Dalgarno sequence allowing protein synthesis to start at the ATG of the adjacent HGH gene. The DNA sequence of the tacI promoter/operator and ribosome binding site is shown in figure 2.

The EcoRI fragments that contain the trp and the tacI promoter also code for the Shine–Dalgarno area of the ribosome binding site. For a systematic analysis of ribosome binding sites it would be very useful to be able to vary the nucleotide sequence in the Shine–Dalgarno area in simple experiments. Therefore, a portable Shine–Dalgarno element was included in the design of the synthetic part of tacII.

```
                        TaqI      HpaI      +1              XbaI EcoRI
Ptrp      GAGCTGTTGACAATTAATCATCGAACTAGTTAACTAGTACGCAAGTTCACGTAAAAAGGGTATCTAGAATTCTATG....HGH
                                             --------------

                                      +1                              EcoRI
PtacI     GAGCTGTTGACAATTAATCAT  CGGCTCGTATAATGTGTGGAATTGTGAGCGGATAACAATTTCACACAGGAAACAGAATTCTATG....HGH
                                                             --------------------

                        TaqI              +1            HindIII  XbaI EcoRI
PtacII    GAGCTGTTGACAATTAATCATCGAACTAGTT TAATGTGTGGAATTGTGAGCGGATAACAATTAAGCTTAGGATCTAGAATTCTATG....HGH
                                                             ------------------

                    HpaII              +1
PlacUV5   CCAGGCTTTACACTTTATGCTTCCGGCTCGTATAATGTGTGGAATTGTGAGCGGATAACAATTTCACACAGGAAACAGAATTCTATG....HGH
                                                             --------------------
```

Figure 2. The DNA sequence of the trp, lac–UV5, tacI and the tacII promoters. The sequence of the DNA strand corresponding to the messenger RNA is shown. The −35 sequence and the Pribnow box sequence of the promoters is underlined. The Shine–Dalgarno and the start codon of the HGH mRNA is overlined. The trp-repressor binding site and the lac–operator is indicated with broken lines. The transcription start site is indicated with +1. In the case of tacI and tacII it is assumed that the transcription start site is the same as that of the lac–UV5 promoter.

The construction of pHGH907tacII is shown in figure 3.
In the trp-promoter the trp-repressor binding site includes
the Pribnow box which in this case also contains a HpaI site
(see figure 4 and ref. 12). Sequences downstream of this
HpaI site were replaced by a synthetic DNA fragment, thus
destroying the trp-repressor binding site. Plasmid
pHGH207-1 was opened with HpaI and XbaI and a synthetic DNA
fragment of 46 bp, whose sequence is shown in figure 3, was
inserted. The sequence of this synthetic fragment is
identical to that found in the homologous area of the
lac-UV5 promoter/operator. Thus the first 2 bp of the
trp-Pribnow box (TTAACTA) are joined to the last 5 bp of the
lac-UV5 Pribnow box (TATAATG) resulting in a hybrid
Pribnow box (TTTAATG). Downstream of this hybrid

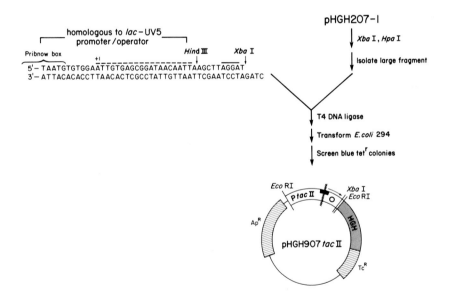

Figure 3. Construction scheme of pHGH907tacII containing
the half synthetic trp-lac hybrid promoter tacII and the
portable Shine-Dalgarno area.

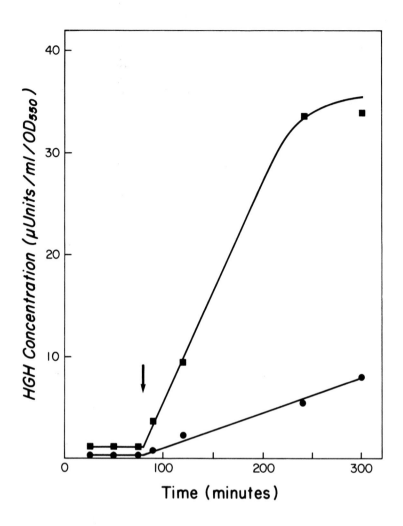

Figure 4. HGH production directed by the lac-UV5, the tacI
and the tacII promoter. Overnight cultures were used to
inoculate 50 ml LB-ampicillin (20 µg/ml) to a cell density
of 0.03 (OD_{550}). After one hour the first sample was
taken (t=40 min). Induction was done by addition of 1.0 mM
IPTG at t=80 minutes. HGH levels were determined in a
radioimmunoassay as outlined previously (1). Symbols:
0-0: E. coli D1210/pHGH107-11 (single lac-promoter); - :
E. coli D1210/pHGH 807tacI (tacI-promoter) and E. coli
D1210/pHGH907tacII (the induction profiles of tacI and tacII
are identical).

Pribnow—box is a 5 basepair sequence which is identical to
that found in the lac-UV5 promoter/operator. This 5 bp
sequence is the left arm of the region of symmetry that
flanks the core of the natural lac-operator (13). This 5 bp
sequence is followed by nucleotides that specify the core of
the lac-operator. The lac-operator is followed by a HindIII
site and a Shine—Dalgarno sequence which has a 4 basepair
homology with 16s ribosomal RNA (11). The synthetic
fragment ends with an XbaI site by which it is fused to the
HGH gene. Thus, the Shine—Dalgarno area is flanked by two
different restriction sites which are unique to the
plasmid. It should be noted that the tacII
promoter/operator lacks the right arm of the region of
symmetry that flanks the natural lac-operator (13).

B. Measurement of the Relative Efficiency of the
Various Natural and Hybrid Promoters

All the natural and hybrid promoters described above are
contained on EcoRI fragments which also encode a
Shine—Dalgarno region but which lack a start codon. In the
case of PtacI, the DNA sequences downstream of position −20
are identical to those of the parental lac-UV5 promoter.
Therefore, it is likely that transcription starts in these
promoters at the same position as that in the lac-UV5
promoter. Thus, the length and nucleotide sequence of the
5' untranslated region of each mRNA is likely to be the same
in these promoters. The 5'-untranslated region contains the
Shine—Dalgarno area, which precedes the start codon of the
HGH gene in the same way in each case. Therefore, the
entire ribosome binding site is similar on the mRNA of the
tacI and the lac-UV5 promoter. Thus, the initial rate of
HGH accumulation after induction of these promoters in cells
of E. coliD1210laci^q must reflect the relative
efficiencies of these promoters. In the experiment of
figure 4, cells of E. coliD1210laci^q containing
pHGH807tacI or pHGH107-11 were grown in M9 medium
supplemented with 0.2 percent casamino acids and induced
with 1.0 mM IPTG. Samples were taken before and after
induction and the HGH level was determined using a standard
radioimmune assay (1). The HGH concentration per optical
density unit of cells was calculated and plotted versus
time. Figure 4 shows that the initial rate of HGH
accumulation after induction of the tacI promoter is about
seven—fold higher than that of the induced lac-UV5
promoter. This means that the tacI promoter is about seven
times as efficient as the lac-UV5 promoter. Before
induction the HGH level in repressed cells with the tacI

promoter is significantly higher than that in repressed cells with the lac–UV5 promoter. This probably reflects the difference in promoter strength.

The tacI and the lac–UV5 promoter could be compared by measuring the expression of the HGH gene since their ribosome binding sites are identical. However, the relative efficiency of these promoters cannot be compared in this way with the trp and the tacII promoter since they are contained on EcoRI fragments that code for different ribosome binding sites. In order to measure reliably relative efficiencies of any promoter we used a plasmid system which has been designed to circumvent such complications. Rosenberg and his colleagues (9,14) constructed a plasmid (pKO–1) that contains the entire galactokinase gene including its natural ribosome binding site and leader mRNA but it lacks a promoter. The galK gene is transcribed only when a promoter is inserted in one of the engineered unique restriction sites located in front of the leader mRNA about 200 bp upstream of the galK start codon (see Figure 5). Between these restriction sites and the galK gene they (9,14) introduced stop codons in all three reading frames. Thus any protein initiation that might occur on the inserted promoter containing fragment is aborted and does not

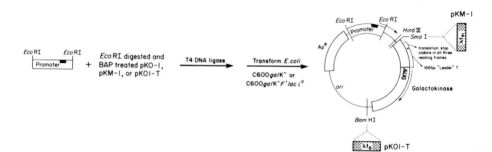

Figure 5. Insertion of the EcoRI fragments containing the various promoters into Rosenberg's (9) promoter probe plasmids pKO-1, pKO1-T and pKM-1. The structure of pKO-1 harboring a promoter containing plasmid pKO1-T is derived (9,14) from pKO-1 by insertion of the terminator λt_0 into a newly constructed BamHI site at the end of the galK gene. Plasmid pKM-1 is derived from pKO-1 by insertion of the rho-dependent terminator λt_{RI} into the SmaI site of pKO-1. This terminator reduces galK transcription by at least 70 percent.

contribute to the galK expression. Thus it is ensured that
the level of galK expression is solely determined by the
efficiency of the inserted promoter and is independent of
the length of the untranslated mRNA and translation signals
therein (for details see refs. 9,14).

Constitutive expression of a gene directed by a strong
promoter on a high copy number plasmid may reduce plasmid
copy number per cell and may disturb balanced growth rate of
the cell which in turn may affect the activity of the
various promoters in different ways. To circumvent such
complications, Rosenberg et al. (9) constructed pKM-1 (see
figure 5) which they obtained by insertion of a
Rho-dependent terminator (λt_{RI}) in the SmaI site of
pKO-1. This terminator reduces the number of transcripts
entering the galK gene by at least 70 percent (M. Rosenberg,
personal communication).

From the plasmids pHGH107-11, pHGH207-1, pHGH807tacI
(figure 1), pHGH907tacII (figure 3) the small EcoRI fragment
containing the lac, trp, tacI and the tacII promoter
respectively, were isolated and inserted into pKO-1 and
pKM-1 (see figure 5). Each plasmid was introduced in
E. coli C600 galK⁻. Cells were grown in minimal medium
containing the required amino acids and fructose and the
galactokinase levels were determined as described (9,14).

The lac-promoter and the hybrid promoters on pKM-1 and
pKO-1 are derepressed since the lac-repressor in the host
cells (E. coli C600) is titrated away by the abundance of
lac-operator sequences on the high copy number plasmids.
The trp-promoter on these plasmids is derepressed due to
lack of tryptophan in the medium (see discussion). The
galactokinase levels in cells with pKO-1 or pKM-1 harboring
the various promoters are shown in Table I. This table
shows that the galK level due to pKO-1trp is 2.2 times
higher than that due to pKO-1lacUV5. This same ratio is
obtained when pKM-1trp is compared with pKM-1lacUV5. These
results show that the trp-promoter is at least (see also
discussion) two times stronger than the lac-UV5 promoter.

Table I shows that the galactokinase levels due to
pKM-1tacI and pKM-1tacII are similar and about seven times
higher than that of pKM-1lac. These results clearly show
that the tacI and the tacII promoter are equally strong and
seven times stronger than the parental lacUV5-promoter.

C. Construction of a Hybrid of the Ribosomal-RNA
      Promoter and the lac-UV5 Promoter

The ribosomal RNA promoters are among the most efficient
promoters of E. coli (15-19). Each rrn-operon sequenced

Table I. Galactokinase activity of the various promoters in
E. coli C600 galK⁻

Plasmid/Promoter	galactokinase units*	ratio relative to Plac
pKO-1 lac	424	1.0
pKO-1 trp	926	2.2
pKM-1 lac	42	1.0
pKM-1 trp	92	2.2
pKM-1 tacI	290	6.9
pKM-1 tacII	310	7.4

*Galactokinase units expressed as nanomoles of galactose phosphorylated per minute per ml of cells at $OD_{650} = 1.0$.

thus far is transcribed by a tandem promoter (16-22). The two promoters are located about 200 and 300 basepairs upstream of the sequences that code for mature 16s rRNA. Both promoters are active in vitro (23) and in vivo (16). It is not known why the ribosomal RNA promoters have maximal efficiency. The relative contribution to the overall efficiency of the first (P_1) and the second (P_2) promoter (16) is also unknown. Since the substitution of sequences upstream of -20 of the lac-UV5 promoter by homologous sequences of the stronger trp-promoter improved the efficiency of the promoter we designed a similar hybrid of the rrnB and the lac-UV5 promoter. If the promoter strength is determined by the DNA sequence upstream of position -20 as seems to be the case in the trp-promoter we wanted to investigate whether the hybrid of the rrn and the lac-UV5 promoter is as efficient as the ribosomal RNA promoter itself.

The construction of such a hybrid promoter (the rac promoter) is shown in figure 6. It involves the subcloning of the -35 area and the AT-rich domains upstream (17,18,20-22) of the -35 area of the promoter of the rrnB operon from plasmid pkk3535 (20,22). This fragment ends at position -28 with respect to the start site of transcription

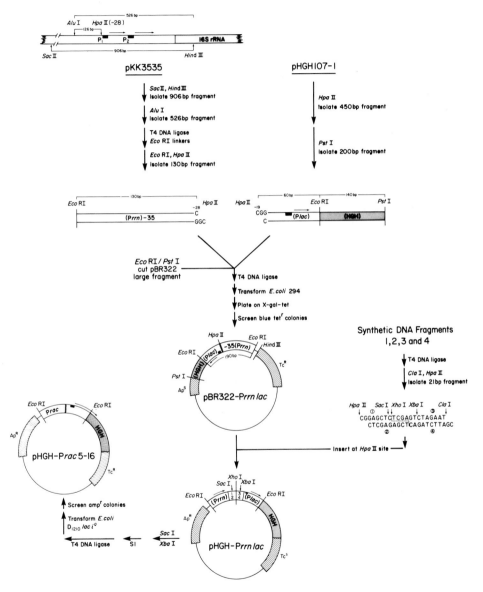

Figure 6. Construction scheme of the rac5-16 promoter and its precursors PrrnlacI and PrrnlacII.

of the first ribosomal RNA promoter P_1 (17,22), and it begins at the AluI site at position -154 which was converted into an EcoRI site using an EcoRI linker. The lac part is the same as described for the construction of the tacI promoter (see figure 1). Thus pBR322-Prrnlac was obtained. The sequence of Prrnlac I is shown in figure 7. The distance between the Pribnow box and -35 area is 10 bp which makes the promoter inactive (data not shown). At the HpaII site at the junction between the rrn part and the lac part of PrrnlacI a synthetic DNA fragment was inserted according to a scheme which will be described elsewhere. This insert increases the length between the Pribnow box and -35 sequence from 10 bp to 31 bp (see figure 5). This promoter (Prrn-lacII) is also inactive. Thus, the assembly of a

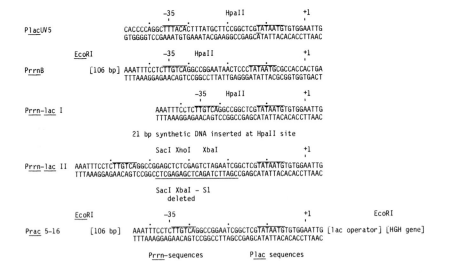

Figure 7. The DNA sequence of the lac-UV5, the rrnB, the shortened (Prrn-lacI), the lengthened (Prrn-lacII) and the activated (rac5-16) hybrid promoter. The -35 sequence and the Pribnow box sequence is overlined. The synthetic insert of Prrn-lacII is underlined. The transcription start site of Plac and Prrn or the nucleotide corresponding to that start site in the case of the Prrn-lacI, the Prrn-lacII and the Prac5-16 is indicated.

potentially lethal promoter was avoided and after each ligation and transformation blue colonies on X-gal plates could be selected and screened. The synthetic insert contains three unique restriction sites, namely SacI, XhoI, and XbaI. In order to activate the rrn-lacII promoter the plasmid (pHGH-Prrnlac) was digested with SacI and XbaI and the protruding ends were removed with nuclease S_1. The plasmid was reclosed with T_4 DNA ligase. At this stage the hybrid promoter was expected to be active. Therefore, cells that are lac-repressor overproducers ($D_{1210}laci^q$) were transformed with this DNA. At this stage the activated promoter was repressed by the lac-repressor. In this activated promoter (Prac5-16) the distance between the -35 sequence and the Pribnow box is 16 bp which is the same as that in the parental ribosomal RNA promoter. The rac-promoter can be induced with IPTG. Preliminary data show that the rac5-16 promoter is at least as efficient as the tacI and tacII promoters.

III. DISCUSSION

This paper describes the construction of three different hybrid promoters. All these hybrid promoters contain the lac-operator at the natural position with respect to the promoter and therefore can be repressed in cells of lac-repressor overproducing strains. All three promoters can be induced with IPTG in such strains.

The data in Table I show that the trp-promoter is at least two-fold stronger than the lacUV5-promoter. The hybrid promoters tacI and tacII are seven times stronger than the lac-UV5 promoter and at least three times stronger than the trp-promoter. This suggests that these hybrid promoters are stronger than either the trp promoter or the lacUV5 promoter. However, we cannot exclude the possibility that in these experiments the efficiency of the trp-promoter is somewhat underestimated. The trp-promoter can be repressed by the trp-repressor and tryptophan. Derepression occurs when the cells are grown in the absence of exogenous tryptophan. Although no tryptophan was added to the growth medium used for the experiments presented here it is still possible that even under these conditions the trp-promoter is repressed to some extent due to the presence of tryptophan synthesized within the cell.

The tacI and the tacII promoter are seven times stronger than the lac-UV5 promoter itself. It is not clear what causes this increased promoter efficiency. Several features in the DNA sequence of a promoter are likely to affect the promoter efficiency: 1) the nucleotide sequence of the Pribnow box; 2) the nucleotide sequence of the -35 area; 3) the distance between these two domains; 4) the AT-richness of areas upstream of the -35 sequence; 5) particular combinations of these features and 6) possibly the nucleotide composition between the Pribnow box and the -35 area.

A consensus sequence for the Pribnow box (TATAAT) and the -35 sequence (TTGACA) has been formulated (for a review see Refs. 2 and 3). The lacUV5-promoter has a consensus Pribnow box sequence (TATAAT) but no consensus -35 sequence (TTTACA). The trp-promoter does not have a consensus Pribnow box sequence (TTAACT) but it has a consensus -35 sequence (TTGACA). Consequently, the tacI-promoter not only has a consensus -35 sequence (TTGACA) but also has a consensus Pribnow box sequence (TATAAT). This correlates well with the observation that the tacI promoter is stronger than the trp promoter and the lac-UV5 promoter. The tacII promoter has a consensus -35 sequence but no consensus Pribnow-box sequence (TTTAAT). Yet the tacI and the tacII promoter are equally strong. Although an A is highly preferred at the second position in the Pribnow-box (3) a T at this position is the second most preferred base (3). If the consensus Pribnow-box sequence of tacI (TATAAT) is a more favorable factor for promoter strength than the tacII Pribnow box (TTTAAT) it follows that another feature in the DNA sequence of tacII must be more favorable than that in tacI promoter. Such a possible feature is the distance between the Pribnow box and the -35 sequence. In tacI this distance is 16 bp whereas this distance is 17 bp in the tacII promoter.

Jaurin et al. (24) showed that the distance between the Pribnow-box and the -35 sequence affects promoter strength considerably. They found a mutant with an insertion of a G:C basepair between -16 and -15 in the promoter of the chromosomal β-lactamase operon of E. coli. This insertion caused a 15-fold increase in β-lactamase synthesis. The distance between the Pribnow box and the -35 sequence thus increased from 16 to 17 bp. A 17 bp distance occurs most frequently in all the sequenced promoters. However, not only the relatively strong ribosomal-protein promoters but also the weak lac i promoter belongs to this group (2,3). On the other hand, in all the ribosomal RNA promoters this distance is 16 bp (16-22), which may be sub-optimal, with respect to the β-lactamase promoter system. In the hybrid tacI-promoter this distance is 16 bp but in the tacII promoter this distance is 17 bp which may compensate for a less favorable Pribnow-box sequence in the tacII promoter. The presence of unique restriction sites in the -20 area of the rac5-16 promoter allows us to vary the distance between the -35 sequence and the Pribnow-box. Mutants with a deletion of one or more basepairs in this region have been made and experiments are in progress to measure the effect of such deletions on the promoter efficiency. Thus the optimal distance between these two domains of the rac5-16 promoter can be determined.

The presence of three unique restriction sites in the synthetic insert of the rac 5-16 promoter also allows us to vary the nucleotide composition in the -20 area without changing the distance between the Pribnow box and -35 sequence. It will also be interesting to see whether the rac5-16 promoter is regulated in a similar way as is the ribosomal RNA promoter whose expression is growth rate dependent and stringently controlled (25).

ACKNOWLEDGMENTS

We thank Drs. Mary Alice Raker, Jurgen Brosius and Harry Noller for donation of their plasmid pKK3535 and Martin Rosenberg for donation of the plasmids pKO-1, pKM-1 and pKO1-T. We thank Jeanne Arch for preparing the manuscript and Alane Gray for preparing the figures. We thank Martin Struble for the purification and sequence determination of the synthetic 42-mer and 46-mer.

REFERENCES

1. Goeddel, D.V., Heyneker, H.L., Hozumi, T., Arentzen, R., Itakura, K., Yansura, D.G., Ross, M.J., Miozzari, G., Crea, R., and Seeburg, P.H. (1979) Nature 281, 544-548.
2. Rosenberg, M. and Court, D. (1979) Ann. Rev. Genet. 13, 319-353.
3. Siebenlist, U., Simpson, R.B., and Gilbert, W. (1980) Cell 20, 269-281.
4. Gold, L., Pribnow, D., Schneider, T., Shinedling, S., Singer, B.S. and Stormo, G. (1981) Ann. Rev. Microbiol. 35, 365-403.
5. Pribnow, D. (1975b) J. Mol. Biol. 99, 419-443.
6. Schaller, H., Gray, C., and Hermann, K. (1975) Proc. Natl. Acad. Sci. USA 72, 737-741.
7. Gilbert, W. (1976) In: RNA Polymerase. Eds. Losick, R., and Chamberlin, M. Cold Spring Harbor Laboratory.

8. Johnsrud, L. (1978) Proc. Natl. Acad. Sci. USA 75, 5314–5318.

9. Rosenberg, M. (1981) Promoters, Structure and Function. Praeger Scientific Publishing Co., M.J. Chamberlin and R. Rodriguez, Eds., in press.

10. de Boer, H.A., Comstock, L.J., Yansura, D., and Heyneker, H. (1981) Promoters Structure and Function, Praeger Scientific Publishing Co., M.J. Chamberlain and R. Rodriguez, Eds., in press.

11. Shine, J., and Dalgarno, L. (1974) Proc. Natl. Acad. Sci. USA 71, 1342–1346.

12. Bennett, G.N., and Yanofski, C. (1978) J. Mol. Biol. 121, 179–192.

13. Heyneker, H.L., Shine, J., Goodman, H.M., Boyer, H.W., Rosenberg, J., Dickerson, R.E., Narang, S.A., Itakura, K., Lin, S.Y., and Riggs, A.D. (1976) Nature 263, 748–752.

14. McKenney, K., Shimatake, H., Court, D., Schmeissner, U., Brady, C., and Rosenberg, M. (1981) Gene Amplification and Analysis. Vol. II: Analysis of Nucleic Acids by Enzymatic Methods (Elsevier–North Holland) Jack G. Chirikjian and Takis S. Papas, Eds.

15. Ellwood, M., and Nomura, M. (1980) J. Bacteriol. 143, 1077–1080.

16. De Boer, H.A., and Nomura, M. (1979) J. of Biol. Chem. 254, 5609–5612.

17. De Boer, H.A., Gilbert, S.F., and Nomura, M. (1979) Cell 17, 201–209.

18. Kiss, I., Boros, I., Udvardy, A., Venetianer, P., and Delius, H. (1980) Biochim. Biophys. Acta 609, 435–447.

19. Hamming, J., Gruber, M., and Ab, G. (1979) Nucleic Acids Res. 7, 1019–1033.

20. Brosius, J., Dull, T.J., Sleeter, D.D., and Noller, H.F. (1981) J. Mol. Biol. 148, 107–127.

21. Young, R.A., and Steitz, J.A. (1979) Cell 17, 225–234.

22. Brosius, J., Palmer, M.L., Kennedy, P.J., and Noller, H.F. (1978) Proc. Natl. Acad. Sci. USA 75, 4801–4805.

23. Gilbert, S.F., de Boer, H.A., and Nomura, M. (1979) Cell 17, 211–224.

24. Jaurin, B., Grundstrom, T., Edlund, T., and Normark, S. (1981) Nature 290, 221–225.

25. Miura, A., Krueger, J.H., Itoh, S., de Boer, H.A., and Nomura, M. (1981) Cell 25, 773–782.

DISCUSSION

D. BOTSTEIN: You mentioned several times the lethal effect of too strong a promoter. Do you have any idea what causes the cells to die? Is it the plasmid that dies and they loose drug resistance, or perhaps the cells actually die from some toxic product?

H. HEYNEKER: I don't know for sure which of these two reasons cause lethality. I can only give some observations that show that a strong promoter can be lethal to the cells. One can clone without problems the tacI promoter into the amp-gene (pBR 322 tacI) such that a nonsense hybrid protein is made.

However, when one attempts to hook up the tacI promoter to the HGH gene in wild type cells, things are less simple. Only two tacI promoters in a row fused to the HGH gene are obtained if wild type cells are used as host. This is not the case when a laciq host is used. In wild type cells selection is apparently against the presence of a constitutively expressing single tacI promoter in front of the HGH gene.

In later experiments with a double and a single tacI promoter fused to the HGH gene in E. coli D_{1210} lacq we found that indeed a double tacI promoter yields after induction with IPTG far less (about half) HGH than a single tacI promoter. These observations show that two tacI promoters in a row together are less efficient than a single tacI promoter. We have evidence that the transcripts that started at the upstream located promoter terminate in the lac-operator region thus interfering with RNA-polymerase binding at the downstream tacI promoter.

D. GELFAND: Your slides indicated a position for transcription initiation in tacI, tacII, and racI promoters. Do you know if alterations in the -10 sequence, or spacing between "-10" and "-35" regions have any effect on the initiating nucleotide?

H. HEYNEKER: No, we have not determined the 5'-initiating nucleotide. However, in the case of the hybrid promoters the nucleotides between the Pribnow box and the presumed starting nucleotide are the same as those of the lac-UV5 promoter itself. Therefore, the starting nucleotide that is in these hybrid promoters is assumed to be the same as that of the lac-UV5 promoter.

BIOLOGICAL ACTIVITIES
OF CLONED GENE PRODUCTS

REGULATION OF HISTONE GENE EXPRESSION
IN HUMAN CELLS

G. S. Stein, J. L. Stein*, L. Baumbach, A. Leza*,
A. Lichtler, F. Marashi, M. Plumb,
R. Rickles, F. Sierra and T. Van Dyke

Department of Biochemistry and Molecular Biology
Department of Immunology and Medical Microbiology*
University of Florida
Gainesville, Florida

Histone genes represent a moderately repeated set of genes in human cells. To study the organization and regulation of human histone genes, we have characterized a series of recombinant lambda Charon 4A phage containing genomic human histone sequences (designated λHHG). Our analyses indicate that human histone genes are clustered but are not organized in a simple tandem repeat pattern as is observed for several lower eukaryotes. For example, several of the human genomic fragments we have isolated contain two each of segments coding for H3 and H4 histones. Of particular interest with respect to organization and expression of the human histone genes is the presence of at least seven different H4 histone mRNAs associated with polysomes of S phase HeLa cells. At least three of the HeLa H4 histone mRNAs are products of distinct genes and two H4 histone genes in the same genomic fragment code for different H4 mRNAs.

We have used our cloned human histone genes to examine the regulation of histone gene expression in human cells. Though it is generally agreed that histone protein synthesis in HeLa cells is restricted to the S phase of the cell cycle, and therefore parallels DNA replication, both transcriptional and post-transcriptional levels of control have been postulated. By probing electrophoretically fractionated, filter-immobilized RNAs with several cloned genomic human histone sequences representing different histone gene clusters, we have assessed the steady state levels of histone RNAs in the nucleus and cytoplasm of G1 and S phase HeLa S3 cells. The representation of histone mRNA sequences in G1 compared with S phase cells was less than 1% in the cytoplasm and approximately 1% in the nucleus. These data are consistent with control occurring primarily at the transcriptional level, but we cannot dismiss the possibility that regulation of histone

FROM GENE TO PROTEIN:
TRANSLATION INTO BIOTECHNOLOGY

gene expression is, to some extent and/or under some biological circumstances, mediated post-transcriptionally. If histone gene transcription does occur in Gl, the RNAs must either be rapidly degraded or be transcribed to a limited extent compared with S phase.

An unexpected result was obtained when a northern blot of cytoplasmic RNA from Gl and S phase cells was hybridized with λHHG41 DNA (containing H3 and H4 genomic human histone sequences). This clone hybridizes with histone mRNAs present in S phase cytoplasmic RNA, but also hybridizes with a Gl cytoplasmic RNA approximately 330 nucleotides in length. This RNA, present in the cytoplasm of HeLa cells predominantly in the Gl phase of the cell cycle, is not similar in size or nucleotide sequence to known histone RNAs.

The possibility of prokaryotic-like organization and regulation of human histone genes is discussed.

I.   INTRODUCTION

Histone genes are represented as a moderately repeated set of genomic DNA sequences in human cells, and compelling evidence points towards a major role for the gene products, the histone proteins, in packaging newly replicated DNA and in modifications of genome structure associated with transcription. In this chapter we will summarize the progress our laboratory has made during the past several years towards addressing the structural and functional properties of human histone genes: First, the structure and organization of human histone genes will be discussed, primarily because of the functional interrelationships between structure, organization and regulation but also because of the utilization of specific regions of genomic histone sequences as probes for analysis of human histone gene organization and regulation of expression. Second, approaches to assessing levels of control of histone gene expression will be considered, and here, variations in levels of control under different biological circumstances and the possibility of prokaryotic-type organization and regulation of human histone gene will be evaluated.

II.   STRUCTURE AND ORGANIZATION OF HUMAN HISTONE GENES

We have isolated a series of twelve genomic clones containing human histone coding sequences, their flanking sequences and noncoding sequences from a λ Charon 4A human

gene library constructed by Tom Maniatis and collaborators
(1). These clones were analyzed by hybridization with heterologous probes as well as by hybrid selection-translation
(Figure 1A), nucleotide sequencing (Figure 1B) and restriction enzyme mapping (39). While the moderately repeated
human histone genes are clustered, they are not arranged in
the form of a tandem repeat such as has been observed in sea
urchin and Drosophila. Rather, human histone genes exhibit
at least three types of arrangements with respect to restriction sites and the order of coding sequences; each of these
arrangements is clearly distinguishable from the others. The
organization of human histone genes is further complicated by
the association of Alu family DNA sequences with some but not
all of the histone coding regions (40). Additionally,
sequences coding for non-histone RNA species, some of which
are expressed throughout the cell cycle and others which are
expressed during specific periods of the cell cycle, are
interspersed among the human histone genes. In situ hybridization studies (2,3) and restriction analysis data from
human-rodent hybrids (4) suggest that in humans the histone
genes may be clustered on the distal end of the long arm of
chromosome 7 in the G-negative band q34. Partial restriction
maps of each of the three types of arrangements of histone
genes, representing information from seven individual genomic
clones, are shown in Figures 1C and 1D.

Of particular interest with respect to the structure,
organization and expression of human histone genes is an
observation we made several years ago of multiple forms of
human H4 and H3 histone mRNAs (5-7). In our initial studies
we demonstrated that two different mRNAs species coding for
H4 histones can be isolated from polysomes of S phase HeLa
cells. The two variants were separated by polyacrylamide gel
electrophoresis of 5-18S polysomal RNA extracted from synchronized cells. Upon elution from the gel both species were
assayed for template activity in vitro using the wheat germ
cell-free protein synthesizing system. Both RNAs were translated into H4 histone. Using several types of denaturing
gels (98% formamide-acrylamide, glyoxal-acrylamide and
methylmercury-agarose), the two H4 mRNAs were found to retain
their distinct mobilities, indicating the two are different
in molecular weight. Characterization of the 3'-terminus of
each H4 mRNA by oligo dT-cellulose chromatography revealed no
significant polyA region in either molecule. After treatment
of the RNAs with T1, T2 and pancreatic nucleases followed by
electrophoresis on DEAE paper it was shown that both species
had a capped 5' terminus (8). Sequence organization was
analyzed by digesting the two RNA species with T1 ribonuclease and subsequently performing two-dimensional fractionation
of the resulting oligonucleotides. Although the majority of

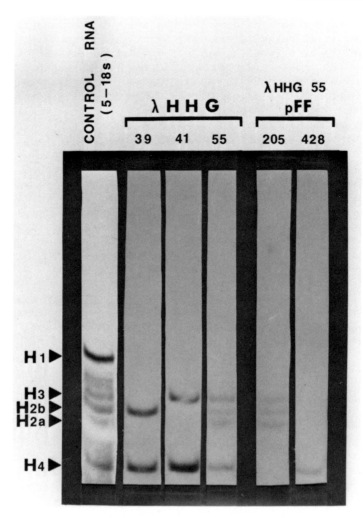

Figure 1A. Hybrid selection–in vitro translation analysis of DNAs from λHHG recombinant phage-containing human
histone sequences and of recombinant pBR322 plasmids containing specific regions of the genomic human histone sequences
subcloned from λHHG phage. Phage or plasmid DNA was immobilized on nitrocellulose filters and hybridized with HeLa S
phase polysomal RNA containing H2A, H2B, H3, H4 and H1
histone mRNAs. Hybridized RNAs were eluted, translated in a
wheat germ cell-free protein synthesizing system, and the in
vitro translated polypeptides were fractionated electrophoretically in acetic acid-urea polyacrylamide gels. Analysis
was by autoradiography. The left lane shows in vitro translation products (all five histone polypeptides) translated
from the RNA preparation used for hybrid selection.

H3 - Histone Sequences

	met	ala	arg	thr	lys	gln	thr	ala	arg	lys	ser	thr	gly	gly	lys	ala	pro	arg	lys	gln	leu	ala	thr	lys	ala	ala
Chicken		GCG	CGT	ACG	AAG	CAG	ACG	GCG	CGT	AAG	TCG	ACG	GGC	GGG	AAG	GCG	CCC	CGC	AAG	CAG	CTG	GCC	ACC	AAG		
Chicken(2.6)	ATG	GCG	CGT	ACG	AAG	CAG	ACG	GCG	CGT	AAG	TCG	ACG	GGC	GGT	AAG	GCG	CCA	CGT	AAG	CAG	CTG	GCC	ACT	AAG		
Sea Urchin		GCA	CGC	ACC	AAG	CAG	ACC	GCT	CGC	AAA	TCT	ACA	GGA	GGG	AAG	GCT	CCC	CGC	AAG	CAG	CTG	GCA	ACC	AAA	GCT	GCC
Mouse	ATG	GCT	CAT*	ACA	AAG	CAG	ACT	GCC	CGC	AAA	TCC	ACC	TGT*	GGT	AAA	GCA	CCT	AGG	AAA	CAA	CTA	GCT	ACA	AAA	GCT	GCT
Human	ATG	GCT	CGT	ACT	AAA	CAG	ACA	GCT	CGG	AAA	TCC	ACC	GGC	GGT	AAG	GCG	CCA	CGC	AAG	CAG	CTG	GCT	ACC	AAG	GCT	GCC

Figure 1B. Partial nucleotide sequence analysis of the human H3 histone gene in the 3.1 Kb Eco RI fragment of λHHG17. Shown for comparison are the nucleotide sequences of several other H3 genes. The asterisk (*) indicates a codon which would result in an amino acid substitution.

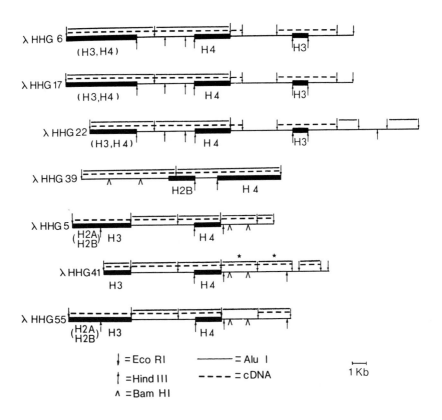

Figure 1C. Partial restriction maps of seven recombinant λ phages containing genomic human histone sequences. Also designated are regions of the phage which contain Alu family DNA sequences and nonhistone coding regions which hybridized with cDNA complementary to S phase polysomal RNAs.

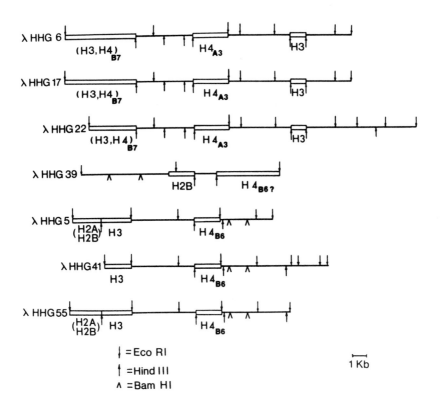

Figure 1D. Partial restriction maps of the λHHG phage in which the variant H4 histone mRNA coding sequences are designated.

the oligonucleotides from the two H4 mRNAs migrated similarly, a few variations were noted, suggesting some sequence
heterogeneity. This is similar to Grunstein's findings for
the C1 and C3 fractions of H4 histone mRNA from sea urchin
(9). Electrophoresis of the in vitro translation products of
HeLa histone mRNAs in an acetic acid urea system could not
separate the independently synthesized products which
comigrate with marker H4. One-dimensional tryptic peptide
mapping of the translation products labeled with several
different radioactive amino acids revealed no differences
between the molecules. Two-dimensional tryptic peptide
mapping has also been carried out and again no differences
were detected (5-7).

    More recently using higher resolution fractionation
procedures, we have observed at least seven different H4
histone mRNA species associated with polysomes of S phase
HeLa S3 cells (41).

    --Electophoretic fractionation under denaturing condi-
tions of in vivo synthesized ^{32}P-labeled S phase HeLa
cell histone mRNAs revealed three major H4 histone mRNA
(Figure 2).

    --Subsequent fractionation of the three H4 histone mRNA
species under non-denaturing conditions, where advantage was
taken of possible variations in secondary structure, resulted
in separation of each band into several additional H4 histone
mRNA species (Figure 3). Each of these mRNA bands was
excised from the gel, denatured in urea and refractionated
electrophoretically in nondenaturing and denaturing gels.
Each of the individual H4 mRNA bands migrated as a single
species under both nondenaturing (Figure 4) and denaturing
conditions, indicating that these RNA bands are not artifacts
of the electrophoresis system and suggesting that each is in
fact a unique H4 histone mRNA species.

    --Each of the numbered bands in Figure 3 (bands 1-10) was
translated in a wheat germ cell free protein synthesizing
system and each coded for only H4 histone.

    --Each of the numbered bands in Figure 3 was also subjec-
ted to T1 ribonuclease digestion and two-dimensional fingerprinting of the resulting oligonucleotides. Variations in
the oligonucleotide maps of bands 1, 2, 5, 7, 8, 9, and 10
(some of which are shown in Figure 5) show sequence differences in at least seven H4 histone mRNA species. The fingerprints also indicate that the oligonucleotides of the smaller
molecular weight H4 histone mRNAs are not a subset of those
generated from the larger RNAs, ruling out a simple
precursor-product relationship.

    --The identity of the H4 histone species was confirmed by
hybrid selection with cloned H4 histone gene sequences of in
vivo labeled S phase HeLa cell polysomal RNA and subsequent

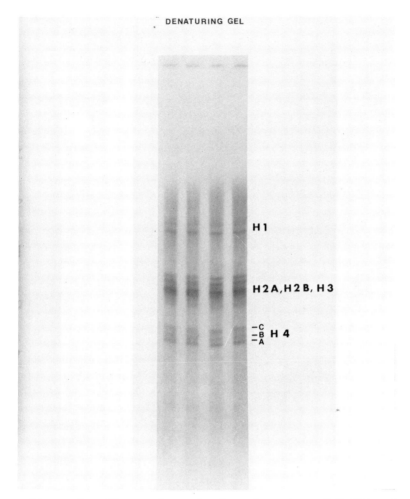

Figure 2. Electrophoretic fractionation of 5-18S poly-somal RNAs under denaturing conditions. S phase HeLa cells were ^{32}P-labeled in vivo and polysomal RNA was isolated in the presence of vanadyl ribonucleoside complex (33). The RNA was fractionated on a sucrose gradient and the region between 5S and 18S was pooled. The RNA was ethanol precipitated and then dissolved in 8.3 M urea – 5 mM EDTA (pH 8.0), heated to 100° and quick cooled before loading on a 6% (w/v) polyacrylamide – 8.3 M urea gel buffered with 50 mM Tris-borate – 1 mM EDTA (TBE). The gel was run at 20 watts which gave a surface temperature of 50-60°C. These conditions are sufficient to denature most RNA molecules (34). After auto-radiography, the bands labeled A, B and C were excised and the RNA was eluted electrophoretically into dialysis tubing.

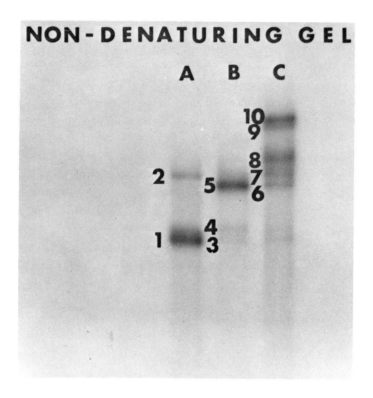

Figure 3. Electrophoretic fractionation under non-denaturing conditions of three molecular weight species of human H4 histone mRNAs. RNAs from the gel shown in Fig. 2 were ethanol precipitated, dried, and resuspended in 4 M urea – 2.5 mM EDTA (pH 8.0). The RNAs were then heated to 100° and loaded onto a 6% (w/v) polyacrylamide – TBE gel with no urea for fractionation under non-denaturing conditions. The numbered bands were excised and the RNAs eluted electrophoretically.

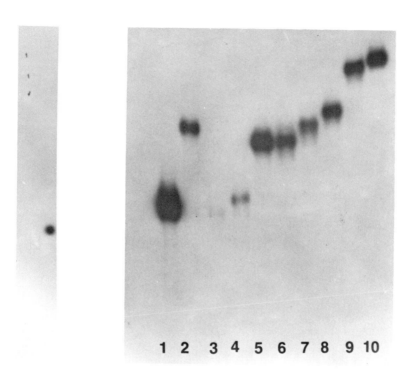

Figure 4. Analysis of fractionated H4 mRNA species. RNAs eluted from the numbered gel bands of Figure 3 were denatured by boiling in 4 M urea (as described in the legend to Figure 3) and re-electrophoresed in the same gel system. Each RNA migrated as a single band with the same relative mobility exhibited in the first gel (Figure 3), eliminating the possibility that the multiple bands in Figure 3 are due to conformational isomers of single RNA species.

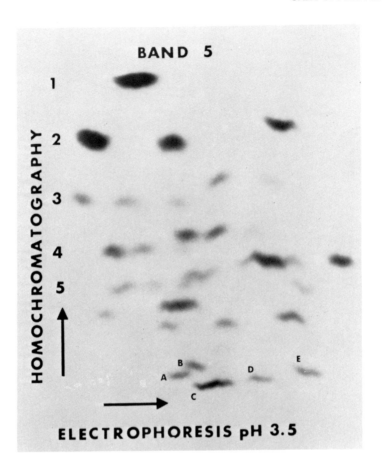

FIG. 5A

Figure 5. Two-dimensional T1 ribonuclease fingerprint analysis of selected H4 histone mRNA species. RNAs eluted from the gel shown in Figure 3 were digested with T1 ribonuclease and the resulting oligonucleotides were fractionated on cellulose acetate at pH 3.5 in 7 M urea in the first dimension, followed by homochromatography on PEI-cellulose thin layer plates in the second dimension. The band numbers correspond to those in Figure 3. Bands 5 and 6 (Figures 5A and B) are identical, as is shown by the correspondence between the lettered oligonucleotide spots. Bands 1 and 10 (Figures 5C and D) are different, as shown by spots a, b and c in 1 which are not found in 10, and spots d and e from 10 which are not found in 1. Also note that both 1 and 10 differ from 5 and 6.

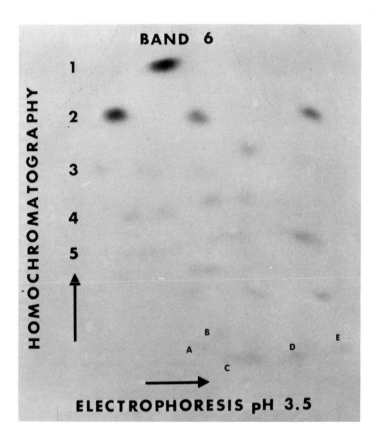

FIG. 5B

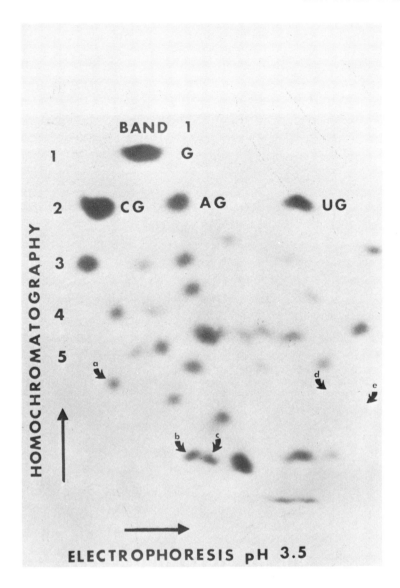

FIG. 5C

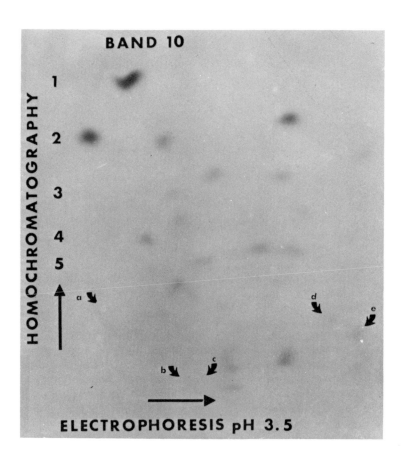

FIG. 5D

electrophoretic fractionation under denaturing conditions of
the selected RNAs (Figure 6).

--Variations were observed in the representation of H4
histone mRNA species when polysomal RNAs from HeLa cells and
WI-38 human diploid fibroblasts were fractionated electrophoretically, immobilized covalently on diazotized cellulose and
identified by hybridization to ^{32}P-labeled, cloned H4 histone DNA sequences (Figure 7). At least one H4 histone mRNA
species that is observed in WI-38 cells is not represented in
the polysomal RNA of HeLa cells, suggesting a possible difference in expression of H4 histone mRNA species in different
human cell types.

--We have begun to assign H4 histone mRNA species to individual H4 histone genes using a modification of the Berk and
Sharp procedure (10). As shown in Figure 8, when H4 histone
mRNA species collectively were annealed with cloned genomic H4
histone sequences, each H4 histone gene formed an S1 nuclease
resistant hybrid with only one H4 mRNA. These results provide
convincing evidence that the various H4 histone mRNA species
do not represent post-transcriptional processing. It is
interesting that where H4 histone mRNA sequences are represented in the genome in close proximity, as in clones λHHG6,
λHHG17, and λHHG22, different H4 histone mRNA sequences are
encoded in the adjacent H4 genes (Figure 1C).

We have recently initiated studies directed toward identification, isolation and characterization of multiple forms
of human H3 histone mRNAs. The approaches being pursued are
similar to those we have been using to examine H4 histone mRNA
species. We have been able to identify at least four distinct
H3 histone mRNA species associated with the polysomes of HeLa
S3 cells and WI-38 human diploid fibroblasts. These H3 mRNAs
can be identified by hybridization of cloned human genomic H3
histone DNA sequences (^{32}P-labeled) to electrophoretically fractionated HeLa and WI-38 polysomal RNAs covalently bound
to diazotized cellulose. As shown in Figure 6, four HeLa cell
H3 histone mRNAs can be isolated from total cytoplasmic RNAs
by hybrid selection with cloned H3 histone sequences.
Additional characterization of the H3 histone mRNAs and
assignment of H3 histone mRNA species to individual H3 histone
genes is being pursued.

The obvious question that arises is the biological
significance of multiple forms of H3 histone mRNAs which are
genetically encoded and serve as templates for the synthesis
of apparently identical histone polypeptides. While here we
can only speculate, our current thinking encompasses a working
model with the following components. All histone genes in a
cluster are coordinately controlled, with their expression
being modulated by common regulatory sequences and/or

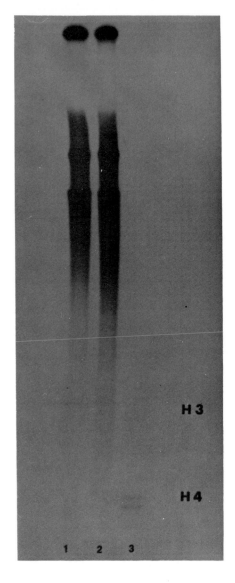

Figure 6. Hybrid selection of _in vivo_ synthesized histone mRNAs. Fifty µg of plasmid DNA (containing H3 and H4 histone genes) were linearized with EcoRI, treated with proteinase K in 2% (w/v) SDS and extracted with phenol-chloroform-isoamyl alcohol (25:24:1). The DNA was denatured with 0.5 M NaOH, neutralized with NaCl- Tris-HCl and passed through 13 mm Sartorius nitrocellulose filters. The filters were baked at 80° in vacuo for 2 hr and then hybridized with ^{32}P-labeled total cytoplasmic RNA from S phase HeLa cells in 1.5 ml Eppendorf tubes in 50% (w/v) formamide – 0.5 M NaCl-10 mM Hepes (pH 7.3) – 1 mM EDTA – 0.2% (w/v) SDS at 47° for 40 h. The filters were washed 10 times with 1 ml of 2 mM EDTA (pH 7.0) at 68°. The hybridized RNA was eluted with two 300 µl aliquots of distilled water at 100° for 2 min, ethanol precipitated, and electrophoresed on an 8.3 M urea denaturing gel (as in Figure 2) (Lane 3) along with total cytoplasmic RNA (Lane 1) and RNA from the hybridization mix which did not bind to the filter (Lane 2).

Figure 7.   Comparison of H4 histone mRNAs from two human
cell lines.   Seventy-five µg of HeLa (Lane 1) and WI38 (Lane
2) cytoplasmic RNAs were preincubated in 50 mM methylmercury
hydroxide and then electrophoretically fractionated on 3%
(w/v) low gelling temperature agarose gels containing 5 mM
methyl mercury.   After a neutralization/equilibration step
(35), the nucleic acids within the gel were electrophoretically transferred to diazotized cellulose as described by
Stellwag and Dahlberg (36).   Prehybridization and hybridization solutions were as described (35) except that 5X Denhardt
solution (minus BSA), 0.7 mg/ml of carrier nucleic acid and
0.1% (w/v) SDS were used.   The paper was hybridized with a
radiolabeled, human H4 DNA fragment (1 x 10^7 cpm/µg, 1 x
10^6 cpm/ml) at 50° for 36 h.   Washing was at 68° using
decreasing concentrations of SSC.   The paper was blotted dry
and exposed to Kodak XAR-5 film at -70° with a Cronex
"Lightning Plus" intensifying screen.

Figure 8.   Identification of H4 histone mRNA species
which are S1 nuclease resistant after hybridization to
histone DNA clones.   10-20 µg of λHHG recombinant DNAs were
mixed with 500-2,000 cpm of ^{32}P-labeled total H4 histone
mRNA isolated from a gel similar to the one shown in Figure 2
and nucleic acids were precipitated with ethanol in a 1.5 ml
microfuge tube.   The pellet was drained thoroughly and
dissolved in 20 µl of redistilled but not deionized formamide
(BRL).   Four microliters of 2 M NaCl-0.4 M Pipes (pH 6.4)-5
mM EDTA were added and the tube was placed at 80°C for 5 min.
Hyridization was at 50°C for 5 h.   After hybridization, 225
µl of ice cold 0.25 M NaCl-25 mM Na Acetate (pH 4.4)-0.45 mM
ZnSO$_4$ were added and the sample was mixed thoroughly and
immediately placed on ice.   One hundred ten units of S1
nuclease (Sigma) in 5 µl were then added, and the sample
incubated at 37°C for 30 min.   Twenty µl of 10% (w/v) SDS, 20
µl 0.2 M EDTA (pH 8.0), 20 µg tRNA, 200 µl phenol, and 200 µl
chloroform/isoamyl alcohol (24:1 v/v) were added, and the
mixture was vortexed, centrifuged, and the supernatant was
ethanol precipitated.   The dried pellet was dissolved in 8.3
M urea-5 mM EDTA, heated to 100°C for 2 min and electrophoresed in an acrylamide gel.
    S1 resistant samples were electrophoresed in an 8.3 M
urea gel (A) or in a nondenaturing gel (B).   Results from two
different autoradiographic exposure times are shown.   Lanes
marked H4A, H4B and H4C correspond to marker RNAs.

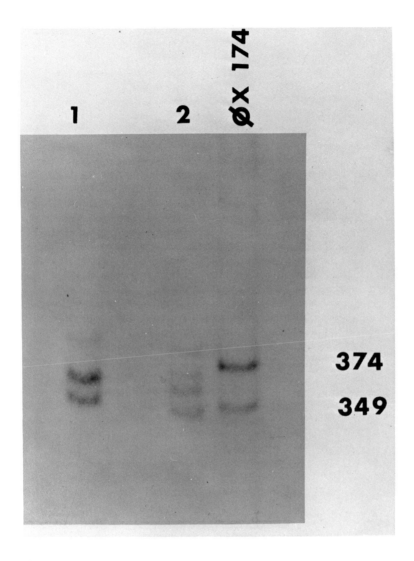

FIG. 7

A DENATURING GEL

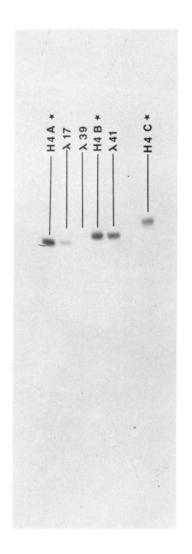
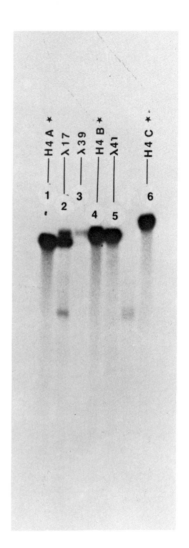

FIG. 8A

B NON-DENATURING GEL

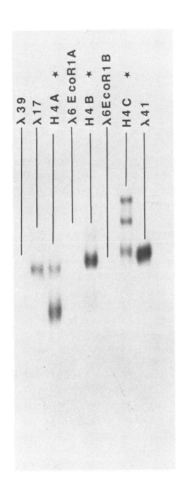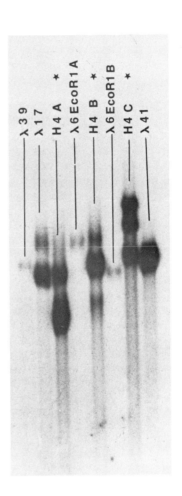

FIG. 8B

regulatory molecules. Only a subset of the reiterated histone genes are expressed, with variations in those histone
genes (gene clusters) expressed in different cells and/or in
different biological circumstances. Selection of clusters to
be transcribed would be based on a requirement for a histone
H2A, H2B, H3, or H1 subspecies where differences in the mRNAs
and in the histone proteins have been observed. The H4 genes
expressed would be predicated on location within a cluster
containing coding sequences for a specific H2A, H2B, H3 or H1
histone subspecies--with the possibility that all genes in
such a cluster contain similar regulatory sequences.

III. LEVEL OF CONTROL OF HUMAN HISTONE GENE EXPRESSION

To understand the level at which regulation of histone
gene expression resides, it is necessary to establish whether
or not histone proteins and histone mRNAs are synthesized in
a cell cycle stage specific manner. Equally important is an
understanding of the relationship between histone protein
synthesis, histone gene transcription and the histone mRNA
sequences in the cell nucleus and cytoplasm.

A. Histone Protein Synthesis During the Cell Cycle

It has been well documented that in most higher eukaryotic cells the synthesis of histone proteins, with the exceptions of H1 synthesis under certain circumstances (37) and
"basal synthesis" of histones in several mammalian cell lines
(38), is primarily if not exclusively restricted to the S
phase of the cell cycle in continuously dividing cells and in
quiescent cells stimulated to proliferate (11-21). And,
while histone protein synthesis has been reported to occur at
a constant rate throughout the cell cycle in S49 mouse lymphoma cells, this appears to be a "special situation" (22).
Early evidence for S phase specific histone synthesis in
human cells, both continuously dividing HeLa Cells and nondividing human diploid fibroblasts stimulated to proliferate,
comes from studies in which cells were pulse-labeled with ^{3}H-
or ^{14}C-labeled amino acids and the specific activities of
nuclear and/or chromosomal histones were determined (11-21).
To eliminate the possibility that histones are synthesized
throughout the cell cycle and become associated with the
genome only during S phase, we recently pulse-labeled G1 and
S phase HeLa cells with ^{35}S-methionine and analyzed
unfractionated total cellular proteins (Figure 9A) or dilute
mineral acid extractable total cellular proteins

(Figure 9B) for the presence of newly synthesized histone
polypeptides. Our ability to detect radioactively labeled
histones only in S phase HeLa cells is consistent with S
phase specific histone protein synthesis or synthesis of
histones predominantly when DNA replication occurs (23). S
phase specific histone protein synthesis is also supported by
the very low levels of histone synthesis and cytoplasmic
histone mRNAs at nonpermissive temperatures in temperature-
sensitive cell cycle mutants (27 ; T. Van Dyke, F. Marashi,
R. Baserga, J. Stein, and G. Stein, unpublished results).

     B.   Representation of Histone mRNA Sequences
           During the Cell Cycle

     To gain a more definitive insight into regulaton of
histone gene expression during the cell cycle, we have used
cloned genomic human histone sequences to examine the representation of histone mRNAs in the nucleus and cytoplasm of G1
and S phase synchronized HeLa S3 cells (42). In agreement
with earlier observations from our laboratory and others
(6,24-31), where histone mRNAs were analyzed by RNA excess
hybridization with homologous histone cDNAs, histone mRNA
sequences are present in significant amounts in the nucleus
and cytoplasm of HeLa cells only during S phase when histone
protein synthesis occurs.
     The representation of HeLa cell histone mRNAs was assayed
by hybridization to cloned H4, H3, H2A and H2B histone
sequences. To standardize conditions for the hybridization,
we took advantage of the longstanding observation that the
inhibition of DNA replication results in a rapid loss of
histone mRNAs from the polysomes of S phase HeLa cells. As
shown in Figure 10, under the standard hybridization conditions employed in our studies, ^{32}P-labeled plasmid DNAs
containing histone sequences anneal with S phase HeLa polysomal RNAs fractionated electrophoretically in methyl mercury
agarose gels and transferred electrophoretically to diazotized cellulose. Consistent with the anticipated loss of
histone mRNA sequences from polysomes following inhibition of
DNA synthesis, a greater than 95% inhibition was observed in
the hybridization of radiolabeled, cloned histone DNAs to
filter-immobilized polysomal RNAs from S phase HeLa cells
when DNA synthesis was blocked by treatment with cytosine
arabinoside. The sensitivity of our hybridization assay is
such that we can detect less than 500 pg of mRNA.
     To address the level at which regulation of histone gene
expression occurs during the cell cycle in HeLa S3 cells, we
assayed the abilities of ^{32}P-labeled (nick-translated)
histone DNAs to hybridize with filter-immobilized RNA from G1

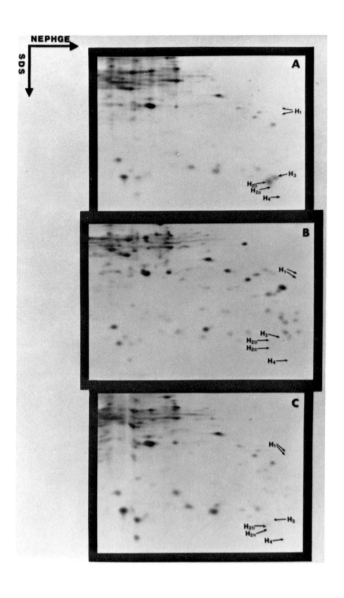

Figure 9A. Two-dimensional electrophoretic analysis of ^{35}S-methionine-labeled total cellular proteins. A – S phase, B – G1 phase, C – cytosine arabinoside-treated S phase.

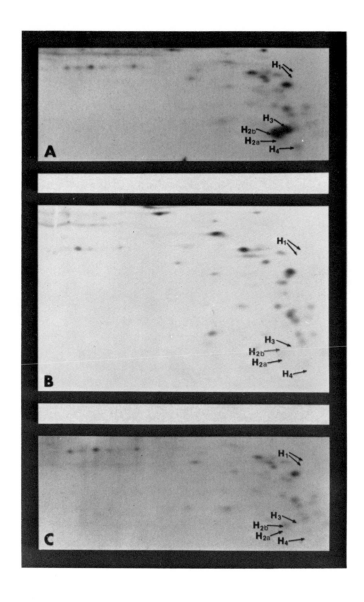

Figure 9B. Two dimensional NEPHGE/SDS electrophoresis of acid-extracted nuclear proteins of A) S phase, B) Gl and C) cytosine arabinoside-treated S phase cells.

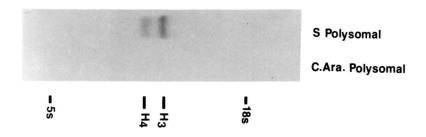

S Polysomal

C.Ara. Polysomal

Figure 10. Hybridization of electrophoretically fractionated polysomal RNA from S phase and cytosine arabinoside-
treated S phase HeLa cells with ^{32}P-labeled human genomic
H3 and H4 histone sequences. One hundred μg of total HeLa
cell polysomal RNAs were fractionated in a 2% (w/v) agarose
gel containing 5 mM methylmercury hydroxide, electrophoretically transferred to DBM paper, and hybridized with
^{32}P-labeled (nick-translated) plasmid DNA containing
human genomic H3 and H4 histone sequences.
Figure 11. Hybridization of human genomic recombinant
phage λHHG 55 DNA containing H2A, H2B, H3 and H4 histone
sequences with cytoplasmic and nuclear RNA from G1 and S
phase HeLa cells. S phase cells were obtained 1 h after
release from two cycles of 2 mM thymidine block while the G1
population was obtained by mitotic selective detachment. One
hundred μg of total cytoplasmic RNA were fractionated in a 2%
(w/v) agarose - 5 mM methylmercury hydroxide gel and equal
amounts (100 μg) of nuclear RNA were fractionated in a gel
containing 1% (w/v) agarose - 5 mM methylmercury hydroxide.
The positions of 28S, 18S and 5S markers were determined
optically after staining the gel with ethidium bromide. RNAs
were electrophoretically transferred to DBM paper and hybridized with ^{32}P-labeled (nick-translated) λHHG 55 DNA.
The blots were analyzed autoradiographically at -70° using
preflashed XAR-5 film (Kodak) with a Cronex "Lightning Plus"
intensifying screen. Autoradiograms were scanned with a
Joyce-Loebel densitometer and quantitated by planimetric
analysis; A) Cytoplasmic RNAs, B) Nuclear RNAs. Note that
under these electrophoresis conditions the mRNAs for H2A, H2B
and H3 histones are not resolved.

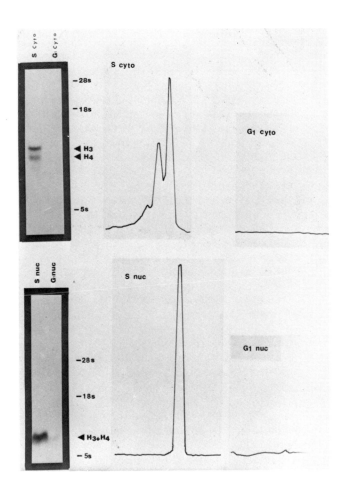

FIG. 11

and S phase cells. The rationale was that because histone synthesis is confined to S phase, hybridization of the histone DNA probes to RNAs from S phase but not from G1 cells would be consistent with nuclear and/or transcriptional control. However, hybridization to both G1 and S phase RNAs would suggest that regulation of histone gene expression resides at a post-transcriptional step.

As shown in Figure 11, hybridization between λHHG55 DNA, a recombinant λ Charon 4A phage containing human genomic H2A, H2B, H3, and H4 histone sequences, and G1 nuclear or cytoplasmic RNAs was barely detectable, while hybridization of λHHG55 DNA with both nuclear and cytoplasmic RNAs of S phase HeLa cells was apparent. In these experiments G1 cells were obtained by mitotic selective detachment, a procedure which yields a G1 population containing less than 0.5% S phase cells. 100 μg of both G1 and S phase RNA were fractionated electrophoretically. Ethidium bromide staining indicated that similar amounts of 18S and 28S RNAs from G1 and S phase cells were fractionated in the gels and transferred to diazotized cellulose (greater than 90% transfer was obtained) in both cases. Because there is an increase in the amount of ribosomal RNA per cell in S phase compared with G1 the hybridization observed is probably an underestimation of the amount of histone mRNA in S phase cells. On longer exposure of the "northern blot" some hybridization of the probe with G1 RNA becomes apparent (approximately 1% the level observed with S phase RNA); this amount of annealing can be explained by a limited number of S phase cells in the G1 population (the biological limits of the system) or by basal level histone synthesis dring G1 (38). When the northern blot of G1 and S phase HeLa cell RNAs shown in Figure 11 was rehybridized with ^{32}P-labeled plasmid DNA containing different H3 and H4 histone coding sequences, S phase specific hybridization was also observed.

An unexpected result was obtained when a northern blot of cytoplasmic RNA from G1 and S phase HeLa cells was hybridized with ^{32}P-labeled DNA from λHHG41, a recombinant phage containing H3 and H4 coding sequences derived from another human histone gene unit (Figure 12). While hybridization with S phase but not G1 histone mRNAs was observed, intense hybridization was also seen with an RNA of approximately 330 nucleotides present predominantly in G1 cells.

The increase in the representation of both nuclear and cytoplasmic histone mRNAs in S phase compared with G1 cells suggests that histone genes are preferentially expressed during the restricted period of the cell cycle when DNA replication occurs. Since synthesis of histone proteins is also confined to S phase, it is reasonable to postulate that nuclear and/or transcriptional level control is operative.

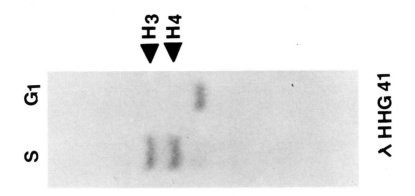

Figure 12.   Autoradiographic analysis of Gl and S phase
cytoplasmic RNAs hybridized with DNA from λHHG 41, a recombinant phage containing H3 and H4 human histone sequences.   The
DBM blot shown in Figure 11A was incubated in sterile water
at 100° to remove ^{32}P-labeled λHHG 55 probe.   After
confirming that the λ55 probe was no longer detectable, the
filters containing fractionated cytoplasmic RNAs were
annealed with ^{32}P-labeled λHHG 41 DNA.

Transcriptional regulation of histone gene expression during
the cell cycle of the HeLa cells is consistent with earlier
observations that histone cDNA hybridized preferentially with
in vivo synthesized RNAs (nuclear and cytoplasmic) and in
vitro chromatin transcripts of S phase but not Gl cells
(6,24-31).
     However an unequivocal demonstration that regulation of
histone gene expression during the cell cycle resides solely
at the transcriptional level requires establishing that: 1)
the very limited presence of histone mRNA sequences in Gl
cells obtained by mitotic selective detachment is attribut-
able to those few S phase cells in the Gl population (approximately 0.5%) and 2) histone gene transcription is initiated
only during S phase.
     Although Melli and coworkers reported the presence of
approximately equivalent amounts of histone gene transcripts
in Gl and S phase HeLa cells (32), these results are not
surprising since histone sequences were assayed by hybridization under non-stringent conditions with a heterologous

probe--sea urchin histone DNA sequences cloned in lambda.
Also in the experiments of Melli and coworkers, cells were
synchronized by double thymidine block which yields a G1
population containing more than 20% S phase cells.

We cannot completely dismiss the possibility that regulation of histone gene expression during the cell cycle may to
some extent be mediated post-transcriptionally. However, the
low steady state level of histone mRNA sequences we observed
in G1 nuclear RNA would indicate that if histone sequences
are transcribed at the same rate throughout the cell cycle
they must be rapidly degraded in G1 cells. Alternatively, it
is possible that histone sequences are transcribed at a much
lower level outside of S phase. Such low level transcription
might be functional or might reflect "leaky" transcription of
some or all of the histone genes. In this context the
persistence of a limited amount of histone transcripts from
the previous S phase into the subsequent G1 should be
considered.

A definitive explanation for the observed hybridization
of ^{32}P-labeled λHHG41 DNA (which contains H3 and H4 human
histone genes) with a 330 nucleotide cytoplasmic RNA species
present predominantly in G1 cells is not yet available.
However, since the G1 cytoplasmic RNA species is encoded in a
genomic sequence in close proximity to histone coding
sequences, a possible regulatory role for the G1 RNA should
not be dismissed. Further analyses of the RNA and the
genomic DNA sequences in which it is encoded are underway.

> C. Regulation of Human Histone Gene Expression
>   Under Various Biological Circumstances

In agreement with the evidence presented above for
nuclear and/or transcriptional level control of histone gene
expression during the cell cycle in continuously dividing
HeLa S3 cells, a parallel situation appears to be operative
when nondividing WI-38 human diploid fibroblasts are stimulated to proliferate. That is, concomitant with the onset of
DNA synthesis, as human fibroblasts make the transition from
the prereplicative to the S phase of the cell cycle, there is
a comparable stimulation of histone protein synthesis and
increased representation of histone mRNAs sequences (29).
However it is unlikely that control of human histone gene
expression is mediated at the nuclear level under all biological circumstances. By analogy with the manner in which
histone gene expression is regulated during early stages of
development in several lower eukaryotes, it is possible that
histone gene expression is controlled post-transcriptionally
under comparable circumstances in human cells. A very likely

possibility, and one which is being pursued experimentally, is that those copies of the human histone genes or those human histone gene clusters expressed concomitantly share similar or identical regulatory sequences, with differences in putative regulatory sequences among those genes whose expression is mediated transcriptionally versus post-transcriptionally.

D. Are Human Histone Genes Organized and Regulated In a Prokaryotic Manner?

Several features of human histone gene organization and the expression of these sequences suggest that it may be realistic to think in terms of "prokaryotic like" gene organization and control. One striking feature of human histone gene organization which appears to reflect that of prokaryotic genes is the apparent absence of intervening sequences. But perhaps the strongest argument for human histone genes being similar to prokaryotic genetic sequences comes from the rapid manner in which newly transcribed mRNAs reach the polysomes and become actively engaged in histone protein synthesis--without an extensive amount of post-transcriptional processing and without addition of poly A to the 3' termini. The "rapid" turnover of human histone mRNAs and the dramatic cell cycle stage specific variations in the representation of histone mRNA sequences in the nucleus and cytoplasm of HeLa and WI-38 cells provide further support for the prokaryotic analogy, i.e., the manner in which histone gene expression parallels the onset of DNA replication is indeed reminiscent of the manner in which lac operon expression in E. coli reflects the levels of specific sugars in the bacterial growth medium. It will be interesting to determine if other eukaryotic genes, whose expression is modulated in a cyclic fashion in response to specific yet changing cellular environments (e.g., inducible metabolic enzymes), are organized and regulated in a manner similar to that of human histone genes. Such organization and expression are apparently strickingly different from those of other eukaryotic genes such as ovalbumin and globin, but it must be considered that the latter genetic sequences are expressed in conjunction with long term rather than cyclic cellular commitments--often being permanent commitments of terminally differentiated cells. It remains to be established whether the "simple" or "prokaryotic type" of organization and control reflected by histone genes or the "complex type" as reflected by globin genes is the rule or special situation in eukaryotic cells.

ACKNOWLEDGMENTS

These studies were supported by grant PCM 80-18075 from the National Science Foundation and Basil O'Connor Starter Research Grant 5-217 from the March of Dimes Birth Defects Foundation.

REFERENCES

1. Lawn, R.M., Fritsch, E.F., Parker, R.C., Blake, G., and Maniatis, T., Cell 15, 1157 (1978).
2. Yu, L.C., Szabo, P., Borun, T.W., and Prensky, W., Cold Spring Harbor Symp. Quant. Biol. 42, 1101 (1978).
3. Chandler, M.E., Kedes, L.H., Cohn, R.H., and Yunis, J., Science 205, 908 (1979).
4. Naylor, S., Stein, J.L., and Stein, G.S., Unpublished results.
5. Stein, G.S., Stein, J.L., Laipis, P.J., Chattopadhyay, S.K., Lichtler, A.C., Detke, S., Thomson, J.A., Phillips, I.R., and Shephard, E.A., Miami Winter Symp. 15, 125 (1978) Academic Press, New York.
6. Stein, G.S., Stein, J.L., Park, W.D., Detke, S., Lichtler, A.C., Shephard, E.A., Jansing, R.H., and Phillips, I.R., Cold Spring Harbor Symp. Quant. Biol. 42, 1107 (1977).
7. Lichtler, A.C., Detke, S., Phillips, I.R., Stein, G.S., and Stein, J.L., Proc. Natl. Acad. Sci. 77, 1942 (1980).
8. Stein, J.L., Stein, G.S., and McGuire, P.M., Biochemistry 16, 2207 (1977).
9. Grunstein, M., Levy, S., Schedl, P., and Kedes, L., Cold Spring Harbor Symp. Quant. Biol. 38, 717 (1973).
10. Berk, A.J., and Sharp, P.A., Proc. Natl. Acad. Sci. 75, 1274 (1978).
11. Spalding, J., Kajiware, K., and Mueller, G., Proc. Natl. Acad. Sci. 56, 1535 (1966).
12. Robbins, E., and Borun, T.W., Proc. Natl. Acad. Sci. 57, 409 (1967).
13. Stein, G.S., and Borun, T.W., J. Cell Biol. 52, 292 (1972).
14. Stein, G.S., and Thrall, D.L., FEBS Lett. 34, 35 (1973).
15. Gallwitz, D., and Mueller, G.C., J. Biol. Chem. 244, 5947 (1969).
16. Borun, T.W., Scharff, M.D., and Robbins, F., Proc. Natl. Acad. Sci. 58, 1977 (1967).
17. Butler, W.B., and Mueller, G.C., Biochim. Biophys. Acta. 294, 481 (1973).
18. Breindl, M., and Gallwitz, D., Eur. J. Biochem. 45, 91 (1974).

19. Borun, T.W., Gabrielli, F., Ajira, K., Zweidler, A., And Baglioni, C., Cell 4, 59 (1975).
20. Jacobs-Lorena, M., Baglioni, C., and Borun, T.W., Proc. Natl. Acad. Sci. 69, 2095 (1972).
21. Gallwitz, D., and Breindl, M., Biochem. Biophys. Res. Comm. 47, 1106 (1972).
22. Groppi, V.E., and Coffino, P., Cell 21, 195 (1980).
23. Marashi, F., Baumbach, L., Rickles, R., Sierra, F., Stein, J.L., and Stein, G.S., Science 215, 683 (1982).
24. Stein, J.L., Thrall, C.L., Park, W.D., Mans, R.J., and Stein, G.S., Science 189, 557 (1975).
25. Detke, S., Lichtler, A., Phillips, I., Stein, J.L., and Stein, G.S., Proc. Natl. Acad. Sci. 76, 1995 (1979).
26. Stein, G.S., Park, W.D., Thrall, C.L., Mans, R.J., and Stein, J.L., Nature 257, 764 (1975).
27. Delegeane, A.M., and Lee, A.S., Science 215, 79 (1981).
28. Detke, S., Stein, J.L., and Stein, G.S., Nucl. Acids Res. 5, 1515 (1978).
29. Jansing, R.L., Stein, J.L., and Stein, G.S., Proc. Natl. Sci. 74, 173 (1977).
30. Parker, I., and Fitschen, W., Cell Diff. 9, 23 (1980).
31. Chiu, I.M., Cooper, D., and Marzluff, W.F., Abstracts of the Second Annual American Cancer Society (Florida Division) Cancer Research Seminar, Abstr., 38 (1979).
32. Melli, M., Spinelli, G., and Arnold, E., Cell 12, 167 (1977).
33. Berger, S.L., and Birkenmeier, C.S., Biochemistry 18, 5143 (1979).
34. Peattie, D.A., Proc. Natl. Acad. Sci. 76, 1760 (1979).
35. Alwine, J.L., Kemp, D.J., Parker, B.A., Reiser, J., Renart, J., Stark, G.R., and Wahl, G.M., Meth. Enzymol. 68, 220 (1979).
36. Stellwag, E.J., and Dahlberg, A.E., Nucl. Acids Res. 8, 299 (1980).
37. Bradbury, M., Personal Communication.
38. Wu, R.S., and Bonner, W.M., Cell 27, 321 (1981).
39. Sierra, F., Lichtler, A., Marashi, F., Rickles, R., Van Dyke, T., Clark, S., Wells, J., Stein, G., and Stein, J., Proc. Natl. Acad. Sci. 79, 1795 (1982).
40. Sierra, G., Leza, A., Marashi, F., Plumb, M., Rickles, R., Van Dyke, T., Clark, S., Wells, J., Stein, G.S., and Stein, J.L., Biochem. Biophys. Res. Comm. 104, 785 (1982).
41. Licthler, A.C., Sierra, F., Clark, S., Wells, J.R.E., Stein, J.L., and Stein, G.S., Nature, in press.
42. Rickles, R., Marashi, F., Sierra, F., Clark, S., Wells, J., Stein, J., and Stein, G., Proc. Natl. Acad. Sci. 79, 749 (1982).

DISCUSSION

P. RECZEK: Are the four mRNA's coding for H4 present
normally in cells or simply an expression of the
transformation process?

G.S. STEIN: Actually, electrophoretic fractionation of H4
histone messenger RNA's first under denaturing and
subsequently under nondenaturing conditions suggest that
there are at least seven distinct H4 histone mRNA species.
We have recently initiated studies aimed at characterizing
the H4 histone mRNA species represented in a series of
normal, transformed and malignant cells. To date our studies
have been limited to only three human tissue culture cell
lines and H4 mRNA's have only been fractionated in denaturing
gel.   We have thus far observed differences in the
representation of H4 molecular weight species but not pattern
reflecting the transformed state.

P. RECZEK: Have you, in fact, sequenced in vitro translation
products of the four H4 mRNA's to determine sequence
homology?

G.S. STEIN: We have not sequenced H4 proteins but two
dimensional tryptic peptide maps of in vitro translated
histone polypeptides do not reveal differences.

M. ADESNIK: Are there not early studies which indicate that
histone mRNA's decay rapidly after DNA synthesis is inhibited
or after cells exit from S phase? Would not such a mechanism
account very well for the low levels of histone mRNA in G1?
In this regard then would it not be crucial to directly
measure transcription rates of histone genes in the S and G1
phases?

G.S. STEIN: Indeed, early studies from several laboratories
including ours indicate that histone messenger RNAs are
"lost" from the polysomes following inhibition of DNA
synthesis and that histone messenger RNAs are associated with
polysomes primarily or exclusively during S phase. Taken
together with "northern blots" of G1, S phase and cytosine
arabinoside treated cellular RNAs, where we recently assayed
the steady state representation of histone mRNAs by
hybridization with cloned genomic histone sequences, the data
are consistent with histone messenger RNAs being present in
cells when DNA synthesis occurs.   We have initiated studies

aimed at directly examining histone gene transcription during the cell cycle -- hybridizing pulse labeled (in vivo) HeLa cell RNAs throughout the cell cycle to cloned human histone sequences. Preliminary results suggest that histone genes are transcribed primarily during S Phase.

B. NELKIN: How many of the histone genes are active?   Do you have any DNA sequence data which can be correlated with differential expression of the H4 genes?

G.S. STEIN: We do not have a "hard figure" for the number of histone genes which are active.   However, in the case of H4 where we have identified a minimum of seven distinct H4 mRNA species with evidence for at least four being encoded by distinct genes (established by formation of SI nuclease resistant hybrids between various H4 mRNA species and specific human genomic H4 sequences under conditions where single mismatched base pairs can be detected), it appears that there are differences in those H4 genes which are expressed in different human cells.

We are in the process of sequencing several human H4 histone genes and their flanking sequences.   At this point I really can't say anything definitive about them and give you reason to suspect that one would be a particularly strong promoter. Representation of mRNAs can reflect properties of the promoters and/or number of gene copies.

CLINICAL INVESTIGATION OF PARTIALLY PURE
AND RECOMBINANT DNA-DERIVED LEUKOCYTE INTERFERON
IN HUMAN CANCER

Jordan Gutterman[1]
Jorge Quesada

Department of Clinical Immunology
and Biological Therapy
University of Texas System Cancer Center
M. D. Anderson Hospital and Tumor Institute
Houston, Texas
U.S.A.

Seymour Fein

Department of Medical Oncology and Immunology
Hoffmann-La Roche, Inc.
Nutley, New Jersey
U.S.A.

I. INTRODUCTION

Interferon, a naturally occurring protein discovered in 1957 (1), has potent antiviral, antiproliferative and immuno-modulating properties (2-4). Clinical experience in treating viral and neoplastic diseases has been limited due to the small quantities of species specific human interferon (5). In February 1978 we initiated a clinical investigation of the antitumor properties of partially pure leukocyte interferon

[1]Supported by NCI grant CA05831 and grants from Hoffmann-La Roche, Inc., the Albert and Mary Lasker Foundation, the Enid Haupt Foundation, the Interferon Foundation, and the Clayton Foundation for Research (Dr. Gutterman is a Senior Clayton Foundation Investigator).

prepared in Finland. At that time, a large body of preclinical data suggested that the interferons might be important
antitumor agents. However, the only significant clinical data
on the effects of interferon in cancer patients existed in
Sweden, where Strander and co-workers suggested that buffy
coat cell-derived leukocyte interferon prolonged the post-
operative disease-free survival of patients with osteogenic
sarcoma (6). It was our feeling that if the interferons were
to play a major role in cancer therapy, they should possess
the capacity to induce regression of established tumor. Thus,
we decided to test whether tumor regression occurred with
leukocyte interferon, despite the fact that there existed no
preclinical tumor model in which tumor regression had been
achieved. We selected patients with metastatic breast cancer
as our first clinical "model".

A.   Initial Study to Determine if Leukocyte Interferon
     Induces Tumor Regression in Human Cancer

The study design, patient selection, dosage schedule and
response criteria have been reported previously (7,8).
Shortly after initiation of this study, we became aware of
preliminary results in Sweden and at Stanford where tumor regression was observed with interferon therapy in individual
patients with multiple myeloma and nodular poorly differentiated lymphocytic lymphoma, respectively. Thus, in July 1978
we extended our clinical evaluation to patients with B cell
tumors (8).
Thirty-eight patients with advanced breast cancer, multiple myeloma, and malignant lymphoma were treated and evaluated in the study. Thirteen of the 38 patients achieved
either complete or partial remission (table 1). An additional
6 patients showed biological improvement or less than a partial remission. Nineteen of the 38 patients showed either
stability or progressive growth of tumor during the study.
Interestingly, prior chemotherapy did not preclude the
ability of interferon to induce regression of tumor. Thus,
only one of nine patients previously unresponsive to chemotherapy had a response to interferon. However, among 18 patients relapsing from previous chemotherapy-induced remissions,
13 showed an antitumor effect with interferon. Eight of these
13 patients previously untreated with chemotherapy achieved an
antitumor effect. Four of these achieved either a complete or
partial remission. The remissions in breast cancer patients
varied from 8 to 78 weeks, with a median of 27 weeks.
Since the patients were treated with either 3 or 9 million
units daily, there was no attempt to determine an optimal

TABLE I.   Evaluation of Leukocyte Interferon in Cancer
           Patients

	Complete Remission	Partial Remission	Less Than Partial Remission or Improvement	No Remission
			no.	
Breast cancer	0	6	1	10
Multiple myeloma	1	2	3	4
Malignant lymphoma	2	2	2	5
Totals	3	10	6	19

dosage schedule in this study. However, several clinical
variables were predictive for a successful antitumor response.
All patients achieving an antitumor effect with breast cancer
experienced a lowering of the leukocyte count to $2,500/mm^3$ or
less and of the granulocyte count to $1,500/mm^3$ or less. In
contrast, only 4 of 10 patients failing to achieve an antitumor response had a decrease in the leukocyte count to
$2500/mm^3$ or in the granulocyte count to $1,500/mm^3$. This was
statistically significant by chi square analysis ($P = <0.05$).
     In our initial study, 6 of 10 patients with advanced
multiple myeloma achieved an antitumor effect. Four of the
10 patients achieved either a complete or partial remission.
One of these patients, a 60 year-old lady with Bence-Jones
proteinuria and extensive osteolytic bone lesions, remains
in complete remission three years after initiating interferon
treatment. She remains on treatment, 3 million units, 3 times
weekly.
     Among the 11 patients with non-Hodgkin's malignant lymph-
oma, six achieved an antitumor effect; 4 of 11 achieved
either complete or partial remission. The two complete remissions were maintained for greater than 1-1/2 years until re-
lapse occurred. During this period of time, Mellstedt, et al
and Merigan, et al reported on the ability of leukocyte interferon to induce remissions in patients with multiple myeloma
and nodular poorly differentiated lymphocytic lymphoma (9,10).

B. Confirmatory Studies of Tumor Regression
with Leukocyte Interferon

After we observed tumor regression in 2 of 5 patients
with breast cancer, we submitted a grant application to the
American Cancer Society for additional financial support to
extend clinical investigation of leukocyte interferon in cancer patients. The American Cancer Society decided to support
phase II studies in four tumors: metastatic breast cancer,
multiple myeloma, nodular poorly differentiated lymphocytic
lymphoma and malignant melanoma. The studies were conducted
by ten institutions. Although the results from these studies
will be presented in detail in separate papers, it is important to point out that the American Cancer Society studies
confirmed the antitumor activity of leukocyte interferon in
patients with metastatic breast cancer (11,12) and multiple
myeloma (13, 14). The overall remission rate was 22% in
breast cancer and 20% in multiple myeloma. The correlation of
antitumor effects and myelosuppression was confirmed in the
breast cancer studies (12). Overall remission rates of 20%
to 25% (15) in these tumors compare favorably with the antitumor activity of many chemotherapeutic agents used today.
However, since the interferons open up an entirely new approach
to the control of malignant growth, it is not particularly
useful, in our opinion, to compare remission rates with standard chemotherapy. The important observation is that a partially pure natural substance can favorably influence the
growth of some metastatic tumors.

C. Additional Studies Demonstrating Tumor Sensitivity
to Leukocyte Interferon

Following these initial studies in which the ability of
partially pure leukocyte interferon to induce regression of
metastatic tumor was demonstrated, we began studies in four
additional malignant tumors and a viral-associated growth,
juvenile laryngeal papilloma, which had been previously demonstrated to be sensitive to interferon by Haglund and co-
workers (16). The malignant tumors studied included metastatic carcinomas of the colon, prostate and kidney (renal cell
carcinoma). We have seen objective evidence of tumor regression in all four malignant tumors. Patients with metastatic
renal cell carcinoma appear particularly sensitive to leukocyte interferon.

1. <u>Colon Cancer</u>. Three of 14 patients with metastatic colon cancer achieved objective evidence of tumor regression. A partial remission occurred in two patients (14%) and improvement occurred in a third for an overall response rate of 21%. In the patient who achieved improvement, an antibody to leukocyte interferon developed during the course of treatment. This almost certainly precluded a more complete response (see below for discussion of antibody to leukocyte interferon).

2. <u>Prostate Cancer</u>. Among 10 patients with metastatic prostate cancer, subjective or objective improvement occurred in four. One patient achieved a complete remission. Three others achieved improvement in pain, a decrease in acid phosphatase, and/or regression of soft tissue disease.

3. <u>Renal Cell Carcinoma.</u> During the last year, partially purified human leukocyte interferon was administered intramuscularly at a dose of 3×10^6 units/day to 19 patients with metastatic renal cell carcinoma (17, 18). Five patients (26%) showed partial responses; two patients (10.5%), objective minor responses, three patients (16%), mixed effects (evidence of biological effect with regression of some lesions but concomitant progression); two patients (10.5%), disease stabilization; and seven patients (37%), progressive disease. All tumor responses were seen in lung or mediastinal metastases. Tumor response significantly correlated with interferon-induced leukopenia and granulocytopenia ($p < 0.04$ and < 0.01, respectively) and with pretreatment performance status ($p < 0.01$). No correlation was found with serum interferon levels. Antibodies to interferon were found in one patient prior to treatment.

4. <u>Juvenile Laryngeal Papillomatosis</u>. Fourteen patients, ages 22 months to 30 years (median 9 years) with aggressive juvenile laryngeal papillomatosis were entered into a non-randomized clinical study. All patients were objectively evaluated and treated by conventional endoscopic microsurgical and laser methods at two month intervals. The dosage of leukocyte interferon was 2 million units/m^2 daily. The treatment has been carried out for at least 6 months in 11 patients. Of these patients, there was one complete remission which has been maintained for over 6 months. Seven additional patients have had at least 50% or greater objective decreases in measurable disease. In at least five of these patients, reduction of dose to 2 mg/m^2 3 times weekly has been associated with slight to moderate regrowth of tumor,

suggesting that a maintenance schedule of 3 times weekly may
not be optimal to achieve control of the disease.

D.   Antibodies to Leukocyte Interferon

During the course of clinical investigations of the buffy
coat-derived leukocyte interferon from Finland, we have found
the presence of neutralizing IgG antibodies to interferon in
three patients (19).   In two of these patients, antibodies to
leukocyte interferon was detected in serum prior to clinical
use of the interferon.   In the third patient, an antibody developed during the course of treatment for colon cancer.
Mr. S.K. was a 46-year-old man in excellent health who
developed severe abdominal pain March 12, 1980.   He was
evaluated and found to have an obstructive lesion of the sigmoid colon.   A diverting colostomy was done on March 12, 1980.
Within three days after the colostomy had been performed, he
had a primary sigmoid resection.   A diagnosis of adenocarcinoma of the colon was made.   At surgery, he was found to have
multiple perioneal implantations as well as metastases on
the surface of the liver.   He was referred to the University of
Texas M.D. Anderson Hospital and Tumor Institute at Houston on
April 16, 1980.   He was found to have liver and splenic metastases on the CT scan of the abdomen.   On the 19th of April he
was started on partially purified leukocyte interferon, 3
million units a day by the intramuscular route.   On May 30th
a followup CT scan of the abdomen demonstrated a substantial
decrease in liver metastases and complete disappearance of the
splenic metastases.   He was considered to have achieved a
definite antitumor response and the dose of interferon was increased to 6 million units a day on May 30th.   On the 24th of
July, the CT scan showed continued regression of the liver
metastases.   At this point, the patient complained of severe
fatigue and the interferon was stopped for a three-week period.
The interferon treatment was reinstituted August 15, at a dose
of 6 million units a day.   Within three days, he developed
abdominal pain.   Toward the end of August, surgery was performed and he was found to have a resolving sub-diaphragmatic
abscess as well as peritoneal implants of the tumor.   During
September, October and November of 1980 he had continued pain
and increasing growth of tumor.   He was given one cycle of 5-
fluorouracil chemotherapy with no success.   During this period
of time, he was taking large doses of morphine for control of
pain.   By the middle of January 1981 the morphine had no
effect on the pain and he died February 4, 1981 with progressive metastatic colorectal cancer.   Because of the patient's
refractoriness to the analogous effect of morphine, and the

recent suggestion of immune cross-reactivity to endorphins
and interferon (20), his serum was evaluated for the development of antibodies to interferon.

The patient achieved a maximal serum interferon titer of
160 units/ml after the initial IM injection of 3×10^6 units.
The maximal serum interferon titer after the initial 6×10^6
unit dose was 320 units/ml. The patient's post-interferon
serum obtained just prior to his death contained neutralizing
IgG antibodies to leukocyte interferon. This is the first
report of the development of an antibody to leukocyte interferon in à patient undergoing treatment. In addition, the
presence of neutralizing antibodies to leukocyte interferon
prior to clinical exposure was observed for the first time in
a patient with breast cancer and a patient with metastatic
renal cell carcinoma. These results will be reported in
detail elsewhere (10).

E. Clinical Studies with Recombinant
Leukocyte Interferon (IFLrA)

On January 15, 1981 we initiated the first clinical study
with a recombinant interferon protein (21). We report on the
pharmacokinetics, single dose tolerance and biologic activity
of IFLrA in cancer patients. In addition, a comparative
analysis of the pharmacology and tolerance of recombinant
interferon and partially purified buffy coat leukocyte
interferon was made at selected doses. The results are also
presented elsewhere.

1. Materials and Methods. The isolation, expression and
purification of IFLrA were done by Hoffmann La Roche, Inc.
and Genentech and have been described (22-24). The plasmid
containing an entire leukocyte interferon gene was identified
and engineered in E. coli in collaboration with Genentech
(22,23). The purification of IFLrA was done using a specific
monoclonal antibody column to leukocyte A interferon (24,25).
The purified protein was homogeneous by sodium dodecyl sul-
fate-polyacrylamide gel electrophoresis. Material was more
than 95% pure; the specific activity was 2 to 4×10^8 units/
mg of protein when tested on AG1732 (human fibroblast) or
MDBK (bovine kidney) cell lines. Molecular weight was estimated to be 19,200 daltons (26). Cultures of the material
for bacteria and mycoplasma were sterile, and a limulus test
for endotoxin was negative.

Several biological activities of this purified recombinant
interferon were seen; the recombinant interferon showed equi-

valent in vitro antiviral and antiproliferative activity compared to crude and purified natural leukocyte interferons (22,27). Pre-clinical in vivo testing showed antiviral activity comparable to human leukocyte interferon and was safely used in rabbits and squirrel monkeys (22). The interferon was provided in ampules of 3×10^6 units/ml and 18×10^6 units/ml. The final freeze-dried preparation contained human albumin and was reconstituted immediately before use.

Partially pure human leukocyte interferon (IFN-C) was prepared as described by Cantell and Hirvonen (28). The specific activity was approximately 1×10^6 units/mg protein.

A total of 16 patients were treated, eight at The University of Texas Cancer Center, M.D. Anderson Hospital and Tumor Institute, and eight at Stanford University Medical Center. Each patient received an initial intramuscular injection of 3×10^6 units of interferon with doses escalating to 198×10^6 units, as tolerated. A minimal interval of at least 72 hours followed each dose to permit clinical and laboratory values to return to baseline and to evaluate each dose independently. At the two lowest doses, there was a cross-over evaluation of IFLrA and IFN-C to permit comparative analysis of pharmacokinetics, single-dose tolerance and biologic effects.

2. Patient Selection and Experimental Design.

Patients in the study had advanced metastatic cancer and were considered incurable. They were ambulatory and had not had anticancer therapy for at least 4 weeks before entering the study.

Recombinant leukocyte A interferon was given intramuscularly into the deltoid muscle with a minimal interval of 72 hours between injections. Individual doses of 3, 9, 18, 36, 54, 72, 90, 108, 144, and 198 million units were given. At the two lowest doses, IFLrA was alternated with 3 and 9×10^6 units of IFN-C. Patients were given alternating sequential doses; thus, eight patients received IFLrA followed by IFN-C, and eight received IFN-C followed by IFLrA.

One of the initial objectives was to treat each patient with the maximal dose tolerated. Patients were treated on Mondays and Thursdays of each week. A minimal period of 72 to 96 hours was used as an interval between doses. Longer intervals were used depending on biologic effects and at the discretion of the senior investigators.

Patients were seen daily and monitored closely. Vital signs including heart rate, blood pressure, respiration, and temperature were monitored at 1, 2, 4, 6, 12, 18, and 24 hours after each dose. Each intramuscular site was examined locally

at 1/2, 1, 3, 6, 12, and 24 hours after injection. A physical
examination was done 6 days a week until the end of the study.
An electrocardiogram and chest roentgenogram were done before
the study and after the final dose. Urinalyses were done at
baseline and every 72 hours thereafter until completion of the
study.

     3. Laboratory Studies. A complete blood count with
differential, leukocyte count enumeration, reticulocyte count,
and platelet count was done every 24 hours. A sequential
multiple analysis of alkaline phosphatase, asparate transaminase (formerly called serum glutamic-oxalacetic transaminase), serum bilirubin, fasting blood sugar, blood urea
nitrogen (Bun), serum cholesterol, total protein, uric acid,
phosphorus, calcium, serum albumin, and lactic acid dehydrogenase (LDH) was done before the study and 24 hours after
each dose. Alanine transaminase (formerly serum glutamic-
pyruvic transaminase), serum creatinine, electrolytes, and
prothrombin time were measured before each dose.
     Quantitative changes in tumor size were evaluated in
physical examination, radiologic examination, and pertinent laboratory studies such as serum protein electrophoresis and
measurement of Bence Jones protein (8). The criteria for
responses have been described previously (8).

     Serum samples were collected at 0, 1, 2, 3, 4, 61, 8,
12, and 24 hours after doses of 3, 9, 36, 72, 108, and 198 x
10^6 units. Urine samples were collected 2 hours before the
dose and at 0 to 2, 2 to 4, 4 to 8, 8 to 12, and 12 to 24 hours
after treatment. Serum interferon titers were measured at
Hoffman-LaRoche, Inc., by a modified bioassay using MDBK cells
as targets and vesicular stomatitis virus as described
previously (29). Interferon titers are expressed as recipro-
cals, with the dilutions producing a 50% reduction of virus
cytopathic effect. All samples were corrected to the standard
GO 23901-527 reagents from the National Institutes of Health,
Bethesda, Maryland.
     An enzyme-linked immunoassay using two monoclonal anti-
bodies to IFLrA was done at Hoffman-LaRoche, Inc. (Spiegel H.
Unpublished data). In principle, the assay depends on
stoichiometric binding to interferon by two monoclonal antibodies (25), one of which is covalently linked to peroxidase.
The peroxidase reacts with o-phenylenediamine in the presence
of peroxide. The color produced was read in a spectrophotometer at 492 mm; details of the methods used will be reported
separately. The development of antibodies to IFLrA and IFN-C
was tested using a modification of the bioassay in which

neutralization of the antiviral effect of interferon containing samples were measured.

4. <u>Results.</u> The patients ranged in age from 32 to 65
years; eight were men and eight women. Five patients had
malignant lymphoma, three had adenocarcinoma of the breast,
four had multiple myeloma, and one each had adenocarcinoma of
the ovary, malignant melanoma, adenocarcinoma of the colon,
and chronic myelogenous leukemia. Twelve of the patients had
previously received chemotherapy or hormonal therapy or both.
Four patients had previously received partially purified
buffy coat leukocyte interferon, and two previously had not
had systemic therapy. The number of IFLrA injections ranged
from 6 to 10, with a median of 8. The maximum dose given
ranged from 72 to 198 million units, and the period of treatment ranged from 29 to 92 days. The cumulative dose of IFLrA
ranged from 192 million units to 750 million units.

The pharmacokinetics and biopharmaceutical measures of
IFLrA were evaluated in 16 patients. Bioavailability of IFLrA
was the primary pharmacokinetic variable evaluated in this
study. Due to the limited sampling and variability of data,
the clearance rate and the volume of distribution could not
be evaluated. Not all subjects received each dose, but five
to 16 patients were included at each dose level. Figures 1
and 2 show the mean serum concentrations as measured by the
enzyme immunoassay (in picograms per millilitre) and the
bioassay (in units per millilitre) respectively, up to 48
hours after representative doses. The enzyme immunoassay
is a research tool and direct correlation with the bioassay
results remain to be proved.

In general, mean maximum observed serum concentration
(c_{max}) and area under curve (AUC) values increased with in-
increasing doses. When c_{max} and AUC were normalized for the
dose given and represented as $pg/ml/10^6$ unit dose or $units/ml/
10^6$ unit dose, there was dose proportionality. The time of
the maximum observed serum concentrations (t_{max}) tended to increase slightly with increasing doses up to 72×10^6 units.
The t_{max} varied from 4.6 ± 2.5 hours at the 3 million units
IFLrA dose to 6.1 3.6 hours at the 72 million unit IFLrA
dose. Above this level, two injection sites were used and the
trend disappeared. This result is consistent with a depot
effect on the intramuscular route of the injection site. The
half-life ranged from 6 to 8 hours regardless of the dose.

Figure 3 shows the arithmetic mean serum concentrations of
IFLrA and IFN-C at doses of 3×10^6 units and 9×10^6 units,
respectively, as measured by the modified bioassay. The only
statistically significant difference between the IFLrA and

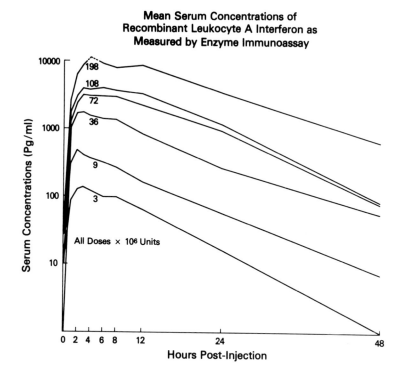

**Mean Serum Concentrations of
Recombinant Leukocyte A Interferon as
Measured by Enzyme Immunoassay**

FIGURE 1. The arithmetic mean serum concentrations of interferon as measured by the enzyme immunoassay. The numbers of patients measured at 3, 9, 36, 72, 108, and 198 million units are 16, 16, 16, 16, 14, and 5, respectively.

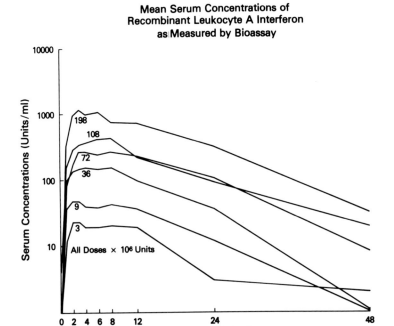

Mean Serum Concentrations of
Recombinant Leukocyte A Interferon
as Measured by Bioassay

FIGURE 2. The arithmetic mean serum concentrations of
interferon as measured by the bioassay with MDBK cells as
target cells. The numbers of patients measured at 3, 9, 36,
72, 108, and 198 million units are 16, 16, 16, 16, 14 and 5,
respectively.

IFN-C value is the AUC at 9×10^6 units, where the IFN-C value is significantly larger than the IFLrA value, $p < 0.05$ level. The half-lives of elimination were calculated at 9×10^6 unit doses and were 7.3 and 8.2 hours for IFN-C and IFLrA respectively.

The maximal tolerated dose varied from 72 to 198 million units. Five patients reached the maximal level of 198 million units. Interferon was discontinued in four patients because of severe and prolonged fatigue and in three patients

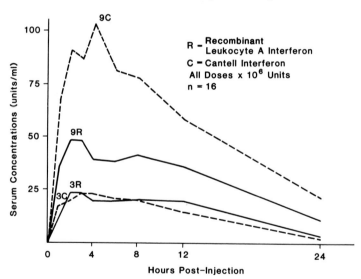

FIGURE 3. The comparative serum levels of 16 patients treated with IFLrA and IFN-C as measured by a modified bioassay at 3 and 9 million units.

because of numbness of the hands or feet, or both. Interferon was discontinued in the other four patients due to
personal reasons unrelated to side effects.

Recombinant leukocyte A interferon and IFN-C at 3 and 9
million units, respectively, produced virtually the same
clinical side effects. The symptoms of fever, chills, myalgias, and headache occurred after almost all doses. Fever
occurred after virtually 100% of interferon injections, but
there was no correlation between increasing febrile response
and increasing interferon doses. Fever generally began 2 to
6 hours after an injection of interferon and peaked at 6 to
12 hours; in almost all patients, it resolved spontaneously
within 24 hours. Some patients received acetaminophen for
fever during the first 8 to 12 hours after each dose of interferon. Headache, chills, and myalgias occurred frequently
after a dose of interferon. These symptoms usually resolved
spontaneously within 24 hours and did not show a relation to
dose response. Antihistamines frequently relieved the
headaches.

Fatigue was a common complaint when interferon was given
in higher doses. With increasing doses, fatigue became more
severe and persisted from 7 to 21 days after individual
interferon doses. In four of 16 patients, fatigue was the
dose-limiting factor, usually after escalation of the dose to
108 million units.

The development of mild numbness and paresthesias of the
hands or toes or both occurred in eight patients. In general,
these effects occurred at doses of 72 million units and above,
although two patients experienced transient symptoms at lower
doses. The experience at The University of Texas, M.D.
Anderson Hospital and Tumor Institute and Stanford University
Medical Center was different. In five patients at M.D.
Anderson Hospital, numbness and paresthesias occurred in the
hands and toes. These symptoms persisted for 2 to 14 days
after individual doses in 4 of the 5 patients. In the other
patient, the symptoms lasted for a few minutes only. In the
three patients at Stanford, numbness and paresthesias were
present only in the hands. Only one patient had this sensa-
tion on more than one occasion, and it lasted for seconds.

Objective decrease in sensory perception of the finger-
tips was noted by a neurologist in two patients. This finding was attributed to a mild sensory peripheral neuropathy.
Nerve conduction studies and electromyograms were done at
The University of Texas, M.D. Anderson Hospital; findings
were normal in all instances. Interferon was discontinued
in three patients because of this side effect. All cases of
numbness and paresthesias resolved completely after discon-

tinuation of the interferon. There was no correlation between these symptoms and previous treatment with agents known to cause neurotoxicity.

Occasional side effects reported in one or two patients with a few doses were tightness of the chest or throat, light-headedness, dizziness, nasal congestion, cold or pale hands, dryness of the neck and face, bad taste in the mouth, and trembling. Gastrointestinal symptoms, including anorexia, nausea and diarrhea, tended to occur at the higher dose levels. Minimal alopecia developed in one patient. Weight loss ranging from 1.4 to 7.3 kg (median value = 3.1 kg) was a common occurrence in this study. Lesions indicative of recurrent herpes simplex developed in three patients after the initial one to three injections.

The patient with chronic myelogenous leukemia was excluded from hematologic analysis. Total leukocyte count in 13 of the 15 patients decreased in comparison to the baseline value measured before each dose. The progressive decrease in the total leukocyte count was analyzed for each patient after each dose was given. A proportionately greater decrease from the baseline value was seen with higher doses. When patients were divided into hematologic and nonhematologic groups, only patients with hematologic malignancies showed this dose response ($p < 0.05$). Because of the design of the study, the decrease could be related to the cumulative effect of many doses or to increasing dose levels or both.

A minimal decrease in the hematocrit values and platelet counts was observed; however, the decrease in hematocrit values (average of 6% per patient during the study) could have been accounted for, in part, by the phlebotomy requirements.

The effects of interferon on blood chemistries were minimal. Seventeen percent of the patients had elevated lactic dehydrogenase (LDH) levels during the study, but these levels returned to their baseline values before the next dose was given. Levels of alanine transaminase rose after only 6% of the injections. Most of these effects were caused by one patient who was abusing alcohol during the study. No effects on serum creatinine and BUN were seen.

Three of 16 patients (patients 1, 7 and 14) developed antibodies to IFLrA. Antibody was not present before the study and was undetected while the patients were receiving interferon. Patient 1 had a detectable antibody level of 1:7.5 10 days after receiving her last dose of interferon. Six months later, the antibody titer had risen to 1:30; however, during this period the patient was treated for 4 weeks with IFN-C. Patient 7 had a titer of 1:20 10 days after the end of the study; this titer has remained stable 2 months after the end of study. Patient 14 developed a titer of 1:30 3 weeks

after her last dose, and this titer has remained stable
1 month after study. Titers of 1:7.5, 1:20, and 1:30 are
sufficient to neutralize in vitro 150,400, and 600 units/ml
of interferon, respectively. Preliminary evidence indicates
that the antibody is of the IgG class.

Seven of the 16 patients showed objective evidence of
tumor regression during this study. Among the five patients
with nodular poorly differentiated lymphocytic lymphoma, one
had a partial remission, three had minor responses, and one
had a mixed response. The partial remission occurred in
patient 15, who had previously had a complete remission using
IFN-C. Two of the minor responses occurred in patients 1 and
2 at The University of Texas, M.D. Anderson Hospital; they had
previously been treated with a 28-day course (total of 84
million units) of IFN-C without response. Patient 1 had shown
progression 1 month before entering the study and the condition
of patient 2 had remained stable for 1 year before the study.
Both patients showed objective evidence of tumor regression on
radiologic examination of abdominal lymph nodes: patient 1
showed a 30% to 40% decrease in tumor-involved lymph nodes;
in patient 2, several pathologically involved lymph nodes
disappeared, while others remained stable or decreased less
than 50%. One patient with chronic myelogenous leukemia
showed a reduction of peripheral leukocyte count from 22,000
to 12,300/mm^3 and a reduction of more than 50% in spleen size.

F. Discussion

Recent clinical investigation has clearly demonstrated
that regression of several types of human malignancy can be
achieved by partially pure preparations of leukocyte interferon (8-10, 12, 18). This work opens up a new area of therapeutic research for human malignancy, namely the use of biological agents that can directly influence cellular growth.
With the exception of the use of corticosteroids in limited
forms of hematologic malignancies and the use of hormones in
sensitive tumors such as breast carcinoma and prostate carcinoma, this is the first demonstration of important biological
activity with natural antigrowth substances. Due to the
rapid progress in recombinant DNA technology, many forms of
interferon will now be available for in depth clinical
investigation.

Recent progress in recombinant DNA technology and
production of monoclonal antibodies has yielded a single
purified species of leukocyte interferon (22-25,30). This
study is the first clinical investigation of a purified recombinant DNA-produced interferon; previous clinical studies

of interferons have used preparations of approximately 1%
purity. With recent knowledge gained from recombinant DNA
techniques, we know that these preparations contained a
mixture of eight or more species of leukocyte interferon
(31-33).

This investigation has described the pharmacologic
profile, single dose tolerance, and biologic effects of
IFLrA. Recombinant leukocyte A interferon is well absorbed
after intramuscular injections, resulting in serum concentrations similar to those previously reported for buffy-coat-
derived leukocyte interferon (5, 8, 10). For the first time
in clinical investigation, an enzyme immunoassay has been used
to measure a single antigenic species of leukocyte interferon with concentrations expressed in a weight per volume
basis. The development of monoclonal antibodies to leukocyte
interferon (25) has led to the production of an enzyme immunoassay for the detection of IFLrA. The enzyme immunoassay is
still a research tool, and a direct correlation between IFLrA
and the bioassay is not possible at this time. However, the
ability to measure serum levels on a weight per volume basis
was useful for monitoring interferon levels in this study and
should improve the specificity, sensitivity, and reproducibility of measuring interferon levels in future clinical
studies.

Blood levels 10 to 20 times higher than those previously
reported for buffy coat-derived leukocyte interferon were
achieved at the higher dose ranges (8). The pharmacologic
profile of the IFLrA was similar to that produced by the mixture of natural leukocyte interferon. Thus, the inability of
E. coli bacteria to glycosylate the interferon protein may
not be an important determinant in the pharmacologic behavior
of recombinant DNA-derived interferon. In addition, recent
evidence indicates that the major, naturally occurring, leukocyte interferon species are largely devoid of carbohydrate (26).

Antibody to IFLrA developed in three of 16 patients. Although no apparent subjective or objective clinical abnormalities were seen that could be attributed to this development, the precise significance of these antibodies must be
evaluated by future clinical studies. The degrees of neutralization detected in vitro in the current study indicates that
the in vivo neutralization of pharmacologic doses of interferon by an antibody is possible. Antibodies to partially pure
fibroblast interferon (34) as well as partially purified leukocyte interferon have been detected in patients treated with
these agents (19).

The clinical and laboratory side effects induced by IFLrA
were similar but not identical to those previously recorded
with buffy-coat interferon (8-10). Hematologic toxicity was

frequent, mild, and not dose limiting. The decrease in circulating leukocytes and granulocytes was rapidly reversible
during interval-dosing periods. The kinetics of leukocyte
recovery indicate that interferon may produce or cause cell
margination or redistribution rather than affect maturation
of myeloid stem cells (35). However, in vitro effects on
myeloid differentiation have been shown with both IFN-C and
IFLrA (36).

In contrast to the frequent increase in the alanine
transaminase in most patients treated with daily doses of
crude leukocyte interferon, the alanine transaminase of
patients in this study rarely rose. The lack of effect on
liver enzymes noted in this study may be due to the intermittent schedule or the qualitatively and quantitatively
different effect of this particular pure interferon species.

Interestingly, the acute side effects associated with
IFLrA are similar to those reported with viral infections.
The preclinical studies done with IFLrA in subhuman primates
did not predict a febrile response (22), suggesting that
other species may have to be evaluated for animal toxicology
studies of interferon. Although we cannot be absolutely
certain that the fever induced by the recombinant preparation
is due to the interferon molecule, there is no evidence
indicating the presence of residual E. coli protein or any
other contaminants. Similar pyrogenic as well as flu-like
responses recently have been reported after intramuscular injection of of a homogeneous interferon preparation purified from
human buffy-coat leukocytes (37). These data, therefore,
support the conclusion that the interferon molecule is responsible for the side effects noted in treatment.

The dose-limiting side effect in several patients was
attributed to fatigue, and further studies are needed to elucidate its mechanism. Although the frequency and severity of
fatigue increased with higher doses, it is not possible to
distinguish a true dose-response effect from cumulative
effects of the interferon. The significance of the numbness
and paresthesias in half the patients also must await further
studies. In at least two of these patients, the symptoms
were felt to represent a mild sensory peripheral neuropathy.
It should be noted that Calvert and Gresser reported that in-
vitro interferon can increase neuronal activation (38).

The similar clinical and laboratory toxicities produced
by IFLrA and IFN-C indicate that purification and synthetic
production methods do not alter the basic biologic activities
intrinsic to the interferon molecule itself.

Previous reports have clearly shown the ability of impure
preparations of leukocyte interferon to induce regression of
metastatic tumor in breast cancer, multiple myeloma, malig-

nant lymphoma and renal cell cancer (8-10,17,18). Although this study was not designed primarily to measure antitumor activity, it is of great interest that patients receiving an intermittent schedule of increasingly high doses had clinical evidence of biologic activity. This activity suggests that the interferon molecule is responsible, at least in part, for the antitumor properties of crude or partially pure interferon preparations.

The development of recombinant technology to produce interferon proteins should greatly increase their use as potential antiviral and antitumor agents. Further Phase I, II, and III studies are now necessary to establish the role of this as well as other species of leukocyte interferon in the treatment of human viral diseases and cancer.

REFERENCES

1. Isaacs A., Lindemann, J. Virus interference: 1. The interferon. Proc R. Soc Lond [Biol.] 1957; 147:258-67.
2. Baron S., Dianzani F., eds. The interferon system: a current review to 1978. Tex Rep Biol Med. 1977; 35:1-573.
3. Gresser I. Antitumor effects of interferon in cancer. In: Becker F., ed. Cancer: A Comprehensive Treatise, Vol.5, New York: Plenum Press; 1977:521-71.
4. Paucker K. Cantell K., Henle W. Quantitative studies on viral interference in suspended L cells: III. Effect of interfering viruses and interferon on the growth rate of cells. Virology. 1962; 17:324-34.
5. Strander H., Cantell K., Carlstrom G., Jakobsson P.A. Clinical and laboratory investigations on man: systemic administration of potent interferon to man.
6. Strander, H.: Antitumor effects of interferon and its possible use as an antineoplastic agent in man. The Interferon System: A Current Review to 1978. Tex Rep Biol Med, 35:429, 1977.
7. Gutterman, J.U., H.-Y. Yap, A.U.Buzdar, R.Alexanian, E.M. Hersh, and F. Cabanillas: Leukocyte interferon (IF) - induced tumor regression in patients with breast cancer and B cell neoplasms. Proc Amer Assoc Cancer Res, 20:167, 1979.
8. Gutterman, J.U., G.R. Blumenschein, R. Alexanian, H.-Y. Yap, A.U. Buzdar, F. Cabanillas, G.N. Hortobagyi, E.M. Hersh, S.L. Rasmussen, M. Harmon, M. Kramer, and S.Pestka: Leukocyte interferon-induced tumor regression in human metastatic breast cancer, multiple myeloma and malignant lymphoma. Ann Int Med, 93:399, 1980.

9. Mellstedt, H., A. Ahre, M. Bjorkholm, G.Holm, B. Johansson and H.Strander: Interferon therapy in myelomatosis. Lancet, 1:245, 1979.

10. Merigan, T.C., K. Sikora, J.H. Breeden, R. Levy, and S.A. Rosenberg: Preliminary observations on the effect of human leukocyte interferon in non-Hodgkin's lymphoma. N Engl J Med, 299-1449, 1978.

11. Borden, E., T. Dao, J. Holland, J.U. Gutterman, and T.C. Merigan: Interferon in recurrent breast carcinoma: Preliminary report of the American Cancer Society Clinical Trials Program. Proc Amer Assoc Cancer Res, 21:187, 1980.

12. Borden, E., J. Holland, T. Dao, J.U. Gutterman, L. Weiner, Y.-C. Chang, and J. Patel: Leukocyte-derived interferon (a) in human breast carcinoma. In Press 1982.

13. Osserman, E.F. and W.H. Sherman: Preliminary results of the American trial of human leukocyte interferon (IF) in multiple myeloma (MM): Proc Amer Assoc Cancer Res. 21:161, 1980.

14. Osserman, E.F. and W.H. Sherman: Human leukocyte interferon (HuαIFN) in multiple myeloma (MM): The American Cancer Society (ACS) sponsored trial. In: The Biology of the Interferon S stem. De Maeyer, E., G. Galasso and H. Schellekens (eds.). Elsevier/North-Holland Biomedical Press, pg 409, 1981.

15. Mellstedt, H., A. Ahre, M. Bjorkholm, H. Strander, G. Brenning, L. Engtedt, G. Gahrton, G. Holm, B. Johansson, R. Lerner, B. Lonnqvist, B. Nordenskiold, A. Killander, A. M. Stalfelt, B. Simmonsson, B. Ternstedt and B. Wadman: Interferon treatment of human plasma cell myeloma. In: Recent Advances and Future trends in Myeloma. Durie, B.G.M., and Seligmann (Eds.). Elsevier/ North Holland. In Press.

16. Haglund, S., P.G. Lundqvist, K. Cantell, and H. Strander: Interferon theraphy in juvenile laryngeal papillomatosis. Arch Otolaryngol. 107:327, 1981.

17. Quesada, J.R., J.U. Gutterman, D. Swanson and A. Trindade: Antitumor effects of partially pure human leukocyte interferon in renal cell carcinoma. Proc Amer Assoc Cancer Res. In Press, 1982.

18. Quesada, J.R., J.U. Gutterman, D. A. Swanson and A. Trindade: Antitumor effects of Leukocyte interferon in patients with metastatic renal cell carcinoma. Ann Int Med, submitted, 1982.

19. Trown, P.W., M.J. Kramer, R.A. Dennin Jr., E.V. Connell, A.V. Palleroni, J.R. Quesada and J. U. Gutterman: Antibodies to natural and recombinant human leukocyte interferons (HuIFN-αs) in cancer patients. Submitted, 1982.

20. Blalock, J.E. and E.M. Smith: Human leukocyte interferon: Structural and biological relatedness to adenocortico-tropic hormone and endorphins. Proc Nat Aca Scie, 77:5972, 1980.

21. Gutterman, J.U., S. Fein, J. Quesada, et al. Recombinant Leukocyte A interferon: Pharmacokenetics, single dose tolerance and biologic effects in cancer patients. Ann Int Med. In Press, 1982.

22. Goeddel D.V., E. Yleverton, A. Ullrich, et al. Human leukocyte interferon produced by E. coli is biologically active. Nature, 1980, 287:411-16.

23. Maeda S., R. McCandliss, M. Gross, et al. Construction and identification of bacterial plasmids containing nucleotide sequence for human leukocyte interferon. Proc Natl Acad Sci USA. 1980; 77:7010-13.

24. Staehelin T., D.S. Hobbs, H. Kung, C.Y.Lai, S. Pestka: Purification and characterization of recombinant human leukocyte interferon (IFLrA) with monoclonal antibodies. J Biol Chem. 1981; 256: 9750-54.

25. Staehelin, T., B. Durrer, J. Schmidt, et al. Production of hybridomas secreting monoclonal antibodies to the human leukocyte interferons. Proc Natl Acad Sci USA. 1981; 78:1848-52.

26. Pestka, S., S. Maeda, T-R.Chiang, et al. Cloning and expression of human interferons in microorganisms. In: Tex Rep Biol Med 41, in press.

27. Evinger M., S. Maeda, S. Pestka: Recombinant human leukocyte interferon produced in bacteria has anti-proliferative activity. T. Biol Chem. 1981; 256:2113-14.

28. Cantell K., S. Hirvonen: Large-scale production of human leukocyte interferon containing 10 units per ml. J. Gen Virol. 1978; 39:541-3.

29. Rubinstein S., P.C. Familletti, S. Pestka: Convenient assay for interferons. J. Virol. 1981; 37:755-58.

30. Nagata S.H.H.Taira, H. Hall, et al. Synthesis in E.coli of a polypeptide with human leukocyte interferon activity. Nature. 1980; 284: 316-20.

31. Nagata S., N. Mantei, C. Weissmann: The structure of one of the eight or more distinct chromosomal genes for human interferon. Nature. 1980; 287:401-18.

32. Goeddel D.V., D.W. Leung, T.J. Dull, et al. The structure of eight distinct cloned human leukocyte inter-feron cDNAs. Nature. 1981; 290:2026.

33. Rubinstein M., W.P. Levy, J.A. Moschera, et al. Human leukocyte interferon: Isolation and characterization of several molecular forms. Arch Biochem Biophys. 1981; 210:307.

34. Vallbracht A., J. Treuner, B. Flehmig, K-E. Joester,
     D. Niethammer: Interferon-neutralizing antibodies in a
     patient treated with human fibroblast interferon.
     Nature. 1981; 289: 496-97.
35. Verma D.S., G. Spitzer, J.U. Gutterman, A.R. Zander,
     K.B.McCredie, K.A.Dicke: Human leukocyte interferon
     preparation blocks granulopoietic differentiation.
     Blood. 1979; 54:1423-27.
36. Verma, D.S., G. Spitzer, J.U.Gutterman, et al. Homogene-
     ous preparations of natural and recombinant human leuko-
     cyte interferon block granulopoietic differentiation;
     manuscript submitted.
37. Scott, G.M., D.S. Secher, D. Flowers, J. Bate, K. Cantell,
     D.A.J. Tyrrell: TOxicity of interferon. Br Med J. 1981;
     282: 1345-8.
38. Calvet M.C., I. Gresser: Interferon enhances the excit-
     ability of cultured neurons. Nature. 1979; 278:558-60.

DISCUSSION

H. YOUNG: At the beginning of treatment did any patients have concurrent viral infections such as recurrent herpes, and if so, did the interferon have any effect?

J. GUTTERMAN: We have not investigated this phenomenon with the cloned material. However, we have observed some interesting and dramatic effects on the concurrent infections in several cancer patients given partially pure interferon. In one side effect that I failed to mention, activation of herpes simplex following the first dose of interferon was observed. This was noticed only when interferon was administered at lower doses. This paradoxical effect may be related to the fever or perhaps to prostaglandin release, etc. At lower doses interferon may not be sufficiently antiviral to control the herpes. Thus, the first lower dose of interferon may actually activate viral infection. Studies are proceeding throughout the country to test the effect of interferon on concurrent viral infections.

P.A. LIBERTI: Why did you choose to administer interferon intramuscularly? I would have thought that I.V. administration might decrease the chances of getting an antibody response as well as providing a more direct route to the tumors.

J. GUTTERMAN: We injected interferon intra-muscularly since the primary experience with the partially pure material has been via this route. You get a much slower clearance. On the other hand, you get very rapid clearance when it is injected I.V. I'm not sure that the I.V. route would be better or would preclude production of antibodies. The anti-proliferative data suggest that a more sustained blood level might be better for the anti-tumor activity of the interferon. Interferon is tolerated quite well when administered I.M. For convenience purposes, it's much better to go I.M. because eventually these patients might self administer interferon much like some diabetics administer insulin. Studies are now proceeding at the Memorial Sloan-Kettering Institute comparing the pharmacokinetics of I.V. versus I.M.

J. CAHILL: The interleukins and the lymphokines have been shown to have a net negative nitrogen effect as part of their general response. Is the fatigue in patients receiving interferon associated with muscle wasting and net negative nitrogen balance?

J. GUTTERMAN: We did see weight loss and we are planning to study the metabolic consequences. There is some decrease in appetite of these patients with concomitant G.I. disturbances. We are now beginning to study the caloric intake, etc. Weight loss has been observed both with the partially pure and pure interferon preparations. I think a lot of attention has to be paid to the effects on muscle, etc. Besides weight loss we have not seen any specific changes in muscle. We've done some muscle testing but the reason for the fatigue is not clear at the moment.

D. BURKE: Experience in Britain has suggested that some of the side effects of interferon are due to the prostaglandin pathway and could be supressed with indomethacin. Have you done, or do you plan to do any trials using indomethacin?

J. GUTTERMAN: You may be familiar with the work by Dr. Cuatrecasas and his group at the Research Triangle Park (Burroughs-Welcome) in which inhibitors of fatty acid cyclo-oxygenase (prostaglandin synthetase) seem to decrease the anti-viral, and possibly the anti-proliferative activity. In addition, an L1210 line of leukemia, resistant to the type 1 interferon, seems to lack this particular enzyme. So use of compounds that inhibit prostaglandin synthesis has to be carefully considered. I know from our previous experience that patients feel better when given agents such as indomethacin. However, we must realize that we may be destroying or decreasing anti-viral, anti-tumor activity of interferon.

V. CHOWDHRY: You mentioned three patients developed low antibody titers. Did these three belong in the set of non-responders or were they part of another set?

J. GUTTERMAN: Two of the three were non-responders. One of the patients with the lowest titer, is a patient with lymphoma who had previously been exposed to the partially pure interferon. Her lymphoma did respond. Based on her serum titers the antibody did not appear until late in the study. We noticed, with the last dose of interferon which was a repeat dose of 108 million units, that her serum titers were dramatically lower in comparison with the previous dose. I think her antibody didn't appear till the end of the study. The other two patients had very good titers up until the end of the study. Therefore, I think that the antibody in these three patients appeared late, toward the end of the study.

THE USE OF BIOSYNTHETIC
HUMAN INSULIN IN MAN

John A. Galloway
Mary A. Root

Lilly Laboratory for Clinical Research
Eli Lilly and Company
Indianapolis, Indiana
U.S.A.

In the family of significant discoveries in human biochemistry and endocrinology, insulin is like a firstborn child. Not only has the birth or identification of this unique substance been the result of the fortuitous combination of research advances, like the firstborn child in a family it has been the source of experience, the model, by which the parents or researchers have acquired skills which have facilitated the understanding of the siblings next to arrive. Thus, insulin was the first substance to be administered as replacement therapy in a deficiency disorder, diabetes mellitus,[1] the first peptide hormone to be sequenced,[2-4] and among the first peptide hormones to be produced by recombinant technology.[5-7] In this paper we discuss insulin as the first product of recombinant research to be studied extensively in clinical tests, presenting clinical pharmacologic data generated at our own institution in normal subjects, with a brief discussion of difficulties encountered in the radioimmunoassay (RIA) for human and pork insulin in the serum of volunteers who received these insulins. Additionally, we present preliminary information on patients who have never received insulin and have been treated with biosynthetic human insulin (BHI) for six months. Finally, we speculate on the therapeutic potential of human proinsulin which soon will be introduced into limited clinical trial. The BHI used in the studies described was all derived from human A and human B chain material produced by recombinant techniques and combined and purified by chemical methods.[7]

FROM GENE TO PROTEIN:
TRANSLATION INTO BIOTECHNOLOGY

We have administered Neutral Regular Insulin
(NRI) and NPH (Neutral protamine Hagedorn,
Isophane) Insulin to normal fasting
volunteers.[8] Purified pork insulin (PPI) has
been used as a control agent. In all studies the
total nitrogen, amino acid nitrogen, and rabbit
hypoglycemic potencies of the BHI and PPI were
virtually identical. The routes of
administration and doses utilized are summarized
in Table 1.

In Figures 1, 2, and 3 are presented the plasma
glucose responses to Neutral Regular BHI and PPI
administered intravenously and subcutaneously.
There are no statistically significant
differences in the glycemic profiles after
Neutral Regular BHI or PPI administered
intravenously or subcutaneously or after human or
porcine NPH Insulins given subcutaneously.

While analysis of the plasma glucose responses
was straightforward, the serum insulin data
generated in these studies contained
inconsistencies which we only recently have begun
to resolve. For instance, although in our
studies the serum insulin concentrations (S.I.C.)
after intravenously administered BHI and PPI were
identical, following subcutaneous administration
the peak insulin concentrations of BHI ranged
from 30 to 100 percent above those for PPI. The
same phenomenon occurred with NPH BHI and PPI but
was less striking. Suspecting that these
differences might be due to the RIA we examined
carefully the standard curves for human and pork
insulins produced with the antiserum used for
these measurements.[9]

Figure 4 presents the mean standard curves for
porcine and human insulins generated by the RIA
in which the S.I.C. were measured for these
studies in normal subjects. The same reagents
were used for all assays in these studies and
each assay contained a standard curve of porcine
insulin and a standard curve of BHI. Although
there were small differences in the relationship

Table 1 - Clinical Pharmacologic Studies with
          BHI Performed in the United States

Study No. 1

    BHI 0.1 U/kg I.V.
    PPI 0.1 U/kg I.V.

    BHI 0.1 U/kg S.C.
    PPI 0.1 U/kg S.C.

    Same experiments repeated using 0.15 U/kg

Study No. 2

    BHI 0.1 U/kg S.C.
    PPI 0.1 U/kg S.C.

    BHI NPH 0.1 U/kg S.C.
    PPI NPH 0.1 U/kg S.C.

    Same experiments repeated using 0.15 U/kg

Figure 1 -

The Plasma Glucose Response of Normal Fasting
Subjects to Biosynthetic Human Insulin and
Purified Pork Insulin 0.1 and 0.15 units/kg
Intravenously

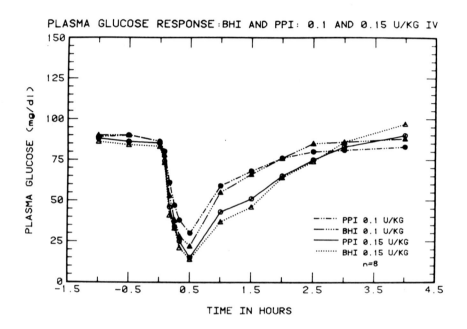

Galloway, J. A., Spradlin, C. T., Root, M. A., and
Fineberg, S. E.: The Plasma Glucose Response of
Normal Fasting Subjects to Neutral Regular and NPH
Biosynthetic Human and Purified Pork Insulins,
Diabetes Care, 4:185-88 (1981)

Figure 2 -

The Plasma Glucose Response of Normal Fasting
Subjects to Biosynthetic Human Insulin and to
Purified Pork Insulin 0.1 and 0.15 units/kg
administered subcutaneously

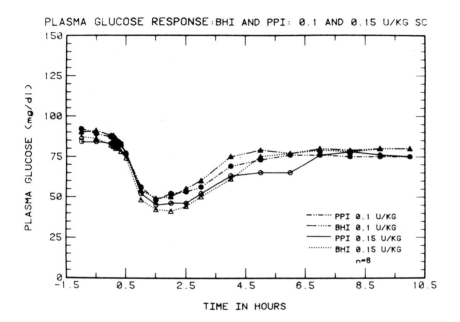

PLASMA GLUCOSE RESPONSE:BHI AND PPI: 0.1 AND 0.15 U/KG SC

Galloway, J. A., Spradlin, C. T., Root, M. A., and
Fineberg, S. E.: The Plasma Glucose Response of
Normal Fasting Subjects to Neutral Regular and NPH
Biosynthetic Human and Purified Pork Insulins,
Diabetes Care, 4:185-88 (1981)

Figure 3 -

The Plasma Glucose Response of Normal Fasting
Subjects to NPH Biosynthetic Human Insulin and
Purified Pork Insulin 0.15 units/kg Administered
Subcutaneously.

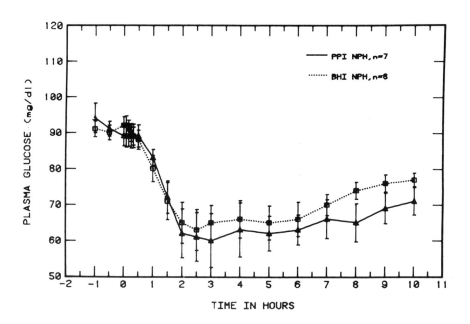

Galloway, J. A., Spradlin, C. T., Root, M. A., and
Fineberg, S. E.:   The Plasma Glucose Response of
Normal Fasting Subjects to Neutral Regular and NPH
Biosynthetic Human and Purified Pork Insulins,
Diabetes Care, 4:185-88 (1981).

Figure 4 -

Standard curves for radioimmunoassay of purified pork insulin and biosynthetic human insulin with a pool of guinea pig antisera to porcine insulin. Percent BOUND represents percent B/B_0. The ^{125}I-porcine insulin used in these studies was monoiodinated at tyrosine A_{19}.

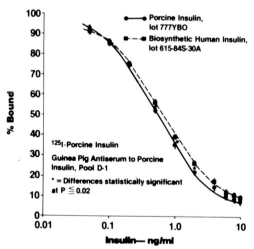

Standard Curves

Mean ± S.E.M. for 28 Assays

The difference in slopes magnifies the difference in unknown samples measured at higher concentrations.

between the two standards (BHI and PPI) from
assay to assay, the curve for BHI always fell to
the right of that for PPI and the differences
were greatest at the higher concentrations.
These differences are clearly demonstrated in
Table 2. Two or more of five different serum
samples were included in each assay as internal
controls. Samples measuring less than 0.5 ng/ml
gave virtually identical concentrations whether
read from the pork or the human standard curve.
Samples containing higher concentrations of
insulin measured significantly higher when read
from the human insulin curve than they did when
read from the porcine insulin curve. This
antiserum pool, D-1, clearly differentiated
between porcine and human insulins to such an
extent that comparison of serum concentrations of
these two insulins in circulating blood is not
possible.

In order to develop an assay which measures human
and pork insulin reliably over a wide range of
serum concentrations, other antisera have been
examined. For example, in Figure 5 the standard
curve for a pooled antiserum from guinea pigs
immunized with pork insulin is demonstrated. It
will be noted that this antiserum at
concentrations of 0.1 to 0.8 ng/ml recognizes
pork better than human insulin. However, at the
higher S.I.C. this pork antiserum actually has a
greater affinity for human insulin. On the other
hand, in Figure 6 it will be seen that a pooled
antiserum developed in guinea pigs immunized with
pancreatic human insulin recognizes human better
than pork insulin over the full range of serum
concentrations. There are two reports[10,11] in
which S.I.C. after BHI were found to be higher
than after PPI, leading to the suggestion that
BHI is better absorbed than PPI. Our findings
highlight the necessity for investigators'
reporting S.I.C. after BHI and PPI indicate the
binding characteristics of the antiserum used in
their immunoassay systems.

After considerable testing we believe we have
identified an antiserum which binds BHI and PPI

TABLE 2

Human Serum Controls

Sample	Number of Assays	Serum Insulin— ng/ml measured by	
		Porcine Insulin Standard	Human Insulin Standard
A	8	0.21 ± 0.02	0.22 ± 0.03
B	19	0.54 ± 0.03	0.64 ± 0.03
C	18	1.00 ± 0.07	1.37 ± 0.11
D	18	0.41 ± 0.04	0.44 ± 0.05
E	18	0.61 ± 0.03	0.75 ± 0.04

Figure 5 -

Standard curves for radioimmunoassay of purified
pork insulin and biosynthetic human insulin with
a pool of guinea pig antisera to porcine insulin.
Percent BOUND represents percent B/B_o. The
^{125}I-biosynthetic human insulin used in these
studies was monoiodinated at tyrosine A_{19}.

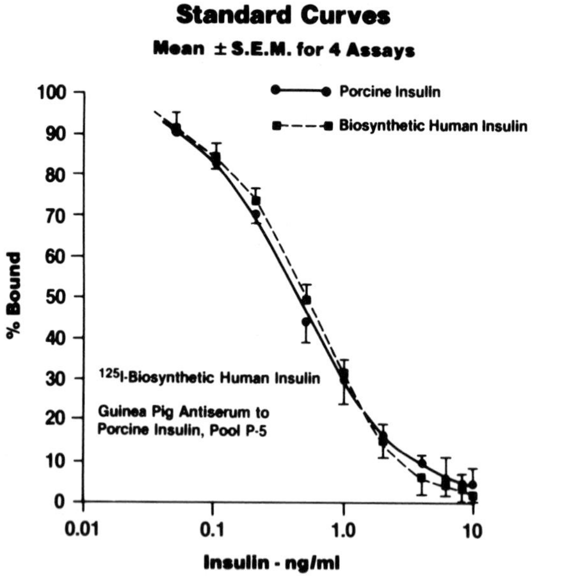

Standard Curves

Mean ± S.E.M. for 4 Assays

The differences in slope with this antiserum produces the greatest differences in
measurements in the upper part of the curves.

Figure 6 -

Standard curves for radioimmunoassay of purified pork insulin and biosynthetic human insulin with guinea pig antiserum to pancreatic human insulin produced in a single animal. Percent BOUND represents precent B/B_0. The ^{125}I-biosynthetic human insulin used in these studies was monoiodinated at tyrosine A19.

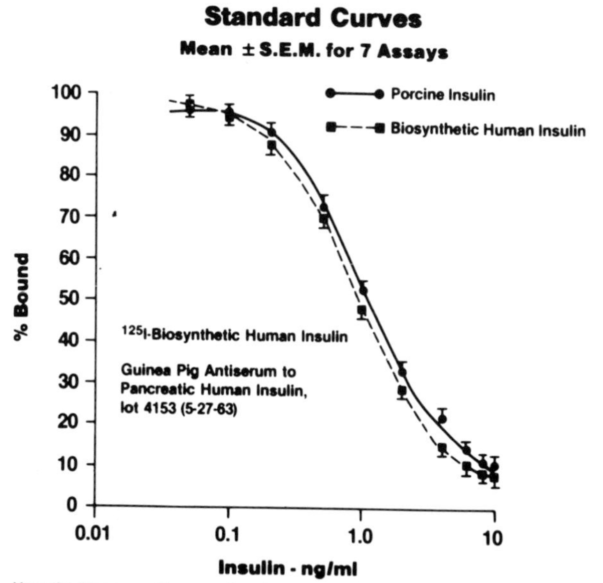

Standard Curves
Mean ± S.E.M. for 7 Assays

Here the slopes are the same but the curves for biosynthetic human insulin fall to the left of that for porcine insulin.

equally. This will permit a complete
re-evaluation of the S.I.C. data we have
generated to date. In addition, we have
requested samples of antisera used by
investigators who have reported on BHI and plan
to compare our standard curves of PPI and BHI
with these antisera. We anticipate that these
two approaches will tell us whether increased
S.I.C. after BHI in comparison to PPI is a real
or apparent difference. We should add that even
if S.I.C. for BHI are found to be higher after
BHI than PPI, the clinical significance of these
findings will not be known until a substantial
number of diabetics have been treated.[12]

New Patient Studies

Ninety-eight diabetics from six private patient
care centers in the U.S. who had never received
insulin have been treated with BHI for six
months. Thirty-eight had Type I diabetes and 60
had Type II diabetes. The breakdown of the study
group with respect to gender is demonstrated in
Table 3. The pretreatment and subsequent mean
glycosylated hemoglobin values and post-treatment
insulin dose in units/kg are presented in
Table 4. It will be noted that insulin treatment
effectively reduced glycosylated hemoglobin.
However, the mean value for both Type I and Type
II patients was still elevated, even after six
months of insulin treatment. These
glycohemoglobin values and insulin doses are,
however, the same as or slightly less than those
we have observed in a similar study with animal
insulins involving over 500 patients.

Of interest is the fact that there were no
reports of local or systemic allergy to insulin
or of insulin lipoatrophy. There were two
patients who developed insulin hypertrophy, or
mounding of subcutaneous tissue at the injection
site. This occurs in patients who receive only
the most pure of the available pork insulin
preparations and is probably due to local
lipogenic effects of insulin. Of the two cases

Table 3 - Type of Diabetes and Gender of
98 Patients Who Had Never Received
Insulin and Who Were Treated With
Biosynthetic Human Insulin

	Type I	Type II	Total
Males	26	27	53
Females	12	33	45
TOTAL	38	60	98

of insulin hypertrophy, one improved after the
patient was reinstructed on rotation of the
injection sites. Serum insulin antibody titers
(IgG class of immunoglobulins) have been measured
before treatment and at two-month intervals
thereafter. The insulin antibody titer data for
this study are not yet available.

Finally, intradermal tests performed before and
after six months treatment with BHI were
consistently negative on both occasions. This
finding mitigates against the presence of
clinically detectable quantities of E. coli
substances or other immunogens in BHI and is
compatible with results of the use of a RIA for
E. coli polypeptides (ECP) in BHI which indicates
concentrations of 1 ppm or less in our BHI.[13]
The assay has a sensitivity of 1 ppm and is
described by Baker et al.[13]

Table 4 – The Pre- and Posttreatment Glycosylated
Hemoglobin* and Insulin Dose (Units/kg)
of 98 Patients Receiving Only Biosynthetic
Human Insulin

	Before Treatment		2 Months		4 Months		6 Months	
	Glyco. Hgb	Dose	Glyco. Hgb	Dose	Glyco. Hgb	Dose	Glyco. Hgb	Dose
Type I	14.2%		11.3%	0.49	10.8%	0.49	10.2%	0.51
Type II	14.3%		11.5%	0.40	10.5%	0.45	10.4%	0.44

*Method of Simon and Eissler[1] (normal 6.0-8.8%)

1. Simon, M., and Eissler, J.: Critical Factors in the
   Chromatographic Measurement of Glycohemoglobin (HgA_1),
   Diabetes, 29:467-74, 1980

A solid-phase immunoassay for antibodies to ECP
in serum has recently been developed. In a
preliminary study, using serum from an ECP
immunized rabbit for a positive control, sera
from a large number of patients who have received
BHI for up to six months, and sera from normal
individuals, we have observed no significant
difference between BHI-treated patients and
normal controls. These results are consistent
with the poor immunogenicity of ECP found in
laboratory animals and the undetectable levels of
ECP in BHI. These studies are continuing.

New patients are now being entered into a
double-blind study in which they are assigned
randomly to treatment with BHI, PPI, or mixed
beef-pork insulin. The study otherwise is the
same as the "open label" study.

A transfer study is also underway. This consists
of switching patients who have been on the animal
insulins in a double-blind fashion either to BHI
or back to their usual beef-pork or PPI. Results
from this study are not yet available.

In addition to the above studies, BHI has been
supplied to three patients in the U.S. with
systemic allergy to insulin. In none of the
three patients who have had marked allergy to the
Lilly and the Danish PPI has lasting improvement
occurred with BHI. One patient treated by us at
the Lilly Clinic reacted equally to an
intradermal test dose of one millionth of a unit
of PPI and of BHI. Another Lilly Clinic patient
requires prednisone 10-15 mg per day to reduce
allergic manifestations to both PPI and BHI. We
ascribe the lack of improvement to BHI in our
patients with systemic allergy to insulin to the
fact that these patients are allergic to the
insulin molecule proper. These findings are
consistent with our experience with highly
purified pancreatic human insulin which we
studied several years ago.[14]

As already indicated, the foregoing studies have
been performed using human insulin prepared from
cloned A and B chains which have been chemically
combined. BHI has also been produced by a
process involving the expression of the
proinsulin gene in E. coli. Human proinsulin is
then enzymatically cleaved to form BHI.[15]
Studies in animals disclose that this material
has the same activity as BHI produced from A and
B chain combination. Therefore, if we expect to
derive unique clinical advantages from
recombinant technology we must look beyond cloned
insulin proper. Based on data from human and
animal studies in which beef or porcine
proinsulin have been evaluated, the availability
of human proinsulin for clinical testing may
represent a significant reward of recombinant
research for diabetic treatment. A brief review
of the physiology and pharmacology of animal
proinsulins may elucidate the potential benefits
which could arise from the availability of human
proinsulin.

Although most investigators agree that serum
proinsulin may constitute from 5 to 40 percent of
insulin-like activity in human plasma,[16-20]

Table 5 - A Summary of Important Biologic
          Characteristics of Proinsulin in
          Comparison with Insulin

---

1. Long T 1/2 (about 6 x insulin)

2. Delayed effect

3. Greater central (hepatic) than peripheral
    activity

4. Possible synergism with insulin

Figure 7 -

A comparison of the serum concentrations and
Hypoglycemic effects of equipotent and equimolar
doses of porcine insulin and proinsulin administered
intravenously to normal fasting subjects.
(Note: MuGm/ml is the same as ng/ml) A comparison
of the glucose responses to porcine insulin
0.1 U/kg and an equimolar dose of porcine proinsulin
(0.02 U/kg) demonstrates that on a molar basis the
effect of proinsulin on the blood glucose is 18-20%
that of insulin.

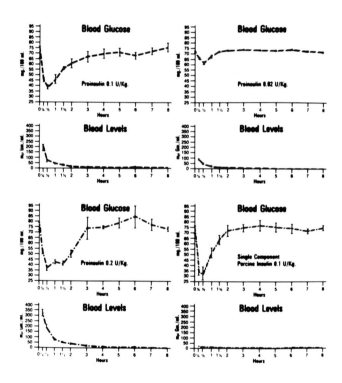

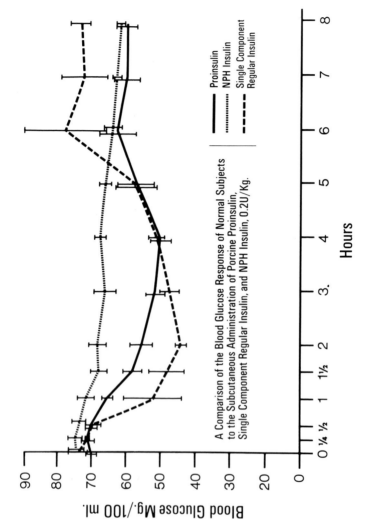

Figure 8 – A comparison of the blood glucose response of normal subjects to the subcutaneous administration of porcine proinsulin, Single Component Regular Insulin, and NPH Insulin, 0.2 U/kg.

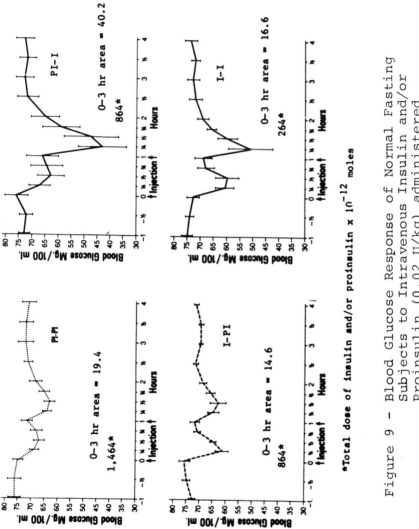

Figure 9 – Blood Glucose Response of Normal Fasting
Subjects to Intravenous Insulin and/or
Proinsulin (0.02 U/kg) administered
60 minutes apart.

409

very little is known about the function of
proinsulin after it leaves the beta cell. An
abundance of in-vitro and in-vivo data developed
with porcine or bovine proinsulin, however,
demonstrate clear differences in activity and
biologic handling between insulin and
proinsulin.[16-20] (Some of these are listed in
Table 5.)

Several investigators (including ourselves[21])
(Figure 7) have demonstrated that serum
concentrations after the intravenous
administration of proinsulin are significantly
higher than after equimolar quantities of
insulin. It will be noted that after proinsulin
the serum levels are higher than insulin and the
effect of proinsulin is delayed. In fact, in
Figure 8 it will be observed that following
subcutaneous administration the blood glucose
lowering effect of proinsulin is intermediate
between that of Regular and NPH insulin. The
reasons for this phenomenon are not clear. It is
important to note that the proinsulin used in
these experiments had a potency of 3 units/mg as
determined by mouse convulsion assay, whereas the
activity of the insulin was 27-28 units/mg.
Thus, proinsulin on a molar basis has 18-20
percent of the activity of insulin (Figure 7).

Another characteristic in which proinsulin
differs from insulin is that its effects on
glucose homeostasis are mediated to a greater
extent by the liver than is insulin. This
feature of proinsulin has been demonstrated by
Tompkins and his associates[16] who performed
glucose turnover studies in anesthetized dogs and
compared the effects of insulin, proinsulin, and
other insulin analogues.

The last item on our list is possible synergism
with insulin, an interaction suggested to us in
studies presented in Figure 9. In these, normal
fasting subjects were administered the
combinations of porcine insulin and/or proinsulin
at 0 and 60 minutes.[22] The dose of each was

0.02 units/kg for a total dose of 0.04
units/kg/experiment. This figure shows the blood
glucose curves, the integrated areas under the
baseline, and the total dose of "insulin" as
insulin and/or proinsulin as picomoles/kg.
Comparison of the upper right hand panel
(proinsulin followed by insulin) and the lower
left hand panel (insulin followed by proinsulin)
discloses that while the treatments are the same
except for the order of administration, the blood
glucose lowering effect of insulin administered
after a dose of proinsulin is nearly three times
that of insulin when injected before the
proinsulin. We would hasten to add that these
experiments were repeated using I^{125} proinsulin
and I^{131} insulin and no effect of one on the
disposal of the other was observed. Thus, if a
synergism exists between insulin and proinsulin,
it apparently is not related to pharmacokinetics
in the plasma. Since isolated perfused forearm
techniques have demonstrated no synergism between
insulin and proinsulin,[23] the site of these
unique effects of proinsulin-insulin combinations
is probably the liver.

Based on this brief review of the pharmacology of
proinsulin, what can we say about its potential
as a research probe and possibly as a therapeutic
agent in diabetes? The questions we will be
addressing are listed in Table 6.

Clearly, we are curious about the immunogenicity
of human proinsulin. While this material is
homologous in humans, the immunogenicity of
animal proinsulins in humans gives cause for
concern in this matter. The second and third
questions deal with the metabolic impact of
replenishing a substance in the insulin-dependent
diabetic which ironically through recent years
insulin purification technology has sought to
remove from commercial insulin preparations.
Consequently, with the availability of human
proinsulin the role of this substance in human
diabetes and its replacement can be studied. If
insulin-proinsulin combinations are found to be

Table 6 - Clinical Questions and Possibilities for Human Proinsulin

1. Safety: Is human proinsulin immunogenic?

2. What are the effects of proinsulin alone or in combination with insulin

   a. On blood glucose

   b. On other parameters

      1. acute

      2. chronic

3. Are proinsulin-insulin combinations physiologic modalities? Will double-barreled pumps be needed to accommodate these two hormones?

4. Is proinsulin uniquely efficacious in patients with massive degradation/sequestration of insulin at the injection site?

necessary, then insulin infusion pump technology may be called upon to add a channel for proinsulin which may make possible selective hepatic effects peripherally administered insulin cannot achieve.

Finally, for the rare but unfortunate patients who destroy or sequester insulin at the injection site,[24] subcutaneous proinsulin as an insulin molecule with a protective shield might be a useful therapeutic alternative to intravenous insulin therapy.

In the foregoing discussion we have demonstrated that in normal fasting subjects BHI and PPI are virtually identical. We have attempted to identify certain problems related to the interpretation of results obtained when one uses a RIA to measure BHI or PPI in the serum. We have presented preliminary data in 98 patients who, receiving insulin for the first time, have been treated with BHI and have failed to demonstrate cutaneous reactivity to BHI or serum antibody titers to E. coli protein that possibly could have been present in the BHI preparation. All available data suggest that BHI is an active, safe insulin. However, in patients who have severe allergy to the animal insulins, unique benefit from BHI has not yet been demonstrated. Finally, we have suggested that possibly one of the most exciting therapeutic benefits arising from the use of recombinant technology in diabetes may result from the availability of human proinsulin.

REFERENCES

1. Banting, F. G., and Best, C. H.: The Internal Secretion of the Pancreas, J. Lab. Clin. Med., 7:251-66 (1921-22).

2. Sanger, F., and Tuppy, H.: The Amino Acid
    Sequence of the Phenylalanine Chain of
    Insulin I, II, Biochem. J., 49:463-81, 481-90
    (1951).

3. Ryle, A. P., Sanger, F., Smith, L. F., and
    Kitai, R.: The Disulphide Bonds of Insulin,
    Biochem. J., 60:541-56 (1955).

4. Nicol, D. S. H. W., and Smith, L. F.: The
    Amino Acid Sequence of Human Insulin, Nature
    (London), 187:483-85 (1960).

5. Goeddel, D. V., Kleid, D. G., Bolivar, F.,
    Heyneker, H. L., Yansura, D. G., Crea, R.,
    Hirose, T., Kraszewski, A., Itakura, K., and
    Riggs, A. D.: Expression in Escherichia coli
    of Chemically Synthesized Genes for Human
    Insulin, Proc. Natl. Acad. Sci. USA,
    76:106-10 (1979).

6. Crea, R., Kraszewski, A., Tadaaki, H., and
    Itakura, K.: Chemical Synthesis of Genes for
    Human Insulin, Proc. Natl. Acad. Sci. USA,
    75:5765-69 (1978).

7. Chance, R. E., Kroeff, E. P., Hoffmann,
    J. A., and Frank, B. H.: Chemical, Physical,
    and Biologic Properties of Biosynthetic Human
    Insulin, Diabetes Care, 4:147-54 (1981).

8. Galloway, J. A., Spradlin, C. T., Root,
    M. A., and Fineberg, S. E.: The Plasma
    Glucose Response of Normal Fasting Subjects
    to Neutral Regular and NPH Biosynthetic Human
    and Purified Pork Insulins, Diabetes Care,
    4:183-88 (1981).

9. Root, M. A., Spradlin, C. T., Galloway,
    J. A., and Chance, R. E.: Factors Affecting
    the Measurement of Insulin in the Blood of
    Human Subjects Following the Administration
    of Purified Pork and Human Insulins,
    Presented at the EASD 17th Annual Meeting,
    Amsterdam. The Netherlands, September (1981).

10. Botterman, P., Gyaram, H., Wahl, K., Ermler, R., and Lebender, A.: Pharmacokinetics of Biosynthetic Human Insulin and Characteristics of Its Effect, Diabetes Care, 4:168-69 (1981).

11. Federlin, K., Laube, H., and Velcovsky, H. G.: Biologic and Immunologic In Vivo and In Vitro Studies with Biosynthetic Human Insulin, Diabetes Care, 4:170-74 (1981).

12. Pickup, J. C., Bilous, R. W., Viberti, G. C., Keen, H., Jarrett, R. J., Glynne, A., Cauldwell, J., Root, M., and Rubenstein, A. H.: Plasma Insulin and C-peptide After Subcutaneous and Intravenous Administration of Biosynthetic Human and Purified Porcine Insulin in Healthy Men, to appear in The Lancet.

13. Baker, R. S., Schmidtke, J. R., Ross, J. W., Smith, W. C.: Preliminary Studies on the Immunogenicity and Amount of Escherichia Coli Polypeptides in Biosynthetic Human Insulin Produced by Recombinant DNA Technology, The Lancet, pp. 1139-42, November 21, 1981.

14. Galloway, J. A.: Potential Clinical Research Impact of Biosynthetic Human Insulin, Presented at the EASD Meeting, Amsterdam. September, 1981.

15. Frank, B. H., Pettee, J. M., Zimmerman, R. E., and Burck, P. J.: The Production of Human Proinsulin and its Transformation to Human Insulin and C-Peptide, Proceedings of the Seventh American Peptide Symposium, p. 729-38 (1981).

16. Rubenstein, A. H., and Steiner, D. F.: Proinsulin, the Single Chain Precursor of Insulin, Med. Clin. North Am., 54:191-99 (1970).

17. Rubenstein, A. H., Melani, F., and Steiner, D. F.: Circulating Proinsulin: Immunology, Measurement, and Biological Activity, Handbook of Physiology: Endocrinology, 1:515-28 (Chapter 33). Endocrine Pancreas, (1972).

18. Horwitz, D. L., Starr, J. I., Mako, M. E., Blackard, W. G., and Rubenstein, A. H.: Proinsulin, Insulin, and C-Peptide Concentrations in Human Portal and Peripheral Blood, J. Clin. Invest., 55:1278-83 (1975).

19. Kitabchi, A. E.: Proinsulin and C-Peptide: A Review, Metabolism, 26:547-87 (1977).

20. Villaume, C., Beck, B., Pointel, J. P., Drouin, P., and Debry, G.: Effects of Liver and Kidney on High Molecular Weight Immunoreactive Insulin (HWIRI), Proinsulin (PI) and Insulin (I) in Man, Horm. Metab. Res., 13:583-84 (1981).

21. Galloway, J. A., Root, M. A., Chance, R. E., Rathmacher, R. P., Challoner, D. R., and Shaw, W. N.: In Vivo Studies of the Hypoglycemic Activity of Porcine Proinsulin, Diabetes, 18:341 (1969).

22. Galloway, J. A., Rathmacher, R. P., Root, M. A., Crabtree, R. E., and Chance, R. E.: The Effects of Low Doses of Intravenous Proinsulin and Insulin Combinations in Normal Fasted Man, J. Lab. & Clin. Med., 78:991 (1971).

23. Fineberg, S. E., and Merimee, T. J.: Effects of Comparative Perfusions of Equimolar, Single Component Insulin and Proinsulin in the Human Forearm, Diabetes, 22:676 (1973).

24. Freidenberg, G. R., White, N., Cataland, S.,
    O'Dorisio, T. M., Sotos, J. F., and Santiago,
    J. V.: Diabetes Responsive to Intravenous
    But Not Subcutaneous Insulin: Effectiveness
    of Aprotinin, <u>New</u> <u>Eng</u>. <u>J</u>. <u>Med</u>., <u>305</u>:363-68
    (1981).

DISCUSSION

S. KING: Over the course of the study the levels of
glycosylated hemoglobin declined from 14% to 10%.   Do you
think this is due to insulin therapy?

J. GALLOWAY: Hyperglycemia, e.g. exposure of the hemoglobin
to elevated glucose levels, results in a transformation of
the hemoglobin to a glycosylated form which can be quanitated
by various methods of chromatography.   Generally, the extent
of elevation of the glycohemoglobin concentration (normal by
the method we use is 6.0 to 8.8%) reflects the average level
of the blood glucose for the preceding 2-3 months.   Very
poorly controlled patients have glycohemoglobins of 18-20%.
We are happy when we can get our patients down to less than
10%.

THE BIOLOGICAL EFFECTIVENESS OF PITUITARY-DERIVED
AND BIOSYNTHETIC METHIONYL-hGH IN ANIMALS AND MAN

S. L. Kaplan+
J. Fenno
N. Stebbing
R. Hintz*
R. Swift

Department of Pediatrics+
University of California San Francisco
San Francisco, CA, U.S.A.

Department of Pediatrics*
Stanford University
Stanford, California
U.S.A.

Genentech, Inc.
South San Francisco, California
U.S.A.

I. INTRODUCTION

The species specificity of pituitary growth hormone limited
its availability for the treatment of growth disorders in
children until 1956 when purification of human growth hormone
was achieved (1,2). Considerable effort was expended
unsuccessfully to identify a "biologically active core" of hGH
(3,4). Structural alternatives including reduction of SH
bonds or substitutions on the COOH terminal do not affect the
biological activity of hGH (5-7). Based on complementation
studies by Burstein and associates (8), maintenance of both
immuno and bioactivity is dependent on the integrity of the
amino terminal of hGH. Other approaches to increase the
available growth-stimulating hormones such as modified forms
of bovine growth hormone (9) or purified human chorionic somatomammotropin (hCS) have not been satisfactory therapeutic
substitutes (10).

FROM GENE TO PROTEIN:
TRANSLATION INTO BIOTECHNOLOGY

419

The advent of recombinant-DNA techniques led to the biosynthesis of human growth hormone (hGH) (11,12). Biosynthetic
methionyl-hGH derived from E. coli has essentially the same
electrophoretic pattern as purified hGH on SDS polyacrylamide
gels; contains less than one percent chimeric hGH; has a
specific activity of approximately two bioassay units per mg
protein and amino acid analyses and tryptic peptides in agreement with data for pituitary hGH (13). The biosynthetic
methionyl-hGH differs from the pituitary hGH preparations by
virtue of absence of other pituitary hormones, the presence of
trace amounts of E. coli proteins rather than human proteins
and the presence of an extra N-terminal methionine which arises
from the method of cloning.

Growth hormone has diversified metabolic effects many of
which are mediated by other growth factors, i.e., somatomedins
(14). Neuroamine regulation of the release of the hypothalamic
peptides somatostatin (SRIF) and growth hormone-releasing factor (GRF) influences the secretion of growth hormone. This
aspect has been reviewed in detail elsewhere (15,16).

The physiologic effects in animals and man of pituitary derived hGH will be reviewed and compared with studies in which
biosynthetic methionyl-hGH was used.

A.   Studies on Animals

A dose-response related increase in body weight is induced
in hypophysectomized rats following administration of either
pituitary-derived or biosynthetic hGH. The other standardized
bioassay for growth hormone in animals is stimulation of the
width of the tibial epiphysis or tibial length in hypophysectomized rats. At equivalent doses of pituitary or biosynthetic
methionyl-hGH, the tibial length was significantly increased
(13).

B.   Studies in Humans

A nonrandomized double crossover study was carried out in
15 normal adult males to assess the safety, pharmacokinetics,
and biologic effects of biosynthetic methionyl-hGH compared to
that of pituitary-derived hGH (Table 1).

At present the biologic effectiveness of biosynthetic methionyl-hGH is under investigation in 20 growth hormone deficient
children in a multi-center national study (Table 2).

TABLE I. Design of Study on the Effect of
Methionyl-hGH in Normal Adults

3 Groups - 5 Subjects Each
Group 1 - Methionyl-hGH (0.25 IU/Kg daily x4 d.)
Group 2 - Pituitary-hGH (0.25 IU/Kg daily x4 d.)
Group 3 - Excipient alone
Time Course
1) Initial phase - 5 days Injections daily for 4 days
2) Wash-out period - 9 days
3) Cross-over period - 5 days reversal of hGH groups

1. Nitrogen Retention

Accretion of protein with nitrogen retention is induced
following administration of hGH in man (10). A decrease in
BUN reflects N retention and was used in the studies by Hintz
and associates (16) to verify this biological action of biosynthetic methionyl-hGH in adult subjects.

2. Linear Growth

Stimulation of linear growth following administration of
hGH provides the definitive assessment of biological activity
of hGH in man. The growth rate of children with growth hormone deficiency before and after treatment with either pituitary hGH or biosynthetic methionyl-hGH (0.2 U/Kg) given by
intramuscular route three times a week is shown in Table 3.
The preliminary data (based on two to four months of therapy)
suggest a comparable effect on growth rate of biosynthetic
methionyl-hGH compared to that described for pituitary-derived
hGH.

3. Generations of Somatomedins

Somatomedin (SM) are peptides which may be the intracellular mediators of growth hormone action. Pituitary-derived
growth hormone stimulates the synthesis and release of SM and
correlates with the bioactivity of growth hormone (14). Hintz
and associates (17) have demonstrated that pituitary and

TABLE II. Design of Study on the Effect of
Methionyl–hGH in Children with Growth Hormone
Deficiency

Subjects

Twenty children (3–11 years of age) with
documented GH deficiency and retardation
of growth documented over at least one year
Control subjects – historical data

Acute Phase – 6-Day Inpatient Study

1) IV glucose tolerance test on Day 2 and
    Day 6

2) Methionyl hGH (0.2 IU/Kg) – 3 times a
    week

3) Routine chemistries, somatomedin-C levels,
    EKG, bone age, antibodies for hGH, and
    E. coli

Chronic Study

1) hGH given (0.2 IU/Kg IM) Monday, Wednesday,
    and Friday by parents

2) Office visits at 2, 4, 6, 9, and 12 months
    after therapy.

3) Height, weight, and complete physical exam

4) CBC, routine chemistries, somatomedin-C
    levels, urinalysis, antibodies to hGH and
    E. coli

---

biosynthetic methionyl–hGH have equipotency in stimulation of
somatomedin release in adult men (Table 4). Studies on somatomedin levels in growth hormone deficient children in response
to biosynthetic methionyl–hGH are in progress.

4. Carbohydrate Metabolism

Growth hormone has both contra–insulin and insulinogenic
effects. The acute effect of pituitary and biosynthetic
methionyl–hGH on glucose disposal and insulin secretion were
compared in the crossover study in normal adult males. A 3-
fold increase in insulin secretion and a lesser increase in
glucose disposal was induced by either pituitary or biosynthe-
tic-methionyl hGH (Table 5). These results are comparable to
previous reports using pituitary hGH (6,10). Similar effects
have been demonstrated in the growth hormone deficient children
to whom biosynthetic methionyl–hGH was administered.

TABLE III.   Growth Rate of Children with GH Deficiency

	Pituitary hGH	Biosynthetic Methionyl-hGH
Pre	<4 cm/yr	<4 cm/yr
During	7-15 cm/yr	8-18 cm/yr

5. Kinetics of hGH Administration

The disappearance rate of exogenous or endogenous hGH is
20 to 30 minutes (16). Following administration of 0.2-0.3 mg/
Kg of pituitary hGH to GH deficient children, peak serum
levels are attained at 1-2 hours and decrease to unmeasureable
concentrations by 18-24 hours (18). In the rhesus monkey peak
serum levels are attained at two hours following IM pituitary
or biosynthetic methionyl-hGH (0.25 U/Kg) and decrease to pretreatment levels by 14 hours in both treatment groups.
(Figure 1).

6. Evaluation of Toxicity of Growth Hormone

High doses of pituitary growth hormone administered chron-
ically to animals induce glucose intolerance and hyperinsulin-
ism, bony over-growth and organomegaly (19). In man, hyper-
secretion of growth hormone by pituitary tumors as in acrome-
galy or gigantism induces similar effects as described in
animals (16). Acute effects of high dose hGH has been studied
previously in adult male volunteers (20). Suppression of
growth hormone response to hypoglycemia, increased fatty acid
levels and glucose intolerance were observed. These effects
were reversed within 48 hours following discontinuation of hGH.
The tolerance of animals and humans to high dose hGH was
re-evaluated using both pituitary and biosynthetic methionyl-
hGH.

TABLE IV.   The Effect of hGH
(16 Units IM Daily X 4 Days)
on Serum RIA Somatomedin-C Levels in Adult Males

	PITUITARY hGH	RECOMBINANT-DNA hGH
Pre	1.11 ± 0.06	1.04 ± 0.07
Post	3.08 ± 0.19	2.95 ± 0.20

Reference No. 17

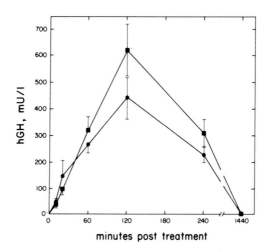

minutes post treatment

FIGURE 1.   Clearance of hGH after intramuscular administration
of 250 µg/kg (■) pituitary hGH; (●) biosynthetic hGH, to
rhesus monkeys. Vertical bars indicate standard errors of
the means.

A.   Rats

In order to reflect clinical usage, the materials were administered intramuscularly and the effects of pituitary hGH
(Crescormon, Kabi AB) and biosynthetic methionyl-hGH were compared in hypophysectomized rats. In addition, the same doses of
the biosynthetic hGH were administered intramuscularly to non-
hypophysectomized rats. Groups of six female Sprague-Dawley
rats at 7 weeks of age were treated with a single intramuscular
injection of 0.05, 0.5 or 5.0 mg hGH/rat. Although the age of
all the rats was the same, the weight of the intact animals
was about 50% greater than the hypophysectomized rats. No adverse effects were noted in any of the rats treated with the
biosynthetic hGH. None of the animals died but adverse reactions were noted in the group of hypophysectomized rats
receiving the highest dose of pituitary hGH. These rats showed
obvious piloerection and in one case edema up to 2-1/2 hours
post-treatment but no abnormalities developed thereafter over
the next 25 days. An additional acute toxicity study involving
groups of ten intact male and female Sprague-Dawley rats who
showed no drug related effects on gross necropsy findings
after intramuscular treatments at doses up to 25 mg/kg.
 Tolerance of female hypophysectomized Sprague-Dawley rats
treated intramuscularly with 10 to 50 µg hGH/rat 3 times weekly
was examined over 8 weeks using both pituitary and biosynthetic

TABLE V. Glucose and Insulin Responsiveness
Following Administration
of hGH (16 Units IM Daily X 4 D) in Adult Males

	Glucose Area Under Curve MG%/HR		Insulin Area Under Curve μU/ML/HR	
	Pre	Post	Pre	Post
Pituitary hGH	333 ± 14	435 ± 25	110 ± 12	324 ± 49
Synthetic Methionyl-hGH	348 ± 16	475 ± 30	133 ± 19	403 ± 55

Reference No. 17

methionyl-hGH. No adverse effects were observed during the
treament period or within four weeks following treatment.

B. Squirrel Monkeys

The effect of repeated injections with hGH was examined in
squirrel monkeys of both sexes administered intramuscularly 3
times weekly for 8 weeks with 250 μg hGH/kg/treatment. Three
groups of five animals were treated with either biosynthetic
methionyl-hGH, pituitary hGH (Crescormon, Kabi AB) or glycine-
phosphate excipient. No adverse effects were observed at any
time during treatment or in the subsequent 4 weeks. Tolerance
to acute treatment with higher doses was carried out only with
the biosynthetic methionyl-hGH using four groups of three animals. Monkeys in each group received intramuscularly at two
sites, total doses of 0.5, 5.0, or 50 μg/kg biosynthetic
methionyl-hGH. No signs of illness and no reactions were observed at any observation time over the first 8 hours post-
treatment or during the subsequent 21 days when twice daily
checks were made.
Low antibody titers to hGH were detectable in 3 of 5
pituitary-hGH treated monkeys and in all biosynthetic methionyl-
hGH treated animals. There was no evidence of antibody formation to trace E. coli contaminants in the methionyl-hGH treated
groups.

C. Rhesus Monkeys

Long-term tolerance of rhesus monkeys to hGH was determined
by a regimen of 250 μg/kg/treatment given intramuscularly 3
times weekly (Monday, Wednesday, Friday) over 8 weeks. Fifteen

young animals, between 15 and 22 months of age, were assigned
to three groups of five. One group was treated with biosynthetic methionyl-hGH, one with pituitary-hGH (Crescormon, Kabi AB)
and the third group with glycine-phosphate excipient. No signs
of illness were observed in any of the monkeys at any time
during the 8 weeks of treatment. Blood chemistries were determined before the first treatment and at 25, 39, and 53 days
thereafter. The liver enzymes (SGOT, SGPT, LDH, and alkaline
phosphatase) were elevated in all groups as treatment progres-
sed. After 39 days both hGH preparations induced an increase
in calcium, significant at the 1% level by analysis of vari-
ance. Blood cell parameters determined before and at 39 days
after commencement of treatment were unchanged. At 21 days
the segmented neutrophils were significantly elevated in the
biosynthetic group compared with the glycine-phosphate group.
Antibody titers to hGH were detectable in four of five biosynthetic methionyl-hGH treated animals at 8 weeks, but at 12
weeks only one monkey had a detectable antibody titer. No antibody formation to E. coli contaminants was found.

D. Man

Two preparations of biosynthetic methionyl-hGH (G1 and G2)
tested in adult human volunteers had similar biological activity. However, IM administration of preparation G1-hGH induced
transient malaise, myalgia, fever, elevated levels of liver
enzymes, decreased iron and other hematologic abnormalities.
Preparation G2-hGH purified by alternative chromatographic procedures was not associated with any adverse effects following
administration to adult human volunteers. Subsequent evaluation of preparation G2-hGH in GH-deficient children was not
associated with any adverse effects either in acute or chronic
studies.

Summary

The biological effectiveness of biosynthetic hGH has been
substantiated in animals and man by these studies. Its biological and in vitro immunologic potency is equivalent to that
of purified pituitary derived hGH. The utilization of hGH for
growth disorders not associated with GH deficiency and for
other metabolic abnormalities (Table 6) will be feasible with
increased availability of biosynthetic hGH.

TABLE VI. Potential Uses of hGH

1. Cachexia associated with burns, surgery and cancer

2. Fractures

3. Osteoporosis

4. Hematologic disorders associated with decreased erythropoiesis

REFERENCES

1. Li, C. H., and Papkoff, H., Science 124, 1293 (1956).
2. Raben, M. S., J. Clin. Endocrinol. Metab. 18, 901 (1958).
3. Li, C. H., Papkoff, H., and Hayashida, T., Arch. Biochem. Biophys. 85, 97 (1959).
4. Niall, H. D., and Treagear, G. W., in "Advances in Human Growth Hormone Research" (S. Raiti, ed.), p. 394. DHEW Publication, No. (NIH 74-612) (1974).
5. Nutting, D. F., Kostyo, J. L., Mills, J. B., and Wilhelmi, A. E., Endocrinology 90, 1202 (1972).
6. Connors, M. H., Kaplan, S. L., Li, C. H., and Grumbach, M. M., J. Clin. Endocrinol. Metab. 37, 499 (1973).
7. Burstein, S., Grumbach, M. M., Kaplan, S. L., and Li, C. H., Proc. Natl Acad. Sci., 75, 5391 (1978).
8. Burstein, S., Grumbach, M. M., Kaplan, S. L., and Li, C. H., J. Clin. Endocrinol. Metab., 48, 1017 (1979).
9. Yamasaki, M., Kangawa, K., Kobayashi, S., Kikutani, M., and Sonenberg, M., J. Biol. Chem., 247, 3874 (1972).
10. Grumbach, M. M., Kaplan, S. L., Sciarra, J. J., and Burr, I. M., Ann. NY, Acad. Sci., 148, 501 (1968).
11. Martial, J. A., Hallewell, R. A., Baxter, J. D., and Goodman, H. M., Science, 205, 602 (1979).
12. Goeddel, D. V., Heyneker, H. L., Hozumi, T., Arentzen, R., Itakura, K., Yonsura, D. G., Ross, M. J., Miozzari, G., Crea, R., and Seeburg, P. H., Nature 281, 544 (1979).
13. Olson, K. C., Fenno, J., Lin, N., Harkins, R. N., Snider, C., Kohr, W. J., Ross, M. J., Fodge, D., Prender, G., and Stebbing, N., Nature 293, 408 (1981).
14. Van Wyk, J. J., and Underwood, L. E., in "Biochemical Actions of Hormones" (G. Litwak, ed.), Vol. 5. Academic Press, NY, (1978).
15. Martin, J. B., N. Engl. J. Med., 288, 1384 (1973)
16. Daughaday, W. H., in "Textbook of Endocrinology" (R. H. Williams, ed.), p. 73. W. B., Saunders Co., Philadelphia, (1981)

17. Hintz, R. L., Rosenfeld, R., Wilson, D., and Bennett, A.,
     Joint Meeting Lawson Wilkins Pediatric Endocrine Society
     and European Society for Pediatric Endocrinology, Geneva,
     Abstract No. 21 (1981).
18. Kaplan, S. L., Savage, D. C. L., Suter, S., Wolter, R.,
     and Grumbach, M. M., in "Advances in Human Growth Hormone
     Research" (S. Raiti, ed.), p. 725, DHEW Publication, No.
     (NIH 74-612), (1974).
19. Savostin-Asling, I., Nakaiye, R., and Asling, C. W., Anat.
     Rec. 196, 9 (1980).
20. Abrams, R. L., Grumbach, M. M., and Kaplan, S. L., J.
     Clin, Invest. 50, 940 (1971).

BIOLOGICAL ACTIVITY OF A CLONED HUMAN ENZYME UROKINASE

P. P. Hung

Genetics Division, Bethesda Research
Laboratories, Inc. Gaithersburg, MD 20877
U.S.A.

ABSTRACT

We have cloned double-stranded cDNA copies of plasminogen activator messenger RNA isolated from human fetal kidney cells. Some of the clones express protein of discrete sizes ranging from 32,000 to 150,000 daltons. These products possess antigenic determinants related to human plasminogen activator from kidney cells, bind to an affinity column specific for serine protease and activate human plasminogen to dissolve fibrin clots.

INTRODUCTION

Acute thromboembolic events, venous and arterial thrombosis, pulmonary embolism, intracardiac thrombosis and systemic embolism are important medical problems. In all of these pathologies, the plasminogen activator urokinase has attractive potential applications, yet it is difficult to isolate in large quantities. We have designed and constructed, therefore, bacterial plasmids which instruct the synthesis of biologically active plasminogen activators in a microbial cell. Moreover, such clones facilitate elucidation of urokinase structure and mode of biosynthesis as well as of urokinase gene regulation and expression.

A plasminogen activator, urokinase, was first observed in urine in 1951 by Williams (1). More recently, fibrinolytic activity in cultures of human kidney cells was demonstrated (2), and it was found that this activity was immunologically indistinguishable from urine urokinase. Based on their sizes, two major types of urokinase have been described. These are the high molecular weight species (HMW) with molecular weight of 54,000 and the low molecular weight species (LMW) of M_r 32,000.

RESULTS

Biochemical properties of Urokinase. Since there are two species of urokinase, it is important to know whether they are derived from one gene or from two genes. If they are coded by the same gene, we need only to clone that particular gene.

FROM GENE TO PROTEIN:
TRANSLATION INTO BIOTECHNOLOGY

Therefore, we studied biochemical relationship between these
two species. Amino acid compositions of the low and high molecular   weight   urokinase   were   determined   and   compared
(Table 1). The individual amino acid content was very similar

Table 1. Amino Acid Composition of Urokinase from Urine and
          from Tissue Culture

	UUK**		TCUK	
	LMW	HMW	LMW	HMW
Asp	19.1*	40.4	19.8	41.6
Thr	19.0	31.6	19.1	29.2
Ser	21.9	34.8	18.7	32.4
Glu	28.3	44.1	26.5	48.4
Pro	16.4	28.7	11.8	23.5
Gly	22.4	41.1	24.5	42.0
Ala	10.1	20.2	10.8	19.8
Cys	9.5	19.4	10.7	5.7
Val	10.9	23.2	9.2	17.5
Met	4.5	7.7	1.6	4.6
Ile	15.6	20.4	13.0	19.0
Leu	20.6	33.6	22.1	32.2
Tyr	12.8	19.6	9.9	13.2
Phe	9.3	14.3	6.9	7.5
His	10.2	19.7	9.1	18.7
Lys	17.3	31.0	15.8	28.2
Arg	14.3	22.1	13.9	23.0

* Data obtained from White, W.F., Barlow, G.H. and Mozen,
  M.M., 1966) Biochem. 5, 2106-2169.

** Amino acid residues per mole of protein.

UUK = urine urokinase
TCUK = tissue culture urokinase

in low molecular weight species from urine (UUK) and tissue culture (TCUK); and similarity also exists in high molecular weight species from both sources. The similarity in amino acid compositions between urinary and tissue culture urokinase strongly suggests that kidney is the common site for urokinase synthesis.

Urokinase is a glycoprotein. Carbohydrate analyses showed that the low and the high molecular weight urokinase from tissue culture contained 7.5 and 8.8 percent of carbohydrate, respectively (Table 2). In both species, there were D-mannose,

Table 2.   Carbohydrate Composition of Urokinase from Tissue Culture

Sugar	LMW		HMW	
	nmole/100μg protein[1]	mol/mol[2]	nmole/100μg protein	mol/mol
D-Man[3]	9.5	3.0	3.97	2.2
L-Fuc[3]	8.7	2.7	7.95	4.4
D-Gal[3]	10.0	3.1	8.65	4.7
D-GalcNAc[4]	ND*	ND	3.39	1.8
Sialic acid[5]	1.6	0.5	7.69	4.2
D-Xyl[3]	ND	ND	0.47	0.3
D-Glc[3]	ND	ND	1.87	1.0
D-GlcNAc[4]	15.3	4.8	11.94	6.5

1. Protein determined by amino acid analysis.

2. For LMW, using molecular weight of 32000g/mol, and for HMW, 54000g/mol.

3. Determined by automated neutral sugar analysis after hydrolysis in 2M TFA at 100° for 3 hr.

4. Determined by automated amino sugar analysis after hydrolysis in 4M HCl at 100° for 6 hr.

5. Determined by the thiobarbituric acid method after hydrolysis in 0.05M H_2SO_4 at 80°.

*   non-detectable

L-fucose, D-galactose and N-acetyl-D-glucosamine. However, the high molecular weight species had significant amounts of sialic acid and N-acetyl-D-galactosamine which were not found in the low molecular weight species.

Edman degradation on the low molecular weight urokinase from tissue culture released two amino acid molecules upon each step of reaction which indicated the presence of two peptide chains in the molecule. Substracting the known N-terminal sequence of urinary urokinase (3), we deduced the partial N-terminal sequences for the two chains (Table 3).

Table 3.   N-terminal Sequence of Low Molecular Weight
           Urokinase

Chain 1. NH_2-Ile-Gly-Ile-Gly-Glu-Phe-Ser/Thr-Ser/Thr-Ile-
           Glu-Asp-Glu-Pro-Try-Phe...

Chain 2. NH_2-Leu-Phe-Lys-Glu-Glu-Gly-Ser/Thr-Ser/Thr-Asp-
           Leu-Gly-Leu-Lys...

Since there are two peptide chains in the molecule, the N-terminals of the molecule were coupled with ^{35}S-phenyl-isothiocyanate and the labelled two chains were separated by reduction and electrophoresis in SDS-poly-acrylamide gradient gel. Two peaks of radioactivity were observed with the low molecular weight species (Fig. 1A). One peak had a molecular weight of about 29,000 and the other about 6000. The high molecular weight species was also analyzed. The results showed that its two radioactive components were similar in size to those observed with the LMW urokinase. In addition, it contained a third component corresponding to a molecular weight of about 19,000 (Fig. 1B).

To show further the close relatedness of the two species, they were digested by trypsin and their fingerprints were compared. The fingerprints from the low molecular weight species had a distinctive distribution pattern with 16 ninhydrin positive spots (Fig. 2A). Fifteen of these were also present in the high molecular weight urokinase (Fig. 2B). However, the high molecular weight urokinase had extra polypeptides not found in the low molecular weight urokinase, which were most likely derived from the third peptide chain. These observations indicated that only one gene was involved in coding for the two species of urokinase.

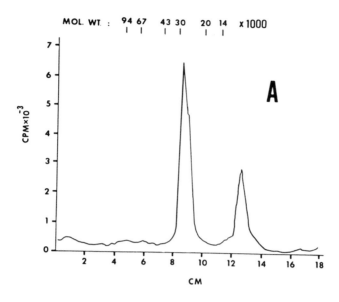

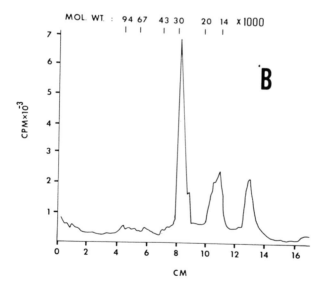

Fig. 1   Peptide chain analyses of urokinase by N-terminal labelling and SDS gel electrophoresis.
(A) Low molecular weight urokinase from tissue culture
(B) High molecular weight urokinase from tissue culture

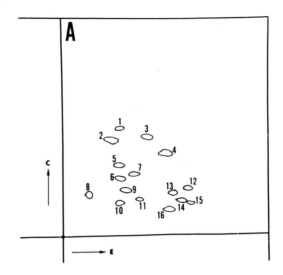

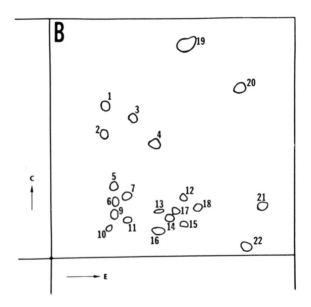

Fig. 2   Tryptic fingerprints of urokinase from tissue culture.
(A) Low molecular weight species
(B) High molecular weight species

   Approaches in Cloning. Experimental strategies for cloning the human urokinase gene and isolating its protein products took the following steps:

   1)   Isolation of RNA from human fetal kidney cells during
        maximal production of urokinase.

   2)   Enrichment of urokinase mRNA by sucrose density cen-
        trifugation and demonstration of its ability to di-
        rect the synthesis of urokinase in a cell-free pro-
        tein synthesis system. Alternatively, immunoprecipi-
        tation of polysomes with anti-urokinase antibody and
        then to demonstrate the presence of urokinase mRNA in
        a cell-free protein synthesis system.

   3)   cDNA synthesis from the mRNA and insertion of the
        cDNA at the Pst I site of pBR322.

   4)   Transformation of E. coli X1776 with the recombinant
        molecules followed by colony selection with tetracy-
        cline.

   5)   Detection by radioimmunoassays of E. coli colonies
        that produce urokinase.

   6)   Isolation and characterization of the protein pro-
        ducts derived from the recombinant DNA.

   Isolation of mRNA. Total RNA from human fetal kidney
cells was isolated by the guanidine thiocyanate method (4).
Poly A-containing RNA was isolated from the total RNA by Poly
U-Sephadex column (5). The RNA was further fractionated by
centrifugation in a sucrose-density gradient. The RNA in different fractions was tested in a cell-free protein synthesizing system (6). Immunoprecipitation of the radioactive products showed that about 7 percent of the synthesized protein
was precipitated when mRNA before fractionation was used
(Table 4). However, only 1 percent of radioactivity was precipitated by anti-urokinase antibody when rabbit globin mRNA
was used. After fractionation, the highest urokinase mRNA activity was located in the fraction corresponding to sedimenta-
tion values of greater than 28.

   mRNA isolated by immunoprecipitation (7) was also studied.
Some preparations of mRNA thus isolated were electrophoresed
in urea gel (Fig. 3). The electrophoretic mobilities corresponded to about 4,900 nucleotides which can code for a protein of approximately 160,000 daltons.

Table 4. Cell-free Protein Synthesis Using Urokinase mRNA

RNA Preparation	Radioactivity Precipitated by anti-urokinase
	(%)
Globin mRNA (control)	1
mRNA before fractionation	7
mRNA after fractionation	
Greater than 28S	39
28S	5
18 to 28S	14
18S	4
Immunoprecipitated mRNA	
Sample No. 1	50
Sample No. 2	90

Fig. 3   Urea-gel electrophoresis of immunoprecipitated mRNA.

In cell-free protein synthesis, one preparation gave 50 percent and the other, 90 percent of immunoprecipitable radio-activity which indicated a high degree of purity of this mRNA species (Table 4).

The immunoprecipitated radioactive products were analyzed by SDS-polyacrylamide gel electrophoresis. A major peak of radioactivity was observed approximately corresponding to a molecular weight of 50,000. Other minor protein peaks corresponding to 30,000 daltons as well as to higher molecular weight materials were also detected (Fig. 4).

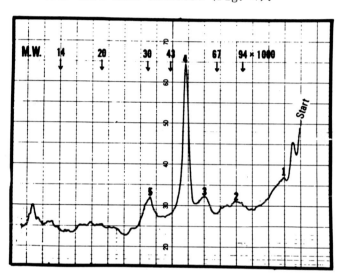

Fig. 4   Polypeptides synthesized by <u>in vitro</u> translation of mRNA.

Recombinant DNA Synthesis

Single-stranded cDNA was synthesized from poly A-containing mRNA with reverse transcriptase. The second strand of cDNA was synthesized by using the large fragment of DNA polymerase I. The double-stranded cDNA was treated with S_1 nuclease and then the DNA was centrifuged in a sucrose-density gradient. DNA of size greater than 2000 base pairs was obtained and was used to attach poly dC tracts. The circular pBR322 DNA was digested with <u>Pst</u> I and tailed with poly dG tracts. The double-stranded cDNA with poly dC tail was annealed to pBR322 DNA with poly dG tail to form recombinant DNA.

Initial Detection of Urokinase-like Material

After transformation of E. coli X1776 with the recombinant
DNA, we observed a total of 32 tetracycline resistant colonies
(Table 5). Of these, four were ampicillin-sensitive and contained inserts in their plasmids. Analyses of Pst I digests
of the plasmids by gel electrophoresis revealed that three inserts were about 4.2 kilobase pairs whereas the fourth was
about 900 base pairs. Two of these transformants, colony
No. 19 and 26 and the negative control, X1776 transformed by
pBR322, were grown and the cell lysates were prepared. Antigens in the lysates were spotted on and covalently bound to
cyanogen bromide-activated paper (8). Detection of urokinase-
like materials was carried out by reaction with ^{125}I-la-
belled anti-urokinase antibody (9). As shown in Fig. 5, the
lysates from both transformants showed weak but positive immunoreactivity as compared to the control, X1776 transformed by
pBR322. Known amounts (0.08, 0.3, 1.25 and 5 ng) of urokinase
(SMW) were also spotted on the paper as the positive control.
The weak positive reaction of the transformants suggested that
they produced small amounts of urokinase. Therefore, the
clones were examined more closely by affinity chromatography.

Table 5.   Selection of Transformants

Clone No.	Tetracycline	Ampicillin	Size of Insert (base pairs)
27-19	Resistant	Sensitive	4,200
27-20	Resistant	Sensitive	900
27-26	Resistant	Sensitive	4,200
27-29	Resistant	Sensitive	4,200
Other 28 clones	Resistant	Resistant	None

Isolation and Purification of Urokinase-like Material.
Benzamidine is an inhibitor of urokinase and other serine proteases and has been used to purify these enzymes by coupling
it to Sepharose for affinity chromatography. Aliquots of the
cell lysate were loaded on a benzamidine- Sepharose affinity
column which was then eluted to collect fractions. Aliquots
from each fraction were assayed by radioimmunoassay in plas-
tic-well microtiter plates (8). Fig. 6 shows the results of

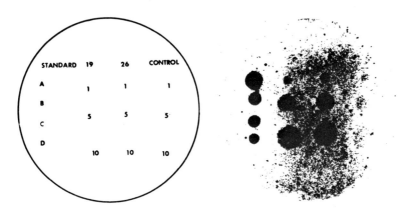

Fig. 5   Detection of urokinase-like material in the lysates
         of transformants.

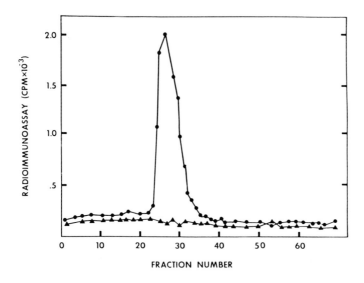

Fig. 6   Benzamidine affinity column chromatography of
         urokinase-like material.

such an experiment. A single positive peak in radioimmunoassays was observed for transformant X1776 (pABB26), which clearly indicates the presence of urokinase-like material. Transformant X1776 (pABB19) showed a similar result (data not shown). As expected, a parallel experiment with X1776 transformed by pBR322 did not show any material which reacted with antibody against the human protein.

Plasminogen Activator Activity of Products from Transformants. To test the products from recombinant DNA expression for functional activity as a human plasminogen activator, the material was subjected to a plasminogen-dependent radioactive fibrinolysis assay (9). Since crude bacterial lysates interfered with the assay, only samples purified by affinity chromatography were employed. Table 6 shows the results of such a study. The negative control, X1776 (pBR322), was devoid of any activity in specific activation of human plasminogen. In contrast, both transformants X1776 (pABB26) and X1776 (pABB19) clearly exhibited functional enzyme activities of a human plasminogen activator. Furthermore, the immunochemical

Table 6. Plasminogen Activator Activity in Bacterial
         Transformants

Sample	Antisera	CPM	Units $(x10^{-3})$	%Activity Remaining
Background	None	798	---	---
Urokinase Std.	None	7,564	35.0	100
Urokinase Std.	Anti-urokinase	1,630	2.0	7
Urokinase Std.	Normal Rabbit Serum	3,458	12.0	35
X1776 (pBR322)	None	952	(0)	---
X1776 (pABB19)	None	18,560	175.0	100
X1776 (pABB19)	Anti-urokinase	1,886	6.0	3.4
X1776 (pABB19)	Normal Rabbit Serum	17,651	156.0	90
X1776 (pABB26)	None	10,887	50.0	100
X1776 (pABB26)	Anti-urokinase	1,479	2.0	4.6
X1776 (pABB26)	Normal Rabbit Serum	5,127	23.0	46

relatedness of urokinase and the recombinant DNA products was studied by the changes in fibrinolytic activity after immuno-precipitation with anti-serum to urokinase. Immunoprecipitation using anti-urokinase and Staphylococcus aureus removed 95 percent of the activity of urokinase or of the enzyme produced from recombinant DNA indicating both molecules share the same antigenic determinants. Normal rabbit serum was also inhibitory, but the extent of inhibition depended on the quantity of plasminogen activator present in the assay. Plasminogen activator activity derived from urokinase or from the product expressed by the recombinant DNA behaved identically in this respect. Serum is known to be inhibitory in this type of assay.

Molecular Species of the Product Expressed by Recombinant DNA. To study the molecular size of the products, samples eluted from the affinity column were subjected to SDS poly-acrylamide gel electrophoresis and then transferred onto cyanogen bromide-activated paper. After reaction with ^{125}I-anti-urokinase antibody and autoradiography, the products were identified and their molecular weights estimated by comparing their mobilities with known protein standards in the same gel. Figure 7 shows that X1776 (pABB26) in Lane 1 produced five

Fig. 7   Electrophoresis and filter affinity transfer analyses of urokinase-like material.

442   P. P. HUNG

bands. Because of diffusion, protein bands were not very
sharp. The positions of the observed bands corresponded to
materials of molecular weight of about 150,000; 125,000;
87,000; 52,000 and 32,000 daltons. The predominant band was
that of the 52,000 species. The smallest two of the products
resemble HMW and LMW urokinase in size (Lanes 2 and 3; visualized with Coomassie blue). The negative control, X1776
(pBR322), in Lane 4 does not show any immunoreactive band
against urokinase antibody.

Characterization of Recombinant Plasmid pABB26. This recombinant plasmid which carries the coding sequence for plasminogen activator in the Pst I site of pBR322, was characterized by restriction endonuclease digestion. Figure 8 shows
the restriction map of pABB26. The total DNA length of pABB26
is about 8,550 base pairs, with an insertion (thick line of
the circle in Fig. 8) of about 4,200 base pairs.

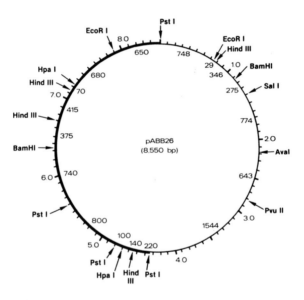

Fig. 8   The restriction map of recombinant plasmid pABB26.

DISCUSSION

The evidence presented in this communication indicates
that we have designed and constructed plasmids containing a
DNA sequence that specifies a plasminogen activator related to

human urokinase, and that E. coli transformed by these plasmids produces a biologically active enzyme. This conclusion is based on the comparisons of the products expressed by the recombinant DNA with urokinase in the following properties: (i) sharing of antigenic determinants as detected by anti-urokinase antibody; (ii) binding to benzamidine, an inhibitor to the active site of urokinase, (iii) similarity in molecular sizes; and (iv) functional enzyme activity as a human plasminogen activator in fibrinolysis. This report represents the first human enzyme gene to be cloned by recombinant DNA techniques, and the products are the largest human proteins engineered and expressed to date.

Inasmuch as the mature mRNA isolated from polysomes revealed a size of 4900 nucleotides, it was not surprising to find some products expressed by the recombinant DNA as large as 150,000 daltons, as well as lower molecular weight species (Fig. 7). We are investigating whether these species are biologically active.

Pulse-labelling of the protein encoded by the constructed plasmid in mini cells showed approximately similar degrees of expression for the plasminogen activator and other gene products encoded in pABB26 DNA sequences. With regard to the question of amounts of plasminogen activator produced, we must recognize that the material initially produced in the cell may be subject to extensive degradation in cultures maintained overnight. Thus, the amount isolated may not be a realistic measure of gene expression. We are improving the degree of expression of the cloned DNA in E. coli.

REFERENCES

1. Williams, J.R.B. (1951) Brit. J. Exptl. Pathol., 32, 530-535.

2. Bernik, M.B. and Kwaan, H.C. (1967) J. Lab. Clin. Med., 70, 650-661.

3. Studer, R.O., Roncari, G., Lergier, W. (1977) In: Thrombosis and Urokinase, Paoletti, R. and Sherry S. (eds), Academic Press, p. 89-90.

4. Ulrich, A., Shine, J., Chirgwin, J., Pictet, R., Tirscher, E., Rutter, W.J., Goodman, H.M. (1977) Science, 196, 1313-1318.

5.   Deeley, R.G., Gordon, J.I., Burns, A., Mullinix, K.P.,
     Bina-Stein, M., and Goldberger, R.F. (1977) J. Biol.
     Chem., 252, 8310-8319.

6.   Pelham, H.R., and Jackson, R.J. (1976) Eur. J. Biochem.,
     67, 247-256.

7.   Rhodes, R.E., McKnight, G.S., and Schimke, R.T. (1973) J.
     Biol. Chem., 248, 2031-2039.

8.   Clarke, L., Hitzeman, R., and Carbon, J. (1979) Methods in
     Enzymology, Vol. 68, pp. 436-442. Academic Press, Inc.
     New York.

9.   Unkeless, J., Dano, K., Kellerman, G.M., and Keick, E.
     (1974) J. Biol. Chem., 249, 4295-4305.

DISCUSSION

H. HEYNEKER: Since you have partial amino acid sequence data on the low molecular weight urokinase, I assume that you have sequenced the cDNA insert. Do these data correlate? Do you know if there is a glycosylation site in the low molecular weight urokinase? I ask this question because bacterially derived urokinase will not be glycosylated; its mobility on SDS-urea gels might therefore be markedly different from the tissue culture derived product.

P. HUNG: The partial DNA sequence has been determined but it is very preliminary and I can't tell you that we found the sequence corresponding to the amino termini of the peptides of urokinase. Regarding the second question---although 75% of the carbohydrate was removed from urokinase by glycosidase, the resulting product had the same electrophoretic mobility as the glycosylated enzyme. The reason why these two forms of urokinase have similar electrophoretic mobilities is not clear at present.

SPEAKER UNKNOWN: What does removal of the carbohydrate do to the activity of the normal enzyme?

P. HUNG: It does not change.

SPEAKER UNKNOWN: How about its half-life?

P. HUNG: I'm not at liberty to discuss it at the moment.

ACTIVITY OF CLONED GENE PRODUCTS
IN ANIMAL SYSTEMS

Nowell Stebbing
Sang He Lee
Benedict J. Marafino
Phillip K. Weck

Department of Biology, Genentech, Inc.
South San Francisco, California
U.S.A.

Kenneth W. Renton
Department of Pharmacology, Dalhousie University
Halifax, Nova Scotia

I. INTRODUCTION

The isolation of therapeutic agents from microorganisms is by no means new, yet the advent of recombinant-DNA derived products has greatly expanded interest in preclinical assessment of the biological properties of such materials. The pharmacological activities of peptides can be determined in appropriate animal systems. However, such studies need to be extended to investigation of any parameter that might affect efficacy of the peptide. In this way optimal treatment regimens may be devised to take account of all known effects both beneficial and adverse. The nature of such studies for products already in clinical use may be obvious. Insulin and human growth hormone (hGH) have been in clinical use for many years and in these cases studies with recombinant-DNA derived materials concentrated on comparison of potency and side effects. Materials not yet in clinical use will present additional problems, particularly development of appropriate animal systems and assessment of potential clinical uses and treatment protocols. Of course, if a recombinant-DNA derived material differs from the natural material, the nature of the

FROM GENE TO PROTEIN:
TRANSLATION INTO BIOTECHNOLOGY

445

difference will require attention. In each of these situations animal systems may be used to assess pharmacological parameters of clinical importance and define a set of desirable properties to be incorporated into any new derivative or second-generation product. The use of such studies for these purposes is here illustrated by reference to recent work with recombinant DNA-derived human growth hormone (hGH), human interferons and thymosin α_1.

II. BIOLOGICAL COMPARISON OF NATURAL AND CLONED GENE PRODUCTS

A. Human Growth Hormone

High level expression of a cloned gene for hGH in E. coli has resulted in production of a form of hGH with 192 amino acids, because the N-terminal initiator methionine is not removed (1). No other chemical differences are apparent between the biosynthetic and pituitary hGH by HPLC and other chemical methods and the two materials are equipotent in growth promoting effects when assayed in hypophysectomized rats (1). These data demonstrate that it is hGH alone which causes growth promotion and rules out synergism with other contaminating bioactive peptides present in pituitary hGH preparations such as FSH, LH, TSH and prolactin. Although most biological comparisons of pituitary and biosynthetic hGH have shown no differences, a sex related difference has been observed recently. In non-hypophysectomized female Sprague-Dawley rats, repeated daily treatments with high doses of biosynthetic hGH cause an increase in growth (Figure 1). This is not observed with pituitary hGH nor in male rats. One explanation of this observation is that growth rate in female animals is normally modulated by some other pituitary factor, such as prolactin, which may be overcome by high doses of hGH free of other pituitary hormones.

Highly purified biosynthetic hGH may also be used to assess other biological properties attributed to hGH. The lipolytic effects observed with clinical preparations of pituitary hGH do not occur with biosynthetic hGH when assayed in the rat fat pad assay or by release of glycerol from rat epididymal adipose tissue (2). Direct effects of pituitary hGH preparations on cells in culture have been reported (3) including stimulation of growth of hematopoietic cells. Using biosynthetic hGH we have confirmed that this effect is due quantitatively to hGH and does not involve other factors present in pituitary hGH preparations.

   Because biosynthetic hGH is an analog of the pituitary material, by virtue of an extra amino acid, it was important to assess the tolerance of experimental animals to the biosynthetic material. The accompanying pre-clinical studies (4) showed no toxicity in acute or chronic studies. However, an effect observed clinically and subsequently in chimpanzees is considered in section III.

   Pituitary hGH preparations have been found to delay glucose clearance in dogs (5) but this was not observed in rhesus monkeys treated with biosynthetic hGH (4). Absence of an effect on glucose clearance in dogs has been found with highly purified pituitary hGH preparations (5) whereas the same material delays glucose clearance after digestion with subtilisin. Thus the "diabetogenic" factor in pituitary hGH preparations is not intact hGH. However, a delay in glucose clearance has been observed in humans treated with biosynthetic hGH (6). Proteinases cleave hGH in the region of amino acid residues 134-150 leaving 2 peptide fragments joined by a disulphide bridge and the growth-promoting effect of this material is comparable to or greater than that of intact hGH (7). Thus partial cleavage of hGH in humans could account for growth promotion accompanied by glucose intolerance effects.

FIGURE 1. The effect on weight gain of pituitary or biosynthetic hGH administered i.m. to male or female Sprague-Dawley rats at 3.125 mg/kg/day for 30 days. Control (saline) animals (●); pituitary hGH (Kabi Crescormon, ☐); biosynthetic hGH (■).

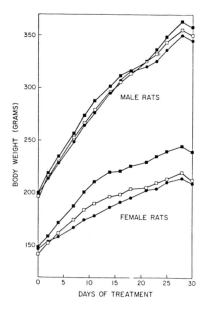

B. Human Interferons

It is now clear that the relationship between antiviral effects in cell cultures and in animals for interferons is superficial because indirect mechanisms are of importance in vivo. Thus destruction of macrophages and probably other cells in mice destroys the efficacy of interferon therapy against EMC virus infection, although the same interferons are effective in mouse fibroblast cultures (8). Moreover, a virus (vaccinia) insensitive to interferon in monkey cell cultures is effective against vaccinia skin lesions in monkeys (9). Interferons have been found to have many effects in vivo which could be important, such as stimulation of NK cells, antibody dependent cell-mediated cytotoxicity (ADCC) and immunosuppression (10, 11).

The importance of indirect effects in vivo for efficacy of interferon therapy means that cloned interferons should not be assumed to be useful simply from cell culture data. However, absence of glycosylation in bacteria has not yet been found to affect in vivo or in vitro activity of interferons. In the case of leukocyte interferons glycosylation does not now appear to occur naturally (12) and in the case of fibroblast interferon, absence of glycosylation does not remove antiviral properties (13). Recombinant DNA— derived human leukocyte and fibroblast interferons, derived from E. coli, have been shown to have antiviral activity comparable to that of natural human interferon preparations in lethal EMC virus infections of squirrel monkeys (14, 15). Studies with HSV-1 infections of the rabbit eye have demonstrated pronounced antiviral effects with cloned leukocyte interferon sub-types (16). In this system treatments from the second day post-infection are as effective as treatments commencing before infection and LeIF-A is as effective as natural buffy-coat interferon despite the fact that LeIF-A shows low activity in rabbit cell cultures (17). These results indicate the importance of indirect mechanisms for antiviral effects against HSV-1 infections of the eye.

Production of a molecular hybrid, LeIF-AD (18), which is active in mice, has allowed examination of antiviral and anti-tumor effects in mouse systems. As in cell cultures, LeIF-AD is highly effective against lethal EMC virus infections of mice and on a weight basis this hybrid is approximately 100-fold more effective than LeIF-D, which is the only non-hybrid human leukocyte interferon sub-type which shows activity in mice and mouse cell cultures (17, 19). Although direct antiproliferative effects may contribute to anti-tumor effects, indirect effects may play an important role in interferon therapy of tumors. LeIF-AD shows little or no inhibition of growth of L1210 cells in culture and NK-cells from LeIF-AD treated mice do not lyse L1210 cells. However BDF$_1$ mice carrying L1210 tumors are protected by LeIF-AD (20) and LeIF-AD is as effective as mouse serum interferon induced by treatment with polyI:C. Treatments before inoculation of mice with L1210 cells are ineffective while treatments commencing 3 to 5 days post-tumor inoculation proved most effective (see Figure 2). Treatments from the third day post-tumor inoculation are most effective and treatments every third day are more effective than daily treatments (21). These results indicate the importance of indirect mechanisms which develop after tumor inoculation and tumor directed antibodies could play a role. Antibodies to L1210 cells, prepared in rabbits, cause pronounced stimulation of L1210 cell lysis in the presence of mouse splenocytes from LeIF-AD treated mice. These observations indicate that a form of ADCC may be important for anti-tumor effects.

The mouse tumor system has also shown that simultaneous administration of cyclophosphamide and LeIF-AD abrogates antitumor effects (20). Cyclophosphamide requires activation by the hepatic P-450 oxygenase system and interferon suppresses this system (22). The data in Table 1 showed that in mice LeIF-AD suppresses specific enzyme and overall cytochrome P-450 metabolism and that interferons without antiviral activity in mice have no effect. However, other factors appear to be involved in the cyclophosphamide/ interferon interaction and effects on parameters of immune responses appear to be important (21).

TABLE 1. Effect of human interferons on hepatic cytochrome P-450 and N-demethylase in mice

Treatment	Units i.p./mouse	Cytochrome P-450 nmoles/mg protein	N-demethylase nmoles HCHO/ mg protein/h
Saline control	---	0.717+0.036	380+17.1
Buffy-coat IFN	5x10^4	0.659+0.015	346+10.8
LeIF-A	5x10^4	0.751+0.029	392+19.6
LeIF-AD	4x10^4	0.533+0.011*	297+28.0*

Asterisks indicate significant difference from control at P<0.05.

Of the various side effects observed with interferons, leukopenia could limit clinical use. Leukopenia has been observed with LeIF-A in rhesus monkeys but not squirrel monkeys indicating that antiviral effects and leukopenia are not inherently linked because antiviral effects with this interferon have been observed in both these species of monkey (14, 15, 23).

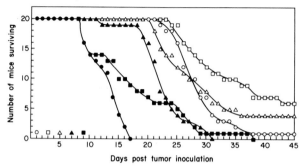

FIGURE 2. The effect of the hybrid interferon, LeIF-AD, 10^5 U/treatment, on L1210 leukemia in BDF_1 mice. All treatments once daily i.p. up to the twelfth day post-tumor inoculation with treatments commencing on day 1 (O); day 3 (□); day 5 (△); day 7 (▲) or day 9 (■). Untreated control group, 10^6 L1210 cells/mouse i.p., (●).

C. Other Cloned Gene Products

Biological activity of bacteria derived thymosin α_1 has been shown to be the same as the natural material and in this case the material without N^α-acetylation shows comparable activity (24). Cloned gene products obtained by cyanogen bromide cleavage of chimeric proteins could have chemical modifications that affect biological activity. However, N-terminal formylation occurs during formic acid chromatography subsequent to cyanogen bromide cleavage necessitating separation of the modified material. In the case of insulin, also produced by cyanogen bromide cleavage of separate chimeric proteins containing the A and B chains, no significant loss of cysteine sulphydryl groups was observed but N-terminal formylation occurred (25).

III. BACTERIAL CONTAMINANTS

Some bacteria derived material will be present in any pharmacological peptide produced in bacteria; the extent of such contaminants will be affected by the nature and difficulty of purification and by the dose of the pharmacologically active peptide to be used. Thus peptides with high biological potency present in fermentation broths at high starting concentrations will present the least problems, provided that the peptide can be separated adequately from bacterial materials by suitable procedures. Clinical use of biosynthetic albumin will present particular difficulties because large amounts are administered (~100 g) and purity of even 99.9 percent would result in administration of ~100 mg of bacterial contaminants. However, the nature as well as the amount of bacterial contaminants is important. Development of host microorganisms which secrete desired cloned gene products and lack endotoxins should prove useful. However, it should be noted that present knowledge of adverse effects of proteins and other materials from bacteria and other microorganisms is limited as are in vitro assays and animal systems for assessing such materials. This may be illustrated by some experiences during development of hGH for clinical use.

An E. coli derived hGH preparation was developed which had greater than 98 percent purity with no detectable endotoxin by the Limulus amoebocyte lysate (LAL) assay (26) and no pyrogenicity in the rabbit test (27). Silver staining of proteins in PAGE revealed minor bands also found in pituitary hGH preparations (probably deaminated derivatives, (7)) and trace amounts of other proteins. This hGH preparation was equipotent with pituitary hGH in various assays (1) and showed no adverse or abnormal effects in acute or chronic tests in rats, squirrel monkeys or rhesus monkeys (4) in which blood chemistry and hematological parameters were monitored. However, during a phase I study, normal adult male volunteers experienced pain at the site of injection accompanied by neutrophilia and a rise in C-reactive protein (CRP), with fever in some cases. These effects have been associated with pyrogens (28) yet no pyrogenicity was observed in rabbits. The adverse signs could not be demonstrated with this same clinical material in guinea pigs nor, on re-testing, in rats, squirrel monkeys and rhesus monkeys at the clinical dose (100 µg/kg) or higher doses. Chimpanzees (Pan troglodytes) were therefore investigated because of their close phylogenetic relationship to man. Temperature was carefully monitored because a contaminating pyrogen was suspected from evidence of fever in the initial phase I clinical study. Calibrated transducers were implanted intramuscularly below the ribs of adult male chimpanzees and after complete recovery of this surgery, temperature was monitored in response to various treatments. The results, in Figure 3, show a rise in body temperature up to 2 h post-treatment, which was significant by t-test analysis compared with untreated animals. However, excipient treated animals showed the same initial temperature response with essentially identical diurnal temperature variations thereafter. Thus the transient temperature effect is attributed to the stress of handling associated with the treatments and not hGH per se. Moreover, the timing of this stress related rise in body temperature is distinct from the pyrexia observed in humans which occurred over a longer period of time.

FIGURE 3. The effect on body temperature in chimpanzees of highly purified bacterial hGH (●) and untreated controls (o). Temperature determined by telemetry from sub-costally implanted AM radio transmitters. Audio beat frequencies on 27MHz receivers were proportional to temperature.

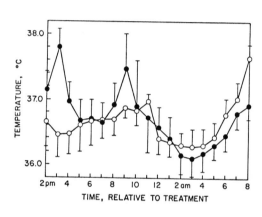

The initial clinical bacterial hGH preparation showed hematological and other effects in chimpanzees also observed in the human phase I study, such as neutrophilia and increased CRP. However, an additional purification step involving dissociation in urea and dextran chromatography produced a material free of any other proteins detectable by silver-staining and this material did not cause significant effects on neutrophils or CRP in chimpanzees. Furthermore, the new hGH preparation has shown no pain or other adverse effects in a repeat human phase I trial (R. Hintz, per. comm.). These results indicate that a bacterial endotoxin was present in the initial batch of hGH but that this caused no treatment related pyrexia in 3 different primates (rhesus and squirrel monkeys and chimpanzees). Moreover, parameters considered to be sequelae to pyrexia, such as neutrophilia and a rise in CRP (28), occurred in chimpanzees but without pyrexia, indicating that the parameters cited are independent effects of endotoxins.

IV. ANALOGS FROM RECOMBINANT GENES

For various reasons cloned gene products may differ in specific features from their natural counterparts. The N-terminal methionine in recombinant DNA-derived hGH is not removed and fibroblast interferon is produced in a non-glycosylated form in E. coli and without the N-terminal methionine present in the natural material (12). In these cases biological activity seems to be unaffected but the cloned gene products are analogs of their natural counterparts. Human leukocyte interferons do not now seem to be glycosylated. However, natural human cell derived preparations are mixtures of various molecular sub-types and the ratio of the constituent sub-types may vary between natural preparations. Thus individual cloned leukocyte interferon sub-types are themselves analogs of the natural mixtures and are known to have distinctive biological properties.

The very process of recombining DNA molecules allows production of new hybrid products from various gene segments and this is illustrated by the hybrid human leukocyte interferons. So far, production of such hybrids has been limited by the existence of common restriction enzyme sites in genes of the leukocyte interferon sub-types. However, the same methods should allow production of interferon fragments either by removal of gene segments between restriction enzyme sites or by terminal deletions. These approaches will allow extension of quantitative structure-activity studies to such high molecular weight pharmacological agents. The method is not limited by the position of restriction enzyme sites because controlled, site specific changes in genes may be achieved using oligonucleotide primers to specific genome sites. Analogs may also be produced by specific amino acid modifications and such studies should indicate the role of particular domains in pharmacological polypeptides.

Elucidation of structure-activity relationships may lead to development of new products having only certain of the desired properties of the natural material. Such materials need not necessarily be smaller "core" structures, as was the case for penicillins, for example. Larger proteins with new functional domains may be produced by recombinant DNA methods and allow incorporation of new features such as targeting to specific tissues or reduced clearance from the circulation. However, in all these cases careful consideration must be given to the biological parameters monitored in structure-activity studies. Derivative or second-generation products should incorporate known

beneficial   properties   and   omit   potentially   adverse
properties.   Appropriate   animal   studies   permit   definition
and monitoring of such properties.

V. DISCUSSION

Assessment of properties of cloned gene products in
animal systems transcends simple confirmation of biological
potency.   Recombinant-DNA derived materials from bacteria
are produced with trace amounts of bacterial contaminants
rather than other human peptides.   Moreover, cloned gene
products may be different by virtue of retention of an
N-terminal methionine or other specific differences, either
a different amino acid sequence or absence of secondary
modifications.   So far absence of secondary modifications
has   not   apparently   affected   biological   activity. N^{α}-
desacetylthymosin α_1 has proved as active, at least in the
properties   examined,   as   natural   thymosin α_1 and   the
bacteria derived non-glycosylated fibroblast interferon is
indistinguishable   in   biological   properties   from   natural
fibroblast interferon.
There is an obvious need for a high degree of purity for
therapeutic agents produced in bacteria and production of
such   materials   will   benefit   from   development   of   new
large-scale   protein   purification   procedures.   Antibody
affinity columns, such as have been used for production of
human interferons (29, 30) could prove generally useful.
However,   the   nature   of   contaminants   is   important.
Identification   of   specific,   potentially   detrimental
contaminants in host microorganisms should allow easier
monitoring of cloned gene products and this will initially
depend on development of predictable animal systems.
Although cloned gene products are a very recent
development, remarkably pure materials have been produced.
This has allowed definitive determination of some of the
varied properties of the interferons and separation of the
properties of hGH from those of other contaminating active
pituitary peptides.   Purified cloned gene products will be
generally useful for investigating the biological properties
of individual species of related proteins and their effects
in combination with existing chemotherapeutic agents.
Without   doubt   some   eukaryotic   proteins   will   not   be
produced in an active conformation in microorganisms and
ultimately this can only be assessed in animal systems.   In
some cases a eukaryotic protein produced in bacteria may
inhibit cell growth and this seems to occur in the case of
human   fibroblast   interferon.   In   other   cases   eukaryotic

proteins may be degraded in bacteria. Although this appears to be the case with the hemagglutinin of influenza virus, fragments positive by antibody binding assays may be readily expressed in fusion proteins (31). Ultimately the antiviral utility of such protein fragments can only be assessed by virus protection studies in animals.

It is now apparent that recombinant-DNA methods may be very effectively applied to production of variants or analogs of natural materials. Complete synthesis of genes (32) or site specific changes to cDNA derived genes may extend classical structure activity studies even to large peptides. The issue now is to devise criteria by which new variant materials are to be selected. In the case of hGH, absence of effects on glucose clearance in man would obviously be desirable although an animal model mimicking this is not obvious. The naturally occurring 20K variant of hGH would seem to be a known candidate material because it does not affect glucose clearance in dogs even after subtilisin treatment (7). Interferons lacking hematological and cytochrome P-450 effects may be desirable and in every case absence or limitation of immunogenicity will be important. As we have seen, studies in animal systems provide the evidence that desirable and undesirable effects may be separated and also the means for assessing whether an analog possesses the desired properties. In some cases, cloning of genes now provides a convenient means for assessing biological properties of distinct molecular species not readily separated from related materials in natural preparations. The distinctive biological properties of particular leukocyte interferon sub-types was not assessed before cloning of the genes. Cloning should prove particularly useful in cases where related materials have agonistic or antagonistic effects, as appears to be the case with the lymphokines. As recombinant-DNA methods reveal the structure of natural pharmacologically active proteins, these materials will pass from the area of poorly defined biological materials to specific, chemically defined agents. Strategies for cloning and high level expression in microorganisms will in part affect the nature of the products and their contaminants, and animal systems for assessing pharmacological properties of these products are likely to become essential for rational drug development.

REFERENCES

1. Olson, K.C., Fenno, J., Lin, N., Harkins, R.N., Snider, C., Kohr, W.H., Ross, M.J., Fodge, D., Prender, G., and Stebbing, N., Nature 293, 408 (1981).

2. Frigeri, L.G., Robel, G., and Stebbing, N., Biochem. Biophys. Res. Commun. 104, 1041 (1982).
3. Golde, D.W., in "Growth hormone and other biologically active peptides." (A. Pecile, E.E. Muller, eds.) p. 52. Excerpta Medica (1980).
4. Kaplan, S.L., Fenno, J.T., Stebbing, N., These proceedings (1982).
5. Lewis, U.J., Singh, R.N.P., VanDerLaan, W.P., and Tutwiler, G.F., Endocrinology 101, 1587 (1977).
6. Rosenfeld, R.G., Wilson, D.M., Dollar, L.A., Bennet, A., and Hintz, R.L., J. Clin. Endocrin. Metab. 54, 1033 (1982).
7. Lewis, U.J., Singh, R.N.P., Tutwiler, G.F., Sigel, M.B., VanDerLaan, E.F., and VanDerLaan, W.P., Recent Prog. Horm. Res. 36, 477 (1980).
8. Stebbing, N., Dawson, K.M., and Lindley, I.J.D., Infect. Immun. 19, 5 (1978).
9. Schellekens, H., Weimar, W., Cantell, K., and Stitz, L., Nature 278, 742 (1979).
10. Gresser, I., Cell. Immunol. 34, 406 (1977).
11. Fridman, W.H., Gresser, I., Bandu, M.T., Aguet, M., and Neauport-Santes, C., J. Immunol. 124, 2436 (1980).
12. Allen, G., and Fantes, K.H., Nature 287, 408 (1980).
13. Harkins, R.N., Hass, P.E., Kohr, W.H., Aggarwal, B.B., Weck, P.K., and Apperson, S., Proc. Natl. Acad, Sci. (in press.)
14. Goeddel, D.V., Yelverton, E., Ullrich, A., Heyneker, H.L., Miozzari, G., Holmes, W., Seeburg, P.H., Dull, T., May, L., Stebbing, N., Crea, R., Maeda, S., McCandliss, R., Sloma, A., Tabor, J.M., Gross, M., Familetti, P.C., and Pestka, S., Nature 287, 411 (1980).
15. Weck, P.K., Apperson, S., Hamilton, E., Estell, D., and Stebbing, N., Virol. (in press).
16. Smolin, G., Stebbing, N., Friedlaender, M., Friedlaender, R., and Okumoto, M., Arch. Ophthalmol. 100, 481 (1982).
17. Weck, P.K., Apperson, S., May, L., and Stebbing, N., J. Gen. Virol. 57, 233 (1981).
18. Weck, P.K., Apperson, S., Stebbing, N., Gray, P.W., Leung, D., Shepard, H.M., and Goeddel, D.V., Nucleic Acids Res. 9, 6153 (1981).
19. Weck, P.K., Rinderknecht, E., Estell, D.A., and Stebbing, N., Infect. Immun. 35. 660 (1982).
20. Stebbing, N., Weck, P.K., Fenno, J.T., Apperson, S., and Lee, S.H., in "The biology of the interferon system" (E. DeMaeyer, G. Galasso and H. Schellekens, eds.) p. 25. Elsevier/North-Holland Biomedical Press (1981).
21. Lee, S.H., Weck, P.K., Moore, J., Chen, S., and Stebbing, N., Symp. Mol. Cell Biol. 25 (in press).

22. Singh, G., Renton, K.W., and Stebbing, N., Biochem. Biophys. Res. Commun. (in press).
23. Schellekens, H., de Reus, A., Bolhuis, R., Fountoulakis, M., Schein, C., Escodi, J., Nagata, S., and Weissmann, C., Nature 292, 775 (1981).
24. Wetzel, R., Heyneker, H.L., Goeddel, D.V., Jhurani, P., Shapiro, J., Crea, R., Low, T.L.K., McClure, J.E., and Golstein, A.L., Biochem. 19, 6096 (1980).
25. Wetzel, R., Ross, M.J., Levy, M.J., and Shively, J.E., Gene (in press).
26. The United States Pharmacopeia, 20th revision, p. 888. The U.S. Pharmacopeial Convention, Inc., Rockville, MD (1980).
27. The United Staes Pharmacopeia, 20th revision, p. 902. The U.S. Pharmacopeial Convention, Inc., Rockville, MD (1980).
28. Atkins, E., and Bodel, P., in "The Inflammatory Process" (B.W. Zweibach, L. Grant and R.T. McClusky, eds.) p. 467. Academic Press, New York (1974).
29. De Maeyer-Guignard, J., Tovey, M.G., Gresser, I., and De Maeyer-Guignard, E., Nature 271, 622 (1978).
30. Secher, D.S., and Burke, D.C., Nature 285, 446 (1980).
31. Davis, A.R., Nayak, D.P., Veda, M., Hiti, A.L., Dowbenko, D., and Kleid, D.G., Proc. Natl. Acad. Sci. USA 78, 5376 (1981).
32. Edge, M.D., Greene, A.R., Heathcliffe, G.R., Meacock, P.A., Schuch, W., Scanlon, D.B., Atkinson, T.C., Newton, C.R., and Markham, A.F., Nature 292, 756 (1981).

DISCUSSION

R. WARD: In the studies you describe on interferon inhibition of virus growth, what virus types have been examined?

N. STEBBING: In cell cultures about 8 viruses have been examined. In monkeys, EMC virus and vaccinia virus have been studied. I did not include HSV-1 studies in my talk but the text cites such studies in herpes induced keratitis of the rabbit eye. These studies were of particular interest because a cloned leukocyte interferon sub-type (LeIF-A) showed activity comparable to natural buffy-coat derived interferon. The LeIF-A has virtually no antiviral activity in rabbit cell culture but is effective in vivo and it is as effective with regimens commencing 2 days after infection as with regimens commencing before infection. These results are obviously encouraging from a clinical point of view and also imply that indirect mechanisms of action are of primary importance in this disease.

L. BAUMBACH: 1) While studying the anti-viral and anti-tumor effects of several IFNs, were any of the tests performed on carcinomas of viral origin? 2) Do you know the primary target of interferons possessing antiproliferative activity?

N. STEBBING: 1) As I stated earlier, three carcinoma cell lines and a control cell line were examined in the cervical carcinoma studies: An in situ, an invasive and an HSV-2 transformed cell line obtained from the control, cervical epithelial cells.
    2) We do not know the primary mode of action in these anti-proliferative studies.

D. GELFAND: Since IFN-α D has a greater activity than - α A in a monkey model system, why have you conducted clinical trials with IFN- α A?

N. STEBBING: This was because work with the LeIF-A (IFN- αA) sub-type was the first to be ready for clinical studies and we did not have the data I referred to at that time. We sometimes have to make decisions of this type before all the data are available!

D. BURKE: Not only do we have the possibility of using interferon hybrids; we also have the possibility of using mixtures of cloned gene products. Do you have any evidence for a synergistic effect?

N. STEBBING: This has been examined in a number of situations and the results generally show additive effects. NK-cell stimulation by individual leukocyte interferon sub-types is not as pronounced as with natural buffy-coat derived preparations. Synergistic interactions between sub-types were therefore anticipated for NK-cell stimulation and these have been observed recently.

STUDIES ON POLIOVIRUS PRODUCED BY TRANSFECTION
WITH CLONED VIRAL cDNA

Vincent R. Racaniello[1]
David Baltimore[2]
Department of Biology
and
Center for Cancer Research
Massachusetts Institute of Technology
Cambridge, MA, U.S.A.

I. INTRODUCTION

Poliovirus is the best-characterized member of the pico-
rnavirus group, which includes positive-strand, non-segmented
RNA animal viruses (1,2). Initially poliovirus was studied
as the agent of a devastating human disease. After polio-
myelitis was controlled by vaccination, research on poliovirus
continued, in part because it possesses a number of intriguing
features. Poliovirus contains a single molecule of RNA
approximately 7.5 kb in length. A small protein, called VPg,
is covalently linked to the 5' end of this molecule and a
stretch of poly(A) is found at the 3' end. A variety of
experimental results have suggested that during infection of
animal cells, the viral RNA is translated into a single continuous polypeptide of 250,000 molecular weight. This polypeptide is subsequently cleaved by proteases to form functional viral proteins needed to replicate and transcribe viral
RNA and to form structural components of newly synthesized
virus particles.

The past several years have seen widespread application
of recombinant DNA technology to the study of RNA viruses.
To use these techniques to better understand poliovirus,
complementary DNA copies of the viral RNA were cloned into
the bacterial plasmid pBR322. Three molecular clones which
together represented the entire viral RNA were used to deter-

---

[1]Recipient of an NIH Postdoctoral Fellowship
[2]American Cancer Society Professor of Microbiology

FROM GENE TO PROTEIN:
TRANSLATION INTO BIOTECHNOLOGY

459

mine the entire 7,440 nucleotide sequence of the poliovirus
genome (3). The nucleotide sequence was correlated with N-
terminal amino acid sequence of poliovirus proteins (3,4),
permitting the construction of a complete genetic map of the
viral genome.

An important question to answer was whether the poliovirus cDNA clones could generate infectious virus in mammalian
cells. To address this problem, the three cDNA clones were
joined to produce a single, full-length copy of the poliovirus genome in pBR322. When this full-length clone was
transfected into CV-1 or HeLa cells, infectious virus was
produced (5). The availability of this infectious cloned
poliovirus cDNA will make possible genetic manipulations
which are not feasible with RNA. For example, it should now
be possible to construct defined poliovirus mutants. Here
we extend our original observations on the infectivity of
cloned poliovirus cDNA by showing that transfection-derived
viruses are identical, by several biochemical criteria, to
the virus originally used to make the cDNA clone. The infectious, cloned cDNA copy of poliovirus RNA will therefore
be extremely useful for performing directed genetic
manipulations.

II. RESULTS

A. Transfection of Mammalian Cells with Cloned Viral cDNA

Previously we reported that transfection of monolayers of
CV-1 or HeLa cells with a full length poliovirus cDNA clone
(pVR106) resulted in production of virus. In those experiments, cells were transfected using the modified calcium
phosphate technique (6). Briefly, a DNA-calcium phosphate
coprecipitate was applied to the cell monolayer, which was
incubated 15' at room temperature, and then covered with
warm medium. After 4 hours' incubation at 37°, the cells
were shocked with glycerol, and then covered with either
fluid medium or with an agar overlay. Using this technique,
it was found that 10 μg of plasmid DNA usually induced 5-70
plaques on monolayers in 10 cm. plastic dishes, a number
which varied greatly from one experiment to another.

When the calcium phosphate technique is performed on cells
in suspension, however, more reproducible results are obtained. In this technique (7), monolayers are removed from
the plastic dish by trypsinization and are centrifuged into a
pellet. The DNA-calcium phosphate coprecipitate is applied
to the pellet and the cells are gently resuspended. After 15'
at room temperature, warm medium is added and the cells are
replated into two 10 cm. dishes. The cells are allowed to

TABLE I.   Transfection of Cultured Mammalian Cells with
           Cloned, Full Length Poliovirus cDNA

Plasmid DNA	Experiment No.	No. of Plaques on Transfected Cell Monolayers
CV-1 cells		
pVR106	1	29,24
	2	24,20
	3	21,26
	4	28,32
pVR106a	1	30,26
	2	31,24
	3	32,35
	4	30,30
HeLa cells		
pVR106	1	15,18

settle for 4 hours at 37° and are then shocked with glycerol.
After glycerol treatment, cells may be covered with either
an agar or a liquid overlay (5).

Using this technique of transfection, from 20–30 plaques
are observed per plate, or a total of 40–60 plaques per 10
µg of plasmid DNA. These results are shown in Table I. Two
independently constructed full-length cDNA clones, pVR106
and pVR106a, were used (5). Each plasmid was transfected
into CV-1 cells on four separate days. After glycerol treatment, cells were covered with an agar overlay, incubated 4
days at 37°, and then stained with crystal violet (5). The
number of plaques observed per dish is much more reproducible than was found previously (5). This technique of
transfection is now routinely used for poliovirus cDNA.

It was also possible to use this transfection technique
with HeLa cells growing in spinner culture (8). Previously
we used a line of HeLa cells adapted to growth in monolayer.
In one experiment, approximately 4×10^{6} cells growing in
suspension were centrifuged into a pellet and transfected with
10 µg of pVR106 as described above. The cells adhered to
plastic dishes within 4 hours at 37°. Cells were treated with
glycerol and then incubated under agar overlay. As shown in
Table I, plaques were observed on these monolayers. This

technique circumvents the trypsinization step and therefore
saves time and eliminates extra manipulation of the cells.

B. Nucleotide Sequences at the 5' End of RNA from
   Transfection-Derived Poliovirus

   We previously speculated on the mechanism by which a cDNA
copy of an RNA virus initiates the infection process (5).
In one scenario plasmid DNA enters the nucleus and RNA
synthesis might initiate at promoter-like regions in pBR322.
Alternatively, the plasmid DNA might integrate next to a
cellular promoter. In either case, the initial plus-strand
transcripts would probably have extra, non-viral sequences at
either end. It is not clear how these initial transcripts
would be replicated. To investigate these problems, it is

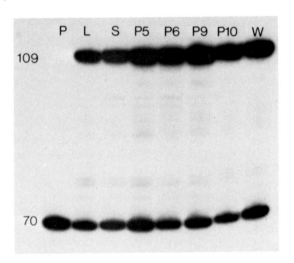

   Figure 1. Electrophoresis of the extension products of
a DNA primer hybridized to different poliovirus RNAs. A
small end-labeled primer, representing bases 39 to 105 from
the 5' end of poliovirus RNA, was hybridized to viral RNA,
and the hybrids were extended with reverse transcriptase,
as described in the text. The products of the extension were
electrophoresed on an 8% polyacrylamide-7M urea gel (12) and
detected by autoradiography. The primer is 70 bp in length;
extension of this fragment to the first nucleotide of the
viral RNA produces a fragment 109 bp in length. P, primer
alone; L, S, P5, P6, P9, P10, display extension reactions
using RNAs extracted from transfection-derived virus; W,
primer extension with wild-type poliovirus RNA.

important to determine the structure of the termini of RNAs
from transfection-derived virus.

Poliovirus was recovered from three isolated plaques in
each of two separate transfections of CV-1 cells with pVR106.
RNA was prepared as described (9) from the six plaque-purified viruses--designated L, S, P5, P6, P9 and P10--as well
as from the wild type poliovirus used to construct the cDNA
clone. The sequences at the 5' end of these RNAs were determined by primer extension. A 70 bp DNA fragment was isolated
which represented positions 39 to 108 from the 5' end of the
viral RNA. This primer was prepared from plasmid pVR105,
which represents bases 1-220 of the viral RNA (5). Plasmid
pVR105 was cleaved with Dde I, and the resulting fragments
were labelled with ^{32}P at the 5' end using polynucleotide
kinase and γ-^{32}P-ATP. The mixture of labelled fragments was
cleaved with Bgl I, and then separated by polyacrylamide gel
electrophoresis. In this wasy a 70 bp primer fragment was
isolated which was labeled with ^{32}P at the Dde I site
(position 108) only.

The primer was hybridized to each viral RNA sample and
the hybrid molecules were extended with reverse transcriptase
as described (3, 10). Figure 1 shows an autoradiograph of
a polyacrylamide-urea gel on which the primer and the seven
extension reactions were fractionated by electrophoresis.
An extension product from the plaque-derived RNAs was synthesized which co-migrated with the extension product obtained when wild-type RNA was used. The length of this extension
product was 109 bases. Apparently RNA from the plaque-
purified virus is identical in length, in this region, to
wild-type RNA.

To determine the sequence of the viral 5' ends, the extension products were purified from the gel, and the nucleotide sequences were determined by the Maxam-Gilbert technique. The sequence of bases 1-38 was identical to the wild-
type sequence (data not shown).

C. Fingerprints of RNA from Transfection-Derived Virus

In the future, the infectious poliovirus cDNA clone will
be used to generate specific poliovirus mutants. To be useful for this purpose, the cDNA clone must give rise to
viruses which are identical and which resemble the wild-type
virus used to construct the cDNA clone. RNAse T1 oligonucleotide fingerprints were used to provide an estimate of
the similarity between transfection-derived poliovirus and
wild-type virus.

The six plaque-purified viruses described above, and wild-
type poliovirus, were grown in monolayers of HeLa cells in the

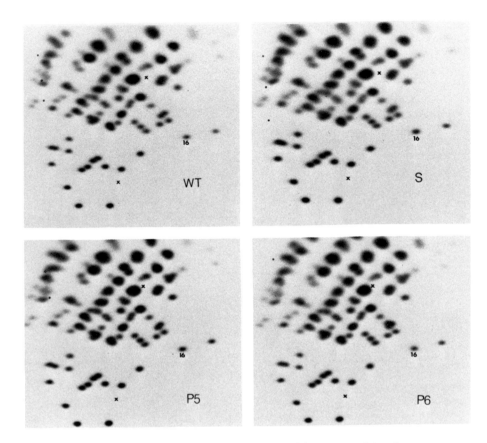

Figure 2. Fingerprints of RNA from wild-type poliovirus type
I(Mahoney) and three transfection-derived polioviruses, S,
P5 and P6. RNA was continuously labeled with ^{32}P in vivo,
digested with RNAse T1, and the resulting oligonucleotides
were fingerprinted in two dimensions on polyacrylamide gels,
as described in the text. The location of oliogonucleotide
#16 is indicated. Separation in the first dimension is from
left to right; separation in the second dimension is from
bottom to top. Lower X, xylene cyanole FF dye; upper X,
bromophenol blue dye.

presence of $H_3{}^{32}PO_4$. At six hours post infection, cells
were scraped from the plastic dish and lysed with NP-40 (9).
The nuclei were removed (9), and the cytoplasmic extract was
fractionated on a gradient of 15-30% sucrose. The peak of
radioactivity representing virus particles was pooled, and
viral RNA was prepared (9). The RNA was digested with ribo-

nuclease Tl, and the digestion products were fingerprinted in
two dimensions on polyacrylamide gels (11).

The fingerprints of RNA from three transfection-derived
viruses as well as wild-type virus are shown in Figure 2.
Examination of the fingerprints indicates complete identity
of all the large oligonucleotides (below the bromophenol blue
dye, the oliogonucleotides are approximately 15 to 37 bases
in length). RNA from wild-type virus was mixed with RNA from
each plaque-purified virus, digested with RNAse Tl, and
fingerprinted. All corresponding spots comigrated, confirming
that the pattern of large oligonucleotides from transfection-
derived virus is identical to that of wild-type virus (data
not shown). In addition, the pattern from three other
transfection-derived viruses was also identical to that of
wild-type virus (data not shown).

Another observation which can be made from the fingerprints concerns the 3' end of the viral genome. Spot number
16 (Figure 2) has been mapped to the 3' end of the viral RNA
(4), representing bases 7419-7434:

          7420   7430   7440
     AGGGGTAAATTTTTCTTTAATTCGGAG-poly(A)
             oligonucleotide #16

This oligonucleotide is present in all six plaque-purified
transfection-derived viruses (Figure 2, and data not shown).
RNA from all the viruses contain poly(A), which migrates as
a smear in the lower left hand of the fingerprint (not
visible in the exposure of Figure 2). Furthermore, experiments indicate that RNA from transfection-derived viruses
binds oligo d(T)-cellulose, indicating that it is polyadenylated (data not shown). These results indicate that the 3'
end of these viruses closely resembles and is probably
identical to the the 3' end from wild-type virus.

C. Polypeptides Induced in Cells Infected with Transfection
                  Derived Poliovirus

Another method for determining the similarity between
transfection-derived poliovirus and wild-type poliovirus is
to examine virus-specific polypeptides in infected cells.
Sequence changes not detectable in an oligonucleotide fingerprint might alter the migration of polypeptides or change
a protease cleavage site.

HeLa cells growing in suspension were infected with wild-
type poliovirus or with each of six transfection-derived

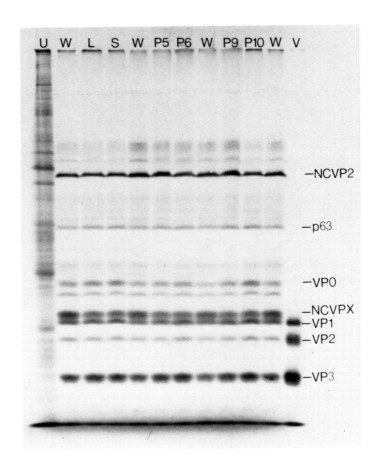

Figure 3. Separation of ^{35}S-methionine labeled polypeptides from uninfected cells and cells infected with wild type and transfection-derived polioviruses. HeLa cells in suspension culture were infected with poliovirus, pulse-labeled with ^{35}S-methionine for 30' at 3.5 hr. post-infection, and cytoplasmic extracts were prepared. Electrophoresis of cell extracts was performed on a 12.5% polyacrylamide-SDS gel(9), and polypeptides were detected by autoradiography. U, uninfected cell polypeptides; W, wild-type polypeptides; L, S, P5, P6, P9, P10, infected cell polypeptides from transfection-derived poliovirus; V, polypeptides from purified wild-type poliovirus labeled with ^{35}S-methionine. Several virus-specific polypeptides are identified at right.

viruses. At 3.5 hr post infection, the cells were pulsed
with ^{35}S-methionine for 30', and then cytoplasmic extracts
were prepared as described (9). Figure 3 shows an autoradiograph of an SDS-polyacrylamide gel on which infected
cell lysates have been fractionated by electrophoresis. The
pattern of virus-specific polypeptides in each of the transfection-derived virus isolates is identical to the pattern
of wild-type polypeptides, both in migration rate and in
number. Therefore viruses derived from transfection appear
to induce the same polypeptides as wild-type virus.

III. DISCUSSION

It was shown previously that a cloned, full length copy
of the poliovirus genome is infectious in mammalian cells (5).
Here we show that, using a procedure for transfection in
suspension, more reproducible numbers of infectious foci are
obtained. In addition, since the cells are split 1:2 after
transfection, there is ample room for the cells to grow for
several days. This condition results in healthier, growing
cells on which viral plaques are more easily visualized.
The RNAs and proteins of viruses which emerge from cells
transfected with viral cDNA have been examined and compared
to those of wild type poliovirus. The sequence of nucleotides 1-38 at the 5' end of RNA from 6 plaque-purified,
transfection-derived viruses was identical to that of wild
type virus. It is likely that when the plasmid DNA enters
a cell, the initial RNA transcripts contain non-viral
sequences at their 5' ends. The result obtained here
indicates that these sequences are lost at some point in
subsequent replication. The mechanism by which these
sequences are lost should be investigated, since it may shed
light on how the viral replicase works.
Oligonucleotide fingerprinting has been used to compare
the RNA of transfection-derived virus with that of wild type
virus. Large RNAse T1-resistant oligonucleotides (n=10-37)
have been mapped on the viral RNA (4), and these are
scattered randomly across the genome. Therefore the RNAse
T1 oligonucleotide fingerprints can be used to compare a
random sampling of the structure of the viral RNAs. We
found that the fingerprints of six transfection-derived
viruses were identical to the fingerprint of wild-type RNA.
From this result we may conclude that these viruses are not
grossly different from wild type virus. In addition, the
presence of a 3'-specific oligonucleotide, as well as poly(A),
in transfection-derived virus, indicates that the 3' end of

the viral RNA has not been detectably altered.

The pattern of proteins induced in infected cells by transfection-derived virus was shown to be identical to that of wild type poliovirus. However, missense mutations would not alter the migration of viral polypeptides, and single base changes might go undetected in an RNA fingerprint. The biochemical comparisons listed here, combined with the observation that transfection-derived viruses appear to have identical growth properties with wild type virus (data not shown) indicate that these viruses are not significantly different from the parent virus. This observation is important because it shows that the full length clone generates identical, "wild-type" viruses. The cDNA clone can therefore be used to produce directed alterations in the viral genome.

The genetics of poliovirus has received only rudimentary attention in the past. The availability of an infectious cDNA copy of poliovirus RNA should permit new genetic manipulations with this virus. For example, it should now be possible to construct a collection of defined poliovirus mutants which will undoubtedly enhance the study of poliovirus genetics. In this way recombinant DNA technology will enable us to continue the fruitful study of poliovirus for many years to come.

ACKNOWLEDGMENTS

We thank Dr. James F. Young for reagents used in fingerprinting, and Philip Hollingshead for technical assistance.

REFERENCES

1. Baltimore, D., Bact. Rev. 35, 235 (1971).
2. Baltimore, D., in "The Biochemistry of Viruses" (H.B. Levy, ed), p. 101. Dekker, N.Y., (1969).
3. Racaniello, V.R., and Baltimore, D., Proc. Natl. Acad. Sci. USA. 78, 4887 (1981).
4. Kitamura, N., Semler, B.L., Rothberg, P., Larsen, G., Adler, C., Dorner, A., Emini, E., Hanecak, R., Lee, J., van der Werf, S., Anderson, C., and Wimmer, E., Nature 291, 547 (1981).
5. Racaniello, V.R., and Baltimore, D., Science 214, 916 (1981).
6. Parker, B.A., and Stark, G.R., J. Virol. 31, 360 (1979).

7.   Chu, G., and Sharp, P.A., Gene 13, 197 (1981).
8.   Baltimore, D., Girard, M., and Darnell, J.E., Virology 29, 179 (1966).
9.   Baron, M.H., and Baltimore, D., Cell, in press.
10. Lamb, R.A., and Lai, C.J., Cell 21, 475 (1980).
11. Pedersen, F.S., and Haseltine, W.A., in "Methods in Enzymology", vol. 34.
12. Maxam, A.M. and Gilbert, W., in "Methods in Enzymology" (Grossman and Moldave, eds.), vol. 65, p. 499. Academic Press, N.Y.

DISCUSSION

R. SAMULSKI: Does the deletion of plasmid pVR106 span the AUG that is translated into polio's polypeptide? Second, do you have any evidence of how this recombinant molecule yields infectious polio virus?

V. RACANIELLO: The deletion shown (pVR106d220-670) does not include the AUG which initiates the polyprotein. That AUG is at position 743.
     We have no evidence concerning the mechanism by which the cDNA clone yields infectious poliovirus.

N. STEBBING: You mentioned that gln/gly was the junction sequence for a protease releasing viral peptides. Are there gln/gly sites elsewhere within the final peptide products?

V. RACANIELLO: There are 13 gln/gly pairs in the polio virus polyprotein; at least 8 of these are sites for proteolytic processing. It is not known whether the other 5 gln/gly pairs are not processed or are processed very infrequently.

D. BURKE: What strain of polio was used, and is there any information about the sequences of different strains?

V. RACANIELLO: The sequence of poliovirus Type I (Mahoney strain) was presented here. Sequences of one other strain (Sabin Type I) are now being determined in Japan, but no comparisons are available.

E. LAMON: I have two questions:
1)   Do you know which gene product turns off host cell macromolecular synthesis?
2)   Have you simultaneously transfected with your incomplete clones to see if you can get complementation?

V. RACANIELLO: 1) The poliovirus gene product responsible for shutoff of host cell macromolecular synthesis has not been identified.

2) Clones pVR105 and pVR104 have been co-transfected into CV-1 cells but no virus was produced.   These two plasmids overlap; pVR105 represents bases 1-220, and pVR104 represents base 116 through the 3' end of the viral RNA.

HORIZONS IN BIOTECHNOLOGY

MARTIN A. APPLE
J. PAUL BURNETT
RALPH W. F. HARDY
DALE H. HOSCHEIT
PATRICK C. KUNG
J. SCHULTZ
STEPHEN C. TURNER
WILLIAM J. WHELAN

Friday Afternoon Panel Discussion

JULIUS SCHULTZ (PAPANICOLAOU CANCER RESEARCH INSTITUTE)
      I would like to introduce the chairman and organizer of
this panel, Dr. Ralph Hardy. Through his early researches at
DuPont, he eventually became director of Life Sciences
Research and an expert in the field of recombinant DNA. Most
of the people who advance to senior positions in science,
especially very busy people, are often called upon to do a
chore for one of the scientific societies. Most of you are
probably unaware that Dr. Ralph Hardy has just retired as
secretary of the Division of Biochemistry of the American
Chemical Society, a position greatly demanding in time and
effort. I think he should be recognized as one of those
people in industry who is willing to devote a great deal of
time to what is essentially an academic activity. Dr. Hardy.

RALPH HARDY (DUPONT)
      Thank you Julius. Let me extend my welcome to the ses-
sion on HORIZONS IN BIOTECHNOLOGY. Some of you may recall
that at the 1977 Miami Winter Symposium, which I believe was
the first one on recombinant DNA, the panel was used to
address the then intense issue of safety and ethics of experimentation in this field. As we all know, the safety issue
has greatly subsided. In its place other issues have sur-
faced. They relate to the multiple approaches in universities and industries to commercialize this science, and the
impact of these commercializations on the future progress of
science. We've assembled in this panel individuals who
represent several of the diverse approaches being used to
commercialize this science.
      We will develop the discussion in the following way:
each member of the panel will "prime the pump" with a five-
minute presentation on his organization's business approach
and comments on specific issues. When all have spoken, we
will open the floor for questions stimulated by either their
comments or your concerns. Dialog between different panel
members with different views will also be encouraged.
      Let me introduce each of the members. Professor Bill
Whelan, who with Julius Schultz has organized these Symposia
for several years. He is Professor and Chairman of the
Department of Biochemistry at the University of Miami. He
will present the view of a university that is seeking appropriate ways to commercialize its inventions. Dr. Martin
Apple is President of the International Plant Research
Institute, IPRI. He will describe the approach of IPRI which
was the first major one specializing in the field of plant

agriculture. Mr. Stephen Turner is President of Bethesda Research Laboratories. His company initiated its activities as a provider of research materials diversified from that base.   Dr. Patrick Kung is Divisional Director of Centocor. His company seeks commercial opportunities based on monoclonal antibodies. Dr. Paul Burnett is Director of Molecular Biology at Eli Lilly Laboratories. He will present the view of a large pharmaceutical company, which in all probability will be the first marketer of a major product based on recombinant DNA technology. Mr. Dale Hoscheit is a member of the legal firm of Schuyler, Banner, Birch, McKie and Beckett. He has spoken broadly on legal aspects in this area and will address a symposium of the American Society of Microbiology on this same subject later this year.

WILLIAM J. WHELAN (UNIVERSITY OF MIAMI)
      Thank you Dr. Hardy.   Ladies and Gentlemen, anything that I could say on the subject of university/industry relationships in the course of five minutes will necessarily have to be rather superficial. Let me recommend to you for an in-depth survey of this topic a symposium that Dr. Hardy and I helped to organize in Rome at the end of September, and deals with the whole issue of university/industrial relationships.   It is "From Genetic Experimentation to Biotechnology--the Critical Transition" edited by myself and Ms. Sandra Black, to be published by John Wiley & Son next month. It was sponsored by the Committee on Genetic Experimentation of the International Council of Scientific Unions and the Italian National Research Council.
      Why am I particularly interested in this topic?   It is said that within every fat man there is a thin man trying to get out and I think likewise that within many a basic scientist there is an applied scientist trying to get out.   I was trained in an era when one was very heavily dependent on industrial funding, because there was very little federal funding of the type that's grown up since World War II.   I have always greatly appreciated the help that I received from the chemical industry that supported the department where I trained.   If there is a controversy about this type of support, one has to say that it really is nothing new. Chemistry and physics have gone through this kind of development. The reason why biotechnology has recently attracted so much similar attention is, perhaps, because the buffer zone of development between the basic discovery and its practical application, which in many cases has been a matter of years or decades, may be, as Nicholas Wade recently said, now down to a matter of weeks. One makes basic discoveries and sees

their practical application exploited extremely rapidly.
Partly because of this many basic scientists have themselves
become involved in this exploitation. If you ask me (and as
a University Professor I ought to declare my self-interest
here), "Should the universities get involved in a practical
way?", I would say "absolutely yes".   Yes, for financial
reasons, for opening new avenues of research for their
faculty, and for opening new opportunities in training and
education for the people who will later exploit this technology.

Of the dangers we often hear about, the obvious ones are
the possible distortion of educational priorities, the constraints on the freedom to pursue any kind of research, to
"mess about" as Milstein said the other day, the secrecy that
may grow up within and between laboratories, and the con-
flicts of interest because of the financial stake of some
faculty members. I have no solution on how to resolve these
problems, only a conviction that they are problems that universities must address. I would rather see this done than
see the universities turn their back and not become involved.
If a university has a $1 billion endowment then it can simply
turn away from these problems, but when a university decides
not to become involved, often the faculty members may start
their own companies. I would rather see this done under the
aegis of the universities.

And if the universities are to become involved, what
should they do? I think two main things, which in my experience most universities presently do badly. The first is to
develop sufficient two-way contact with industrial sponsors.
This is a whole different ball game from federal funding
where, for example, the workings of the NIH are well understood. Industrial companies often don't advertize the fact
that they are interested in sponsoring certain types of
research and almost every project requires its own unique
contract. There ought to be an entrepreneurial, efficient
mechanism within the university to develop those contacts.
The second thing is to improve patenting practices. Universities do a very poor job of helping their faculty to patent
discoveries. They could do better, to the mutual profit of
the university and the faculty. But I am much more interested in the university's gain than in the gain of the individual faculty member. Perhaps the best example of how to do
this is the Wisconsin Alumni Research Foundation which is
currently putting $4.5 million a year back into basic
research in the University of Wisconsin. I was very taken
the other day by César Milstein's suggestion of some kind of
tax or royalties from patents that could be ploughed back

into basic research. I have no idea how one would do this in a practical way but it is an excellent idea that should be given careful consideration.

It has a parallel in scientific publishing. In the last ten years or so I have helped to found five journals. They are all published by commercial publishers, but I have always insisted that royalties come back to the scientific organization that sponsors the journal. Those five journals are now bringing in a half million dollars a year, not to me, but to the biochemical organizations that help foster the growth of the science.

I simply have a firm conviction that universities should not turn their back on this kind of commercial exploitation. It is to everybody's good thus we should be positive about it and work out the best ways to do it. We know, surely, only one small fraction of one percent of what is going to be possible as a result of the new biotechnology. Most of the new knowledge will come out of the universities and it will be exploited by industry. That's good. That's the way our capitalist economy works. We should be working toward developing a dialogue and a mechanism that will create a partnership that will be of mutual benefit to academia and industry.

MARTIN APPLE (INTERNATIONAL PLANT RESEARCH INSTITUTE)

Let me introduce you to the International Plant Research Institute which was started in 1978. It is located primarily in San Carlos, California. It is a private, for profit corporation which has approximately 150 staff members, 65,000 square feet of laboratory space, 90,000 square feet of greenhouse space and farm acreage all over the world. We combine laboratory biotechnology with classical plant genetics. We conduct careful market analyses and add intuitive anticipations to determine what proprietary products we should develop. We work through flexible, product-specific multidisciplinary teams. These teams are drawn from our staff specialists in molecular genetics, cell biology, chemistry, plant pathology, plant breeding, phyto products and marketing. We've focused on the major food crops--wheat, maize (called corn in this country), potatoes, casava, tomatoes, etc. Our second focus is on commercially valuable substances which can be manufactured by plant cells. The end products we produce are special chemicals, food and seeds. We undertake commercial contracts for other companies or countries in the fields of energy, chemicals, foods, timber and agribusiness. The contracts are for the custom design of specific-end products. We are developing joint ventures where our expertise meshes with those of a commercial partner to our mutual advantage. We have experimental crops now growing in Europe, Africa, South America, as well as in a variety of North American locations. Our food crops are being developed to yield more, to resist disease, and to tolerate more stressful environments. We are working on the scale-up of new processes for manufacturing of commercially important substances which can be produced from plant cells. Our revenues should increase over 400% in 1981, to several million dollars. We expect substantial growth this coming year. We utilize such technologies as those involved in recombinant DNA, protoplast fusion, meristem culture, microanalytic and synthetic chemistry, and others in collaboration with plant breeding.

What are the major food crops? In hundreds of millions of metric tons, the major food crops of the world are wheat 360 to 400, rice about the same, maize and potatoes, the next two big crops about 300-350 and the others decreasingly significant in terms of total amounts produced in the world annually. To understand the commercial significance of these numbers, if the annual rice harvest were sold at median prevailing prices it would bring approximately $125 billion.

The changing pattern of world grain trade has great significance. If we look at trends over the last fifty years, they indicate that North America has become the key supplier for the rest of the world, most of which has become a net importer of food. The political and other ramifications of this are enormous. The major dialog of the 1990's will be between food exporting and food importing countries on the topic of food. The constraint on the future growth of agriculture imposed by limited availability of arable land points to the need for new technology. For example, the amount of available arable land per capita in different countries is, in hectares: Japan, .05; Taiwan, .06; South Korea, .07; Egypt, .08; and India, .30. One cannot imagine growing all the food that will be needed on so little land.

World grain reserves are down to approximately a one month supply. In the event of disastrous crop failures throughout the world, we no longer have the luxury of reserves sufficient to supply ourselves for a year. Approximately 34% of the potential value of an annual world harvest of crops is lost from the ravages of insect pests, diseases and weeds. The extent and causes of crop losses are different in different continents, but the solution of each different problem represents an opportunity for the new technology. An important aspect in future food development is the energy output/input ratio of what we develop. That is, we must ask the question, "How many calories do we get back for each one we put in?" Data from Pimentel at Yale and others indicate an output/input ratio of between 15 and 20 in the more primitive agricultural systems of the world. Compared to this, how well do we do in this country, in our highly effective, super productive system? Ten times better? A hundred times better? No, we do worse. We get back between two and three calories for each one we put in, and in some instances, such as for my local, home-grown California tomatoes, we get back 0.6. We get back less than we put in.

The $\underline{E.\ coli}$ of our plant biotechnology industry are plant protoplasts. What Rorvick always wanted to create, "clones", in our case are plants regenerated from the protoplasts. We can do this now. We are on the threshold of the transformation of plants into novel varieties with important new properties. By adding new genetic information to plants and by altering the genetic information they already contain, the possibility of creating food plants which can overcome the barriers we now see as constraining our food supplies is evident. The use of plant cells as solar energy driven manufacturing units for selected chemicals also looms in our future. It will be several years before these all come to pass, but many cautious observers see them on the horizon. As genetic engineering brings its impacts to our world, that world will be changed forever and this decade will be viewed as a turning point in human history. The more resources we devote to this genetic technology now, the sooner will its benefits and results become widely available; it may require substantial resource commitments to achieve all the goals we envisage now.

STEPHEN TURNER (BETHESDA RESEARCH LABORATORIES)

The title of this panel was HORIZONS IN BIOTECHNOLOGY and I'd like to focus just for a moment on the word "horizon" since it's a very powerful word. I took the time to look in a dictionary and it states that the definiton is "the limit of one's perception". Most of you as scientists spend a lot of your time focused on the limit of your perception of the future; on your definition or working hypothesis of your horizon. As businessmen we have a similar problem. We have to view the horizon, plan and prepare for it. Biotechnology is forcing us to come up with, what I consider, novel approaches to organizing the business of science. This is why BRL has been able to achieve the rather dramatic growth that it has achieved up to now. My current view of the horizon obviously has changed a great deal since the company began back in 1976.

Let me focus on just one or two points for this five

minute period. I think we all now realize the depth of the financial and human resources that will have to be committed to bring to fruition the type of scientific insights and inroads that have been made up until now. In the areas that have been touched on in the last week such as eukaryotic gene expression, cell biology or plant molecular biology, I am awed by the resources and commitments that will be required in order to achieve the industrial promises and potentials that have been dreamed of. The second point in my view of the horizon is that commercial realization, the vision of developing science for commercial purposes, is an essential engine for the advancement of science over the next ten years. Whether you blame the environment in Washington, or whoever your favorite villian is, it is a fact of life that industry will be an increasingly essential part of the driving force in the realization of the life sciences. Going from there to BRL, let me say that I strongly believe that the goal of business planning, particularly if you're a small company, is survival. Not dominance, not dreams, just meeting payrolls and hopefully being in business by 1985 or 1990. With that in mind, we think we've done a good job of developing organizational teams of scientists and product development managers.

Let me briefly describe how our organization works. I am asked all the time how it works. I am even asked that by my employees! Basically we work through four divisions. The research product division has as its primary project the application of current science and the development of current methods, for the rapid marketing of high quality, cost effective reagents, of which most of you are familiar (and we are most grateful for your support). Three basic areas involved are molecular biology, protein biochemistry and cell biology. As molecular biology and cell biology start to fuse together you will see more and more BRL products that relate to the enhancement of cell propagation, in addition to reagents relating to DNA methods. The second division is involved with clinical diagnostics or, as we call it, molecular diagnostics. In this we use the technical base that we've built, and essentially recycled back to you in the form of research products, to develop a second generation of clinical products. For example, we recently published a rapid diagnostic method for detecting hepatitis B virus in serum within 24 hours, using modifications of the Southern hybridization techniques. Hopefully, developments of this type will revolutionize clinical virology as we know it.

The third division, which is BRL genetics, is more of a genetic engineering company, except that, we rely on the

support of the scientists in the other divisions to provide
the geneticists with support in biochemistry and methodology.
In the genetics area, our major thrusts and contributions
have been in microbial genetics and bacterial physiology.
The fourth division, BRL instrumentation, supports the other
divisions in the development of instruments for those methods
of molecular biology that have become routine and, therefore,
suitable for automation in your laboratories. This week we
displayed our first contribution from this division, a digi-
tizer which allows one to absorb, identify and analyse in
various ways the information on an autoradiographic gel or a
simple DNA fragment gel. That's a very quick overview of our
company. Within our various divisions there is a great deal
of activity by many scientists who feel they have remained on
the cutting edge of science, and yet enjoy the excitement of
direct application of their science and, hopefully, participation in, building an ever broadening and ever more exciting
scientific company.

PATRICK KUNG (CENTOCOR)
    Three technologies that have been mentioned are recom-
binant DNA, hybridoma, and the third, which is not yet
clearly in focus but is very important, methods of cell culture. We can now culture some cell types almost indefinitely, notably lymphocytes (T & B cells). I think cell
culture is equally important as a technology as are the other
two mentioned.
    One has to be aware that, if biotechnology is a tool, we
have to think about how to use it, and in these terms the
horizon is vast indeed.
    At Centocor laboratories, based in Malvern, we are
interested in developing these three technologies for applications to diagnosing liver and heart diseases, cancers, and
therapeutic developments for treating immune disorders.
    How do we operate in Centocor? We are investing money
in supporting outside research, but this is rather targeted
in the sense that we only support specific projects in which
our own scientists can interact with the outside scientists
at the same level of sophistication.
    The last point I want to make concerns development.
There is a strong misconception that, once a discovery is
made, the development and manufacture of the product is a
simple process. This definitely is not the case. It is a
major effort. In terms of expense it usually runs from a ten
to a hundred or even to a thousand—fold greater investment
than was needed for the original discovery made in the laboratory.

PAUL BURNETT (ELI LILLY)

Eli Lilly and Company is an ethical pharmaceutical company with a world-wide organization. We have a total of about 25,000 employees and about 2,700 persons are involved in research and development. Although a pharmaceutical company such as Lilly is not primarily a biotechnology industry, we do use biotechnology in our industry. The current excitement for the pharmaceutical industry stems from two possibilities for the practical application of molecular genetics. One is that a cell may be endowed with the ability to produce a new type of protein, that otherwise it might not normally produce, by the incorporation into the cell of a piece of new genetic information. The second is that the amount of protein that a cell will produce (either a protein that it would produce normally or a new protein) can be dramatically increased by manipulation of DNA regulatory sequences or by an increase in the gene copy number within the cell. You have heard already at this meeting that, using these techniques, the pharmaceutical industry now has the ability to produce, in a culture of E. coli, proteins such as insulin, human growth hormone and interferon. Insulin and human growth hormone are proteins with a known therapeutic value, and therefore, they will have an assured market. Interferon, on the other hand, is an example of a protein for which one can postulate a therapeutic value but which, heretofore, has not been available in amounts sufficient for adequate clinical testing. Molecular genetics has provided the means of obtaining these types of proteins in amounts that allow clinical trials.

A second thing to keep in mind, when we think about the impact of biotechnology on the pharmaceutical industry, is that proteins are not, as yet, extensively used within clinical medicine, with the exception of a few relatively well known products. Primarily there are three types of protein products used today: hormones such as insulin and growth hormone, biologicals such as vaccines or blood proteins and, finally, enzymes, although enzymes are not used extensively in this country. By and large, proteins presently represent a relatively small product in the pharmaceutical industry. On the other hand a number of proteins may be useful in the future, and recombinant DNA technology and molecular genetics now provide us with a mechanism for determining the usefulness of these proteins. A good example is the group of proteins we collectively refer to as cytokines. Interferon is one example of a cytokine, but there are a hundred or so other cytokines whose biological functions have been characterized, but which have not been isolated in sufficient amounts in order to allow their clinical utility to be tested. These may have potential uses in cancer, autoimmune disease, tissue transplantation and allergies, just to name a few. I think that it is fair to say that the medical uses and, hence, the pharmaceutical production of proteins are likely to continue to increase.

Looking further within the pharmaceutical industry for possible applications of biotechnology, you are immediately struck by the enormous potential for the antibiotic field. Antibiotic production is basically a fermentation business, and antibiotics are the second largest class of therapeutic agents used today, second only to those agents that are used in the central nervous system. They represent a 6 to 7 billion dollar market world-wide. Historically, genetics has been used for strain improvement in the antibiotic field, and selection of mutants to increase yield of antibiotics has made the production of antibiotics a commercial reality. In the future, molecular genetics and recombinant technology may be used to improve the yield of existing antibiotics by affecting gene dosage, by manipulating regulatory sequences and by other techniques. Remember also that most antibiotics used are chemical derivatives of the fermentation product isolated from the culture medium. Thus, there is the potential of an in vivo modification of antibiotics by incorporating into the fermentation microorganism an enzyme that will carry out that modification. Also, there is the possibility of making novel antibiotics by constructing new gene combinations within organisms.

Enzymatic synthesis may become possible for low molecular weight drugs used in a wide variety of pathogenic conditions. The cost of enzymes has prevented the wide-scale use of enzymatic syntheses of small molecules; it is simply not cost competitive with synthetic chemical reactions. Molecular genetics may change this in the future. Finally, I think one of the most important things that molecular genetics may bring to the pharmaceutical industry is a new way of studying the mechanism of action of drugs. For example, there is the potential for cloning and isolating genes which code for cellular proteins that interact with drug substances. Receptor molecules, for instance, could be prepared in large amounts for study of their interactions with known agents. Design for new and better drugs might arise from such a study.

I would like to end by saying that, as exciting as are the present applications of molecular biology in our industry, we must keep in mind that even more important applications, that we can't even imagine today, will really come from those fundamental studies that are presently being carried out on the basic processes of cellular growth and function, and on gene regulation and differentiation.

DALE HOSCHEIT (SCHEYLER, BANNER, BIRCH, MCKIE AND BECKETT)

Ralph was very generous and gave me five minutes in which to summarize all patent law for you. I have resolved, therefore, to reciprocate. I'm going to ask Ralph to come to the next bar meeting we have in Washington, and tell us all about biotechnology in five minutes.

Seriously, a good deal of the biotechnology that is occurring today in universities has an impact in terms of patents. Patents can be the only vehicle by which a professor, or his university, can realize a return on a discovery that leads to commercial activity a year or more later. Realize that, if publication continues at universities, and indeed it should be encouraged, then patents are the only way to protect a development. Trade secrets are simply not a rational alternative.

Let me give you two principles and then, very briefly and generally, summarize some of the options that we have in the patent area. Number one, a patent does not have to be antithetical to publication. It is a question of timing. A patent application that is filed establishes a date for the claim of a property right; publication can thereafter occur. I think it's important to understand that. Number two. Particularly in the area of biotechnology, good patent protection requires the combined input of the scientist and the lawyer. A lawyer cannot know the potential of a development if the scientist isn't alert enough to realize the full impact of what he is doing.

The major premise that I would like to develop within the short time this afternoon is that a little knowledge about patent law is not a dangerous thing for scientists. Quite the reverse; it is very much to be recommended. Let me very briefly describe what a patent is. It is, as perhaps some of you know, a limited 17-year monopoly. The sine qua non for a patent is that the development must be (1) new and (2) non-obvious to one of ordinary skill in the art. You realize immediately that this requires that each development be analyzed in terms of how and what it accomplishes in the light of the background of the art--what has been published and what is generally known in the art. And that is a role you play, to place a development in context.

The patented development also ought to be analyzed in terms of its potential commercial impact. A patent can be a very expensive publication. The difference between a broad patent and a narrow patent may, or may not, have much to do with the breadth of claim language. After all, a patent to a single compound can be a very broad patent if that is the only compound that will achieve a certain commercial end.

What kind of things can be afforded patent protection? You are all aware of the Chakrabarty case which says we can now patent a unicellular microorganism. There was a companion case that did not go through appeal, but the Patent Office adopted the policy espoused by the CCPA in that case, and the Patent Office also allows claims to a pure culture. Thus, if you find an unusual organism in nature, the organism may be old and you cannot claim the organism as such but, if the claim is framed in a context that does not read on the natural habitat of the organism, the Patent Office may grant your claim, e.g. to a pure culture. Patent claims to purified enzymes also are granted. The Patent Office has granted a claim, for example, to a purified endonuclease that was discovered and purified. The Patent Office has always treated attenuated viruses as non-living and has granted patents on attenuated viruses and vaccines for years. The Patent Office grants claims directed to RNA or DNA if it is new or if it's in a form that is not in nature. It is evident from these examples that there are many strings to the patent bow. It takes a good deal of insight and cooperation for the scientist and the lawyer to determine which of those strings ought to be used to best protect the development in question.

And now I'd like to turn to the issue of timing. Let me assure you it is neither criminal nor sinful to think of a patent before you get the page proofs of your publication. But all too often in the university environment the question of patenting or protection arises only days, or at the most, weeks before an article is going to be published. What is the effect of that? It is entirely possible and likely that the submission of a publication may deny you patent rights in countries other than the United States. Therefore, protective steps should be taken early. Publication will not foreclose patenting in the U.S. for at least a year but if a patent application is not filed within the year, then the right is lost. Thus, early publication before the matter is fully considered can lead to loss of rights in most countries, with the exception of the U.S. and Canada. Remember, a world does exist beyond the boundaries of the continental United States.

As I said at the outset, I am not trying to take you deeply into the patent law in the space of these five minutes. However, let me end very much as I began. If I could leave one message, it is--and I'm being quite serious--a little knowledge about patent law is not a dangerous thing for those of you working in biotechnology. I think it is almost essential.

HARDY (DU PONT): I'd like to thank the panel members for their concise and stimulating presentations. The meeting is now open for your questions. Please address your questions generally to the panel and we will identify the appropriate responders.

STEVE COLLINS (BRL): What lessons can we learn from the long and deeply-rooted relationship between the chemical and allied industries and academic chemistry?

BURNETT (ELI LILLY): It is difficult to give any pat answers to that sort of question. It is very important to realize that the previous relationships that you referred to, and Dr. Whelan referred to earlier, evolved over a very long period of time, and that in biotechnology today we don't have the luxury of that span of time. Everything has been dramatically compressed. The relationships that have proven most successful in the past have been those where there has been a true spirit of collaboration between industry and academia, as opposed to a giving or a taking situation in one direction or the other. It is imperative for industry to recognize those things that it can do well, for academia to recognize those things it can do well, and for the two to collaborate and complement each other as much as possible.

WHELAN (UNIV. OF MIAMI): I asked to respond immediately after Dr. Burnett because of something I am reminded of that concerns his company. One positive benefit of help for universities at relatively little cost to industry is that it generates enormous good will. I remember the 1970 Lynen Lecture given by George Wald who told us that when he was a young faculty member on a very low salary and his wife was pregnant with their first child, he received the Eli Lilly award. He said they were so grateful they decided if it was a boy they would call it Eli and if it was a girl.........

APPLE (IPRI): I believe a spectrum of relationships between different universities and different industrial concerns will evolve over the coming years. Within this spectrum experimental types of interactions will develop and, over a period of time, some will emerge as good models for the appropriate types of interaction that will be mutually productive.

MARY ANN DANELLO (SCIENCE/TECHNOLOGY COMMITTEE, HOUSE OF REPRESENTATIVES): What type of involvement would you like the federal government (Congress) to have in the biotechnology areas, in particular with the problems evolving from increased university/industry relations?

TURNER (BRL): We are now seeing the development of a healthy industry structure which is founded on a strong academic base. The federal government has played a constructive role in the long-term funding of life sciences and I think all of us would like to see this commitment to basic research continued. Industry will never be able to provide the total financial support for basic research and it will be in the best interest of our society for the government to maintain its support at the highest level possible. With regard to legislation concerning this emerging industry, I think that it has been handled well through NIH guidelines and there has been a strong scientific representation and peer review and, I think, a healthy overall approach to date.

HARDY (DUPONT): Dale are you willing to respond further to that?

HOSCHEIT (WASHINGTON): Only to underscore what Steve has already said. I think everybody on this panel agrees that the essential role of the university is and will remain far reaching, long-term basic research. It is simply that in this kind of technology there are going to be near-term spin offs from that work.

HARDY (DU PONT): Let me add a general comment to the question. The U.S. spends about $70 billion a year on R&D, split approximately equally between federal and industrial support. There is no way that industry is going to pick up a significant amount of any decrease in the federally funded part. There will be examples of additional support for varying reasons, but one shouldn't expect industry to become a major benefactor beyond its present level of support.

APPLE (IPRI): The mentality of operation of any profit making-organization tends to focus on the eventuality of market success, and future goals probably cannot be as easily set nor as long-term as those made in government funding of basic research.

SCHULTZ (PAPANICOLAOU CANCER RESEARCH INSTITUTE): I'm glad that last question was brought up because it's a question that I'd like to pursue. In view of Dr. Hardy's recognition that industry could not possibly replace the government's investment in the universities and basic research and recognizing that todays biotechnology was developed from university basic research conducted in the '50s and '60s, it seems to me that it would be in the interests of industry to lobby in Washington and work hard to convince the government to maintain adequate levels of funding for basic research.

LAMON (BIRMINGHAM, ALABAMA): I would like the panel to address the issue that Professor Milstein raised, specifically: to whom does the patent or royalty belong? For example, I understand that many monoclonal antibodies are now being patented. Shouldn't the profits from these be shared with Professor Milstein or his institution, and indeed, with Professors Barski in France and Harris and Klein in Sweden, who constructed the first somatic cell hybrids, and with the many others who contributed to the research that led up to the production of hybridoma monoclonal antibodies?

HARDY (DU PONT): There are some contracts between industries and universities that require the patents be held by the industry. This is true in the case of limited and general partnership arrangements because of the tax situation. However, in most cases of industry/university contractual relationships, the patents can reside, and I think they should reside, in the university or with the individual who did the work, depending on the nature of the agreement between the researcher and the university.

HOSCHEIT (WASHINGTON): The question underlines the issue that I raised earlier, which was, whether we like it not, an individual in our society has a choice. If he chooses to publish without attempting to protect, he has made a donation to the public. Protection of a meaningful development really has to come, if we're looking at the academic side, through a patent. I gather the individuals you mentioned did not have patents and they chose to publish their results. Whether the choice was conscious or not, the legal consequence was that they waived their rights.

APPLE (IPRI): One has to realize that it's extremely difficult to commercialize an invention which is published. The reason is that the unit that wishes to commercialize has to put up venture capital, and possibly put at risk a substantial amount of money. If there is no chance to protect a successful development from a commercial competitor before a fair return is received on the investment, then that development is not likely to be undertaken. So there is a great need to patent those developments that have a potential for commercialization.

BURNETT (ELI LILLY): I would like to add to what Dr. Apple has just said. After the initial discovery, marketing of an average pharmaceutical product takes about eight years and a cost of about 50 to 60 million dollars. Thus, development and marketing of a product that is beneficial to society is a very expensive venture.

HARDY (DU PONT): The Patent Restoration Bill is relevant to Dr. Burnett's comment. Patent protection would start after the discovery had received regulatory approval for marketing rather than the date the patent issues. Products such as agrichemicals, pharmaceuticals, etc., would retain the full seventeen years of protection for marketing vs. the nine to ten years that are available under existing laws. The Patent Restoration Bill will stimulate greater R&D since companies will see greater opportunities to meet their expenses and make a reasonable profit.

SCHAEFFER (UNIV. OF VERMONT): Please comment on the relative role of copyright vs. patenting.

HOSCHEIT (WASHINGTON): Copyright has been suggested as an alternative means for protecting a DNA sequence. There are some who say that copyright may provide some protection. Let's look at the name--it means the "right" not to "copy". It doesn't necessarily mean that it will prevent somebody from discovering the DNA sequence independently, for example. My own judgement is that it is very secondary. There are lawyers who view it otherwise but I know quite a number who have my outlook.

HARDY (DU PONT): Might one copyright, as well as seek conventional patent protection on the same invention?

HOSCHEIT (WASHINGTON): Yes, there are those who are suggesting that that be done.

THOMAS MYSIEWICZ (BIOENGINEERING NEWS, SAN FRANCISCO): Could you equate the cost of filing and defending the typical patent with the cost of trade secret protection. I refer in particular to small companies and to process patents, which are generally harder to defend than product patents? Also please comment on copyright protection of DNA sequences. As far as we know the current position of the copyright office is that no DNA sequence copyrights will be granted.

HOSCHEIT (WASHINGTON): Let's deal first with the relative costs. Please realize that because developments vary in terms of their scope, so then costs for patent applications also vary. The process of writing a patent application and making the preliminary search, might cost from $1,500 to $5,000, but it could be well outside those limits, depending on the complexity you have to deal with. On the other hand a trade secret does not cost anything in out-of-pocket expenses but it is valuable to you only so long as you can restrict its disclosure. An employee who has one too many martinis at a bar can put an end to a trade secret. Therefore, trade secrets have a very dubious value in the long term. There are companies that make that choice. Indeed, there are companies in the area of microbiological fermentation that use their own microbial strains and treat these strains as a trade secret. However, in many areas it is not feasible. Certainly, in academia it is not feasible because publication is antithetical to a trade secret. If there is any possibility that there will be publication, then secrecy is not a reasonable alternative to protect a development.

Patent litigation can be very expensive and can run into six figure sums of money very easily. That does not mean a patent is not worth while because very often the development that's being protected may be worth a good deal more than the patent expenses.

MANETTE DENNIS (NEW YORK PUBLIC AFFAIRS TRIBUNE): In this meeting we heard talks and saw posters which for the most part were focused on research and testing done in the U.S. Yet we continuously hear of developments in Japan and Europe. What progress is being made overseas and how much collaboration is there between U.S. workers and workers in other countries?

WHELAN (Univ. of Miami): I'll comment on one aspect of this question that goes back to how the government can help to develop biotechnology by continuing to increase support for basic research. When we complain about decreases in government funding, I have a feeling that members of Congress might react in a positive sense only if they sense that we have another kind of "Sputnik" situation. They will not be persuaded that they should put more money into basic research simply because this is a good thing to do, but they are like-

ly to be persuaded if they become convinced that say, West
Germany, Japan, France, and other countries are forging ahead
of the U.S. in areas of basic research on which progress in
biotechnology is dependent. Only then may we see more funding of basic research.

APPLE (IPRI): I'd like to pursue Dr. Whelan's comments
because I think they're highly germane at this point. I
think the general answer to your question is: Yes, there are
other developments in biotechnology in other parts of the
world and they have a very interesting common history. Their
common history is that there has been a great deal of support
to develop scientists and research facilities capable of providing the expertise that has allowed the generation of biological science into a major technology. You will find that
those centers in Europe, in Asia, and in the Americas that
have a history of this type of background support are now
emerging in the forefront of biotechnology.

TURNER (BRL): I'd like to answer very directly. There are
some areas of biotechnology in which we can identify European
or Japanese scientific groups as leaders. Biotechnology is
a global phenomena; it is an international technological
development which no nation can hope to capture within its
boundaries, just as no single company can hope to capture it
within its boundaries. I think that someday, international
teams of scientists will do much of the work in this area and
that this should be welcomed rather than resisted. This is
one reason we began our European subsidiary company this
year.

APPLE (IPRI): The members of the staff of the International
Plant Research Institute come from about 18 countries.
Obviously, we don't believe the United States has a monopoly
of talent.

BOTSTEIN (MIT): I'd like to make several comments. The
first comment concerns patents. I think it should be understood that the research laboratory doing biotechnology
oriented basic research is not a business concern, and hence
it cannot and, most likely, will not patent everything that
might have use. The sum of $1,500 to $4,000 needed to obtain
a single patent is too large a sum in relation to the money
available for research. There is no source for that kind of
money. Furthermore, for the university to attempt to protect
everything in the hope that it will eventually receive income
is a strategy which is unlikely to be followed and certainly

one which I would not recommend. The legal implications of not patenting are rather unfortunate from the point of view of the university. I believe it is very important for the industrial people to understand that the university is not a profit-making business and that it cannot deduct the cost of defending itself and it should be accorded more reasonable treatment.

My second comment concerns, I think, a more important issue. I agree with the view that industry is unlikely to spend the amount of money which the government has quite properly allocated to basic research in the past. However, one should realize that there are areas in which the drain on the university is substantial. Universities are in the business of training young people to become scientists, and that is expensive. The government, for some reason, has decided to spend as little as possible toward training graduate students and post-doctoral fellows, an area of common interest to both academia and the industry. I believe it is in the interest of the industry to ensure that support for such training continues on a basis consistent with the number of students who are willing and able to learn. Finally, I think that it's very important to recognize that the University is a fragile enterprise and that it cannot exist on a purely competitive basis with all of its constituents.

(Thunderous audience applause)

TURNER (BRL): I'd like to start with the training aspect because I think the industry has to make up the deficiency of funding created by the government. For example, BRL runs DNA workshops conducted essentially on a break-even basis. The tuition pays for radioisotopes and the enzymes. Thus far we have trained over 250 people all over the world in an important molecular biology method. The same thing can be done for other basic laboratory methods. I think that's a legitimate logistical activity that industry can do as well if not better than academia. Secondly, I believe everyone is interested in survival, even a scientist in the university. It is my experience that when an industry is in danger of losing its raw material (scientists in this case) it adopts remedial measures in order to survive. I expect the same thing to happen in the biotechnology industry.

WHELAN (Univ. of Miami): I'd like to respond to Dr. Botstein's comments on the cost of patenting. I was interested and impressed by the Office of Patenting and Licensing at Stanford University, run by Neils Riemers. He has his faculty really attuned to making disclosures. I was there at the end of December when he had just received their 134th disclosure of the year. The way in which he handles patents at minimum cost to Stanford is to find a sponsor to share in the development and marketing of the invention and let him pay the cost of patenting and other expenditures. With minimal office costs he has built up a $1 million plus royalty income per annum not counting the "big hit", the Cohen-Boyer patent.

BURNETT (ELI LILLY): I agree with Dr. Botstein. I think it should be noted that even in industry we don't attempt to patent everything discovered in our own laboratories. Our company, for instance, operates through a review committee made up of scientists representing the research community who recommends whether a particular discovery should be patented or simply published. I think universities will establish the same types of committees. It may very well be that they would want some expert advice from industry representatives serving on that type of committee. Secondly, I think that industry is beginning to shoulder an increasing amount of support for training. We invest a substantial number of dollars, and that amount is increasing every year, to support research within the academic community. Most of this money goes for support of post-doctoral and student salaries.

HARDY (DU PONT): Let me add a comment on training. The contribution of major corporations has been growing over the years. In our case we contribute about $5 million a year in unrestricted funds as grants in aid to education. A large part goes to chairmen of university departments in areas in which we have a current or future business interest. The money can be used for training, for supporting visitors, for equipment, etc. As industry's interests change and become more aligned with life sciences, like Du Pont's currently, then more of these dollars will be allocated to life science research departments.

NORGARD (SOUTHWESTERN MEDICAL SCHOOL, DALLAS): Can a new entity, which is clearly new and has identifiable usefulness, be patented even though the usefulness has not yet been developed to the point of being defendable by appropriate experimentation?

HOSCHEIT (WASHINGTON): Unfortunately the law requires utility to file a patent application. There is precedent that states that identifying usefulness of a compound as a mere possible intermediate in a chemical reaction did not, in that circumstance, establish utility. Therefore, let me address this in two parts if I may. First of all, there is the requirement of utility. We cannot file on something and say here it is; it is new and we hope someday it will be useful. There is an alternative. After all, we generally make our inventions specifically, and then at times we stand back and determine the generic impact of what we have done. A scientist is entitled to use a reasonable scientific judgment to determine what the generic impact is. He is also entitled to use his scientific judgment to determine what the utility will be. However, the Patent Office has the right to challenge the utility and to require a showing of the utility that is disclosed. The patent application has to identify a utility.

NORGARD: I think perhaps I didn't make myself clear. If the uses are clearly spelled out and are identifiable, but it has not yet been demonstrated to the point that they can be defended, is it still patentable? Or does the defense of those uses need to be clearly shown at application of the patent?

HOSCHEIT (WASHINGTON): Something doesn't have to be commercial to be patentable and, therefore, if there is a use and you are satisfied that that use is real, that will satisfy a utility requirement. Utility does not have to be commercial utility.

HARDY (DU PONT): I would like to comment on the lack of knowledge of patent procedures among academicians. Perhaps a one credit patent seminar should be included in a Ph.D. program. Familiarity with patent procedure is clearly important for individuals contemplating careers in industry. It is also becoming increasingly important to academic researchers as they increase their involvement in commercial aspects.

TURNER (BRL): I want to correct the impression that all business is going to have to be transacted through patents. I think it would be a tragedy if you all left here thinking that you have to learn all patent law. In 90% of the transactions we engage in with researchers, whether they concern a new method, a new application, a new apparatus or whatever, patents are not relevant. If the idea is good and we can see a way to commercialize it, we figure on a useful life of probably one to three years anyway before it is superceded by a better idea. I can count over 20 agreements with no prospects for patents, in which we pay royalties and reimburse the laboratory for expenses.

NANETTE NEWELL (CONGRESSIONAL OFFICE OF TECHNOLOGY ASSESS-MENT): The U.S. Congress is concerned that domestic biotechnology industries should remain competitive in international markets. Does the panel feel that legislation could promote international competitiveness?

APPLE (IPRI): I will venture an opinion that will not make us friends and that is--the less legislation the better.

(Audience applause)

WHELAN (UNIV. OF MIAMI): I think it was Dr. Botstein who mentioned training grants. One thing Congress should do is to continue to increase support for scientific training grants. It is sad to see the meataxe approach to training grants that is used from time to time when funds are short. Extremely good programs built up over the years then go out of existence. I think that's a national tragedy.

BURNETT (ELI LILLY): I think the government could assist the pharmaceutical, and other highly regulated industries, to compete on a worldwide basis by getting away from an adversarial situation and interacting in a more positive manner in order to get a product to the marketplace and society as rapidly as possible. Japan is a good example of a country where there is less of an adversarial situation between the government and industry.

HARDY (DU PONT): Certain skills are obviously going to be necessary as biotechnology develops. One of those skills is bioengineering. The U.S. is not training enough bioengineers to meet future needs. If you compare us with, say Japan, we are deficient. Anything that can be done to facilitate training in these areas will be beneficial.
    There is growing excitement about the potential impact of biotechnology in agriculture but the number of trained scientists is very deficient. The problem is compounded by the lack of a National Institute, or its equivalent, in the agricultural area which would fund both internal and external research in this area. A National Plant Molecular Biology Institute is an example.

ALTOSAAR (UNIVERSITY OF OTTAWA): I was glad to see Dr. Apple brought to our attention data showing the importance of plant science research and development. The chairman's last comments on manpower training in the agricultural sector are thus an excellent introduction to my question. A recent issue of <u>Trends in Biochemical Sciences</u> ran an editorial entitled: "Plant Biochemistry--A Cinderella Too Long?". Does the panel feel that government will soon turn into a fairy godmother and start funding plant molecular genetics at a level commensurate with its intrinsic global impact? Research funding is biased against plant research. Similarly, research training is biased against using plant systems or orientations. I feel the problem is rooted deeply in the educational system, in both primary and secondary school teaching. High school students seem very interested in plant natural product chemistry, for example, the use of marijuana. But when it comes to considering important questions in research, the interest is gone. I feel that we, the educators, are largely to blame. Could the panel members please address this pressing problem?

HARDY (DU PONT): In 1940 about 40% of the R&D funds in the U.S. were in the USDA. In 1981 it is something like 2%. The support for agricultural research clearly has gone downhill and there is no evidence that suggests that it will recover.

APPLE (IPRI): I believe somebody on this panel quoted a figure of $70-$80 billion annually for research in the United States. I don't have the exact figure (and I am open to correction), but it is my recollection that last year the commitment of the United States government to plant genetics was about $10 million. This is similar to the R&D budget of some of the very small new companies with which I associate.

WHELAN (UNIV. OF MIAMI): The last question has given me the opportunity to remind you next year's Symposium will be on genetic experimentation in agriculture. When we talked about the details of it early this week we thought that next year's panel meeting would be discussing precisely that topic: how to get more funding for basic research into plant biology.

SIROTKIN (KNOXVILLE): If one of the tests for a patent is one of obviousness, or lack thereof, and if the basic approaches seem obvious albeit arduous, for making and using a plasmid to direct protein production in bacteria (insulin production for example), how is one to view patents on such plasmids?

HOSCHEIT (WASHINGTON): I think you'll understand that we

prefer not to address any specific patent. Let me address the issue a little more generically. It really becomes an interaction between the patent lawyer and the scientist. The ultimate arbitrator is the U.S. Patent Office which makes a judgment that this is or is not patentable. If you and I were to collaborate in a patent application, we have together a duty to provide the U.S. Patent Office with the most relevant prior art of which we are aware. If the Patent Office is going to issue a monopoly, then the least we can do is let them be in the best possible position to make the determination as to whether or not that particular development is obvious. And then, the Patent Office will make the final judgment. Again, it comes back to a case by case basis. You simply have to work with an attorney, look at the prior art, at what is disclosed, and go from there.

HARDY (DU PONT): It is important to obtain valid patents-- making major investments on invalid patents is not good business. The courts decide validity. The newness of patents in the biotechnology area have not allowed time for any significant court decisions. There may be some suprises.

HOSCHEIT (WASHINGTON): More than that---it may be called fraud on the Patent Office, and fraud is a very bad word particularly when it's directed at you.

MARY PATER (NEW JERSEY MED SCHOOL): One of my concerns, also voiced by Dr. Whelan, is that as a result of the developments in biotechnology there is a tendency toward secrecy which may impede rapid dissemination of information among the scientific community of academia. This problem has not been discussed. I would like to hear the views of panel members representing industry.

TURNER (BRL): I think it's a matter of policy for the individual company. The best approach is to participate in the free flow of information with academia since there is a decreasing difference between the research being carried out in academic and industrial laboratories. In my previous statement I mentioned that, in view of the rapid developments in biotechnology, it is rarely feasible to go through the time-consuming process of patenting. I know that this view is unlikely to be shared by others present here. However, I'd also like to focus on a positive aspect. The companies, like BRL, involved in the distribution of technology have played a positive role in the transfer of information. For example, laboratories lacking expertise in say nucleic acid sequencing can easily duplicate published work in a relatively short period of time because of the commercial availability of both the chemicals and the requisite methodology. Thus there exists the potential of an accelerated transfer of information into numerous laboratories all over the world, achieving the exact opposite of what you are concerned about. Recent knowledge and technical skills are also being transferred more rapidly into schools, especially those that have been somewhat distant from the front line of research.

KUNG (CENTOCOR): Sometimes, a detail, which might be regarded as trivial in terms of publication (for example pH 4.0 instead of pH 5.0) could make a big difference in a production process. That detail might not be published but held as a trade secret. On the other hand, the high technology companies as a rule encourage their own scientists to work with outside scientists whose expertise is helpful. The companies often work with outside research scientific advisors for exchange of ideas and reagents. From this point of view secrecy, especially as it relates to the pioneering research aspects, is actually an impediment to a company.

APPLE (IPRI): I hope we at IPRI continue to enjoy good relations with universities all over the world. There are occasions in which ideas that originated in industry are transferred to universities. There seems to be a mistaken concept that the idea-flow is always the other way. It really is not a one-way flow and many fascinating ideas are developed out of interactions. As I said at the beginning of this discussion on university/industry relations, there will always be a spectrum of opinion, a spectrum of activity and a spectrum of results. As these spectra develop, we will find the types of interaction that are the most productive and those will be the ones to prosper.

BURNETT (ELI LILLY): I would like to point out that one of the advantages of patents is that, once the patent has been applied for, the need for secrecy is precluded. As a matter of fact a patent is a publication. You are required to teach how to do the invention that you are describing. So patents do have a beneficial effect in that regard.

HARDY (DU PONT): Let me make a general comment which expands on Mr. Turner's statement. There are some products and processes where it is not essential to have proprietariness--those in which you make a very small investment before you bring them to the marketplace, those that have a short life-span. Others, where there is a major requirement for invest-ment--either as a capital investment or in meeting regulatory requirements--have to be protected. There is no way that anyone is going to bring the unprotected product to the marketplace and then find everyone else able to compete immediately.

Secrecy in filing a patent is not much greater than it is in doing an exciting new piece of research where you don't want your scientific competitors to know about it prior to submittal to a journal. Patents need to be carefully planned but in general they can be filed about as rapidly as a very good publication is written.

SCHUTZBANK (NEW BRUNSWICK, NJ): I have a comment rather than a question with regard to government legislation. What I feel we need is not new legislation but a correction of past mistakes which block U.S. companies from competing abroad. For example, the FD & C Act blocks export for commercial purposes of new drugs not approved in the U.S. when these drugs may, in fact, be approved in many foreign countries. PHS Act 351 also blocks sale, barter or trade in biologics for human use if they are not licensed in the U.S. The 1976 medical device amendments to the FD & C Act does improve the situation somewhat with regard to export of medical devices, but even that provision, in some cases, inhibits the ability of U.S. companies to export medical devices not approved in the U.S. but acceptable abroad. In all three cases, I feel the U.S. government should not regulate products manufactured for export only, since it blocks or greatly hinders U.S. companies from manufacturing in the U.S. and exporting abroad. In fact, many U.S. companies have had to set up foreign production facilities to get around this problem, which, of course, takes money and jobs out of the U.S.

APPLE (IPRI): I would like to respond with respect to my own and my company's area of activity and to the comment made by Nan Newell. If, as someone has predicted, agricultural genetic engineering may develop into a $50 billion-plus industry by the turn of the century, it seems that we have enough interests in common for us to examine how we can work together in the most beneficial way. One expects the first question of a legislator to be: "What legislation do we need?" Perhaps the real question ought to be: What do we need? If the answer is legislation or non-legislation, perhaps the latter also should be considered.

WADE (NEW HAVEN, CONN.): I would like to hear the panel address a question that is somewhat confusing to me. If public money has been used in the development of a product which can later be patented, what obligations do the patent holders now have to the public? What are the guidelines existing now for that and are they sufficient?

HOSCHEIT (WASHINGTON): As of last June a public law went into effect saying that even though government funds were used to support research at a university or non-profit organization, that organization could own the patent rights. If there is a need for an initial large investment then, obviously, the investor has to have a reasonable expectation that he can recover that investment. Therefore, the law allows the organization to grant an exclusive license under the patent. However, there is a fall-back provision in the legislation that allows the government walk-in rights. The walk-in rights provision says that if the public is not being served, the government can step in and see that those patent rights are put out for the better use of the public.

APPLE (IPRI): Fundamentally, what you are asking is what is the benefit to the public. If a patent gets used, presuming that it had some utility by the definition presented earlier by Mr. Hoscheit, it means that it is providing a product that is useful to the public. This utility is the ultimate benefit that the public can always derive when the patent gets used. I'm not sure that public tax money is always used with the clear understanding that there is an ultimate benefit, universally accepted as such, as the result of its use.

CHOWDHRY (DU PONT): We have heard the panel expound on where they see the horizons of biotechnology as they impact on control and well being of cells; that is the realm of biology. Would the panel care to speculate on the interface of biotechnology with the inanimate world? The close dependency on chemistry of the revolution in biotechnology would have been hard to predict from an examination of the history of the chemical industry.

WHELAN (UNIV. OF MIAMI): If you would include an enzyme as an inanimate object, in other words working outside a living cell, I think all you have to do is to look at the list of enzymes that presently are being exploited industrially. What you will be looking at is the index of a textbook of biochemistry of the 1920s. This demonstrates how far we are behind in commercially exploiting the enzymes of which we are already aware. Of course, the enzymes that are easiest to use are those that carry out degradation reactions. What we have to do is to learn how to control synthetic reactions. If you want a prediction, I think we are going to see that the enzyme biotechnology of the next two decades will be derived from very sophisticated chemistry based on what we've learned about natural catalysts.

KING (NEW YORK CITY): A couple of panelists made the important point that it is possible, and in fact happens quite often, that a scientist can do important science within an industrial context. I'd like you to address the flip side of that. That is, when you take someone whose experience so far through life has been school, up to and including post-doctoral fellowship, and then move them to an industrial setting, what new skills do they learn that are of benefit to them there, and how does that benefit the way they do science?

TURNER (BRL): I'd like to answer this, because it is a good question. Obviously we all have a different perspective on what we'd like to learn and why, so these are generalizations, but having brought about 400 persons into industry over the last six years I do have my own perspective.   I think that there is a large component of the Ph.D. training which forces one to be self-reliant in problem solving and in trusting ones own ideas. There are many departments of life sciences that are departments of highly motivated and highly competent individuals. Often times the skills learned in industry are inter-personal skills. By learning how to be effective through other people you are not sublimating or degrading your own capabilities as an individual. Scientific organization and collaboration is a necessity for progress to be made in a number of scientific areas and requires that the scientist involved should possess interpersonal skills that often times are not obvious to a person just out of a Ph.D program.

APPLE (IPRI): Having experienced both, I do not think that the difference in a university and a commercial environment is very significant for an individual. Several important skills: problem solving, interpersonal relationships, etc. are practiced in both environments to a great degree. It may be necessary to develop some skills to a greater or lesser extent, but they probably will be the same in most respects. In both cases the ultimate goal is similar though some immediate goals may differ. Clearly industries are more focused on profits and Universities are more focused on generation of knowledge and its transmission to the next generation of individuals. Nonetheless, they are interdependent. Universities cannot exist in our society without profits, because without profits there would be no taxes, and without taxes we could not have established our schooling systems. In industry we have to solve problems and create new knowledge, and so we have a common purpose and are not working at cross-purposes with the Universities. The things that individuals in both environments need to learn and develop are really quite similar.

KUNG (CENTOCOR):   When a scientist stays in an academic institution he is confined to teaching and research. When he goes to industry, his career possibilities are broadened to include administration, marketing and sales. Maybe one day he can run the whole company.

His career goals should be reviewed every three to five years. If he decides to stay in research then he should be sure that what he wants to do and what the company wants him to do overlap very well.

HARDY (DU PONT):   I think it's important that the academic community and specifically students should become better informed about industry. About ten or fifteen years ago, we accepted post-doctorals with the in retrospect somewhat naive idea that the majority would go back to academia and so be in good position to advise students as to the industry's strengths and weaknesses. Unfortunately, it didn't work out that way. About 70% of our postdoctorals ended up in other industries. Currently we use visiting scientists. University professors still directly involved in the laboratory spend a year with us. They are committed to return to academia and, hopefully, carry the message of the nature of an industrial laboratory back to the students.

KNECHT (UNIV. OF WISCONSIN):   We scientists consider ourselves to be a generally ethical group. Industry had shown itself to be more concerned with making money and cutting costs than with society and ethics. With the inevitable scale-up required for industrial production and the general lack of regulation currently controlling recombinant DNA application, can we trust industry to continue to handle recombinant DNA and new biological materials in a safe manner?

BURNETT (ELI LILLY): At least two issues are involved. One is the fundamental question of whether or not there is any danger to the process at all. To a certain extent it is a mistake to consider "scale" a factor in risk. That is, if one scales up from a laboratory experiment of less than 10 liters to a full scale production of 10 or 50 thousand gallons, does that really necessarily imply an added risk? I think the answer to that is, no. All the risk assessment experiments and all the safety studies that have been done, have indicated that this technology is basically safe. The ethical question you brought up is perhaps a question of the openness of industry. There is a preconceived notion on the part of some people that industries are very secretive and people don't know what they are doing. In the case of the insulin scale-up in our own situation, we have had TV camera crews taking pictures of our pilot plant and throughout our facilities. We have a Bio-Safety Committee that is represented by members of the outside community, including the mayor of the city, and I think that we have done as much as possible to reassure the public, and certainly our local public in Indianapolis, that ours is a very safe procedure.

SCHULTZ (PAPANICOLAOU CANCER RESEARCH INSTITUTE): We are getting close to closing and I would like to leave a message that most of you could take home. Although basic biomedical research has been supported in the past on the level of one to two billion dollars a year, currently $45 billions worth of business is being generated each year based on findings made, in the last 25 years, in basic biomedical research laboratories. I think this is a message that you could let your congressmen and friends of influence know because the horizons of biotechnology are going to depend a great deal on what happens in the university and research institute laboratories.

HARDY (DU PONT): It is time to conclude our panel session. We appreciate your very active participation up until 4 o'clock on a Friday afternoon. I also thank the panel. It was an exciting discussion. What will the session five years from now address? We've gone from safety to economics to question mark.

WHELAN (UNIV. OF MIAMI): It's time to close the meeting. This was the largest ever Miami Winter Symposium, and I think it has been one of the most successful, if not the most successful. We owe our grateful thanks to our panelists and all our speakers. The high note of the proceedings was set by the stirring Lynen Lecture given by César Milstein on Monday evening and the standing ovation he received should have told him how much we appreciated it.   And to you, our audience, about 300 of you who have stuck it out to the end, we are extending our gratitude for your participation and we look forward to seeing you again.   From the financial aspect we'd like to thank the exhibitors whose fees contributed to our coffers and helped to keep down the registration fees.   Of our many sponsors, too numerous to mention, it is not inappropriate to single out one name, Bethesda Research Laboratories, because it gives us an opportunity to thank in person Stephen Turner, the president of the BRL, for taking over the sponsorship of the Lynen Lecture.

FREE COMMUNICATIONS

GENES FOR CYTOCHROME P-450 AND THEIR REGULATION

Michael Atchison, Eugene Ryvkin, Andrea Lippman, Cynthia Raphael and Milton Adesnik. Department of Cell Biology, New York University School of Medicine, New York, N.Y.

Treatment of rats with various specific drugs and xenobiotics leads to marked increases in the hepatic levels of different distinct molecular forms of cytochrome P-450 with different but overlapping substrate specificities.The extremely broad substrate specificity of the P-450 family of enzymes in a wide variety of oxidative reactions contributes greatly to the important role that these mixed function oxidases play in drug detoxification and carcinogen activation. Using recombinant DNA methods we have begun to investigate the genetic mechanisms which determine the size of the P-450 protein repertoire and account for the many-fold increase in liver microsomal levels of different specific molecular forms of this enzyme after treatment of rats with various inducing agents.

We previously demonstrated that treatment of rats with phenobarbital (PB) leads to a substantial increase in hepatic levels of translatable mRNA coding for the PB induced form of cytochrome P-450. DNA-RNA hybridization measurements with cloned PB P-450 complementary DNA indicate that after a single injection of phenobarbital there is up to a 100-fold increase in the liver cytoplasmic levels of PB P-450 mRNA and a rapid commensurate increase in the transcription rate of the corresponding gene.

Southern blot analysis of rat liver genomic DNA and preliminary restriction mapping of fourteen distinct genomic clones isolated from a rat liver-Charon 4A genomic library after screening with cloned cDNA suggest the existence of a multigene cytochrome P-450 family containing at least three, and perhaps up to five, different related genes. Only three of the genomic clones, containing overlapping segments of a single gene, hybridized to the cloned PB P-450 cDNA under conditions of very high stringency. This gene, whose fine structure restriction map in several distinct regions corresponds exactly to that of the cloned cDNA, is more than 10,000 base pairs long and contains at least five intervening sequences as determined by restriction nuclease mapping and electron microscopic examination of R-loops. The cloned PB P-450 cDNA and the corresponding natural gene do not hybridize under conditions of low to moderate stringency to the mRNAs, which code for two different distinct forms of cytochrome P-450 which are induced by 3-methylcholanthrene.

Work supported by American Cancer Society Institutional Grant I-N-14-U and NIH grants AG 01461 and GM 20277.

FROM GENE TO PROTEIN:
TRANSLATION INTO BIOTECHNOLOGY

LOCALIZATION OF A HEPARIN BINDING SITE OF FIBRONECTIN WITH
THE AID OF MONOCLONAL ANTIBODIES

Blair T. Atherton, Elizabeth V. Hayes and Richard O. Hynes.
Center for Cancer Research, Massachusetts Institute of
Technology, Cambridge, MA

We have produced monoclonal antibodies to fibronectin
in order to study structure-function relationships. Our
analysis, so far, has concentrated on the C-terminal portion
of fibronectin. With the aid of these antibodies the location of proteolytic fragments of fibronectin in the intact
molecule has been determined. In particular, a nested set
of heparin-binding fragments ranging in size from 23-200kd
has been identified which extend into the C-terminal region
of fibronectin.

A disulfide bonded dimer of approximately 40kd
containing the C-terminal end of fibronectin and a site of
structural difference between the cellular and plasma forms
of fibronectin does not bind to heparin.Monoclonal antibodies
which bind this fragment do not bind any of the heparin-binding fragments. Therefore, the heparin binding site is at
least 23kd from the C-terminus of fibronectin. Furthermore,
the small size of some of the heparin-binding fragments and
their C-terminal position indicates that the heparin binding
site is located in the C-terminal one-fourth (60kd) of
fibronectin. These results locate a heparin-binding site
within a region 35kd long, lying between 23 and 60kd from the
C-terminal of the protein.

One of our monoclonal antibodies is of particular
interest because it binds to the nested set of heparin-binding
fragments and blocks the binding of heparin to fibronectin.
This antibody should provide a useful probe to determine the
function of the heparin binding site and its role in the
various biological activities of fibronectin.

FROM GENE TO PROTEIN:
TRANSLATION INTO BIOTECHNOLOGY

NUCLEOTIDE SEQUENCE OF THE glg C GENE CODING FOR ADP-GLUCOSE SYNTHETASE FROM E. COLI

Preston A. Baecker,[*] Clement E. Furlong[**] and Jack Preiss[*]
Department of Biochemistry and Biophysics, Univ. of California, Davis[*] and Department of Genetics, Univ. of Washington, Seattle[**]

The majority of the nucleotide sequence of the glg C gene in the glycogen operon of E. coli has been determined by the method of Maxam and Gilbert. The sequenced regions are depicted below as horizontal arrows.

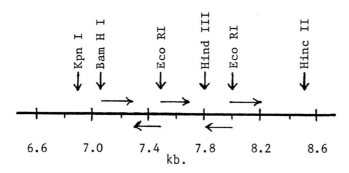

The sequence, which codes for ADP-glucose synthetase, has been used to predict the primary structure of this allosteric protein. The previously determined amino acid sequences of four cyanogen bromide peptides from ADP-glucose synthetase have aided in confirming portions of the nucleotide sequence. One of these amino acid sequences has previously been identified as a portion of the active site of the enzyme, while another has been shown to be involved with the binding of allosteric activators. An ultimate objective of this study is to determine the organization of the coding regions from the nucleotide sequence for the relative positions of the effector and substrate sites of glg C gene product.

Supported by USPHS Grant AI 05520

FROM GENE TO PROTEIN:
TRANSLATION INTO BIOTECHNOLOGY

ISOLATION AND CHARACTERIZATION OF A SOYBEAN GENE FOR THE
SMALL SUBUNIT OF RIBULOSE-1,5-BISPHOSPHATE CARBOXYLASE

S. L. Berry, D.M. Shah and R.B. Meagher. The Department
of Molecular and Population Genetics, University of Georgia,
Athens, Ga. 30602.

Ribulose-1,5-bisphosphate carboxylase (RuBPcase)
catalyzes the fixation of CO_2 via the carboxylation of
ribulose-1,5-bisphosphate. It is one of the most abundant
proteins in nature. The native protein is composed of two
different subunits. The small subunit is encoded by the
nucleus, the large subunit by the chloroplast. The mature
polypeptide for the small subunit contains 123 amino acids.
We are interested in examining the control sequences for
the nuclear genes encoding the small subunit of soybean
RuBP carboxylase.

A cDNA clone for the small subunit of pea RuBPcase
(pSSU160, Bedbrook et al., 1980, Nature 287,692-697)
contains a major portion of the polypeptide coding sequence.
The purified HindIII fragments from this clone were used
to probe the soybean genome for RuBPcase genes. Southern
blot analysis of the soybean genome suggests a small
multigene family encodes the small subunit. Approximately
ten genomic EcoRI fragments ranging from 0.85 to 5kb
hybridize to the probe. Genomic fragments hybridizing to
this probe were isolated from a soybean lambda DNA library
(Nagao et al., 1981, DNA 1, 1-9.). One of these clones,
λSRS1, contains an insert composed of 6 EcoRI fragments
totaling 13.5 kb. Two EcoRI fragments of 2.1 and 0.85kb,
contained in this lambda recombinant and which hybridize to
the pea probe, were subcloned in the plasmid vehicle pBR325.
These bands correspond in size to two of the fragments
observed in the genomic Southern blot of soybean DNA.
Detailed restricion maps were constructed of these two
cloned fragments. Partial sequence data suggests that the
polypeptide coding region contained on these two fragments
is interrupted by at least three introns. These introns
appear to have classic eucaryotic intron-exon junction
sequences. The nucleotide sequence of a major portion of
the processed peptide for this soybean small subunit gene is
at least 80% homologous with pea. A detailed analysis of
the 5' and 3' flanking sequences will be made.

Supported by USDA research grant 78-59-2133-1-1-068-1.

FROM GENE TO PROTEIN:
TRANSLATION INTO BIOTECHNOLOGY

STUDYING POLYOMA VIRUS GENE EXPRESSION AND FUNCTION BY SITE-
DIRECTED MUTAGENESIS USING SYNTHETIC OLIGONUCLEOTIDES

Gordon G. Carmichael[1], David Dorsky[2], Brian Schaffhausen[2],
Donald Oliver[3] and Thomas L. Benjamin[2]

[1]Department of Microbiolggy, University of Connecticut Health
Center, Farmington, CT, [2]Department of Pathology, Harvard
Medical School, Boston, MA and [3]Department of Microbiology,
Harvard Medical School, Boston, MA

We have constructed a transformation defective polyoma
virus mutant (1387-T) which directs the synthesis of a normal
small T antigen, a functional large T containing a single ala
to val change and a truncated (51K) middle T antigen which lacks
37 amino acids at its C-terminus. This shortened middle T
polypeptide is missing the hydrophobic "tail" thought to be
responsible for the anchorage of this protein into the plasma
membrane and, in fact,is found in cytosol fractions. This
deleted middle T is also inert in an in vitro protein kinase
assay.

Mutant 1387-T has a single base change from wild type
virus: a T-A base pair instead of a C-G base pair at nucleotide 1387. This change was introduced into viral DNA by using
a synthetic oligonucleotide 11 bases long as the specific
mutagen. Polyoma DNA was rendered single stranded by molecular
cloning into coliphage M13. The oligonucleotide, which anneals
to polyoma DNA in a single place but with a mismatch at the
site to be altered, was used to prime the in vitro synthesis
of double stranded closed circular DNA by E. coli DNA polymerase and phage T4 DNA ligase. This material was then used to
obtain progeny M13-polyoma recombinants which were screened
by DNA sequencing for the desired base change. Infectious
polyoma containing the mutation was reconstructed from
recombinant phage replicative form molecules.

This methodology is being followed in order to construct
other mutants affecting early viral protein function, as well
as mutants affected in controlling signals for RNA synthesis
and processing.

FROM GENE TO PROTEIN:
TRANSLATION INTO BIOTECHNOLOGY

CHARACTERIZATION OF FOUR DIVERGENT HUMAN GENOMIC CLONES HOMOLOGOUS TO THE TRANSFORMING P21 GENES OF HARVEY AND KIRSTEN MURINE SARCOMA VIRUSES

Esther H. Chang,[+] Matthew A. Gonda,[*] Ronald W. Ellis,[#]
Edward M. Scolnick,[#] and Douglas R. Lowy,[+]
[+]Dermatolgy Branch and [#]Laboratory of Tumor Virus Genetics,
National Cancer Institute, Bethesda, MD, and [*]Biological
Carcinogenesis Program, Frederick Cancer Research Center,
Frederick, MD

Harvey (Ha) and Kirsten (Ki) murine sarcoma viruses are highly oncogenic retroviruses whose transforming gene product is p21. Low levels of a related p21 is found in normal cells. The viral sequences which encode p21 (v-Ha-ras and v-Ki-ras, respectively) are related to each other, but under stringent hybridization conditions there is little hybridization between the two v-ras genes. When the two v-ras genes are used as probes to analyze restriction endonuclease digested genomic DNA from human and other species, each probe hybridizes to distinct "unique sequence" fragments, suggesting that the p21 genes of normal cells are a gene family.[1] Using v-Ha-ras and v-Ki-ras probes, we have now confirmed this hypothesis directly by the molecular cloning from normal human DNA of four different ras containing restriction endonuclease fragments.

Two of these human c-ras clones hybridized preferentially to v-Ki-ras and were designated c-Ki-ras clones; both contained about 0.9 kb of homology with v-Ki-ras. In one c-Ki-ras clone this segment was colinear with v-Ki-ras, while the other contained one intron. The other two clones hybridized preferentially to v-Ha-ras and were designated c-Ha-ras clones. One of these c-Ha-ras clones hybridized poorly to v-Ha-ras under stringent conditions but hybridized to 0.7 kb of v-Ha-ras under less stringent conditions. The other human c-Ha-ras clone hybridized to 0.9 kb of v-Ha-ras and contained three introns. Its structure was very similar to a previously cloned rat c-Ha-ras gene which has been shown to be capable of transforming cells when ligated downstream from a retroviral long terminal repeat.[2]

We conclude that in contrast to other c-onc genes, human cells contain several c-ras genes. These c-ras genes are heterogeneous in their homology and in their number of introns.

1. Ellis et al., Nature 292 506-511 (1981)
2. DeFeo et al., Proc. Nat. Acad. Sci. 78 3328-3332 (1981).

FROM GENE TO PROTEIN:
TRANSLATION INTO BIOTECHNOLOGY

IN VITRO SYNTHESIS OF ENZYMES INVOLVED IN MEMBRANE BIOSYN-
THESIS IN E. COLI

Yang Chang Chen and William Dowhan, Department of Biochem-
istry and Molecular Biology, Graduate School of Biomedical
Sciences, The University of Texas Medical School, Houston,
Texas 77025

Phosphatidylserine synthase (pss gene locus) catalyzes
the committed step in the synthesis of phosphatidylethanolamine in E. coli. This enzyme has been shown to be ribosomal bound in crude extracts of E. coli, but this association appears to be an artifact of isolation (1). The properties of the purified enzyme suggest a membrane location
for the enzyme in vivo (2).

In order to study the processes of transcription and
translation of the pss gene locus and the subsequent assembly of the gene product into a functional enzyme, we have
cloned the pss gene using pBR322 as the cloning vector.
The resulting plasmid (pPS4017) directs the synthesis of
phosphatidylserine synthase in an E. coli S30 protein synthesizing system. The newly synthesized polypeptide has
an apparent subunit molecular weight of 54,000 daltons
(which is identical to the isolated enzyme), is immunoprecipitable by antibody directed against the purified enzyme,
and is enzymatically active. The presence of the protease
inhibitor TLCK during protein synthesis results in the synthesis of a polypeptide of approximately 2,000 daltons
larger than the isolated enzyme. These results indicate
either that the enzyme is made as a higher molecular weight
precursor which is later processed during assembly or that
the previously isolated enzyme of 54,000 daltons is a degradation product of the true in vivo form of the enzyme.
The latter possibility may be the basis for the apparent
artifactual ribosomal association of the enzyme in crude
extracts of E. coli.

Supported by USPHS Grant #GM 25047

(1) Louie, K. and Dowhan, W., J. Biol. Chem. 254, 8391-
     8397 (1980).

(2) Carman, G. and Dowhan, W., J. Biol. Chem. 255, 1124-
     1127 (1979).

STUDIES ON EXPRESSION OF THE PHASEOLIN GENE OF FRENCH BEAN
SEEDS IN SUNFLOWER PLANT CELLS

Prabhakara V. Choudary*, John D. Kemp†, Dennis W. Sutton†,
and Timothy C. Hall
University of Wisconsin, Departments of Horticulture and
Plant Pathology†, Madison, WI   53706

An intermedidate integration vector for plants, pKS 111,
was derived from the broad host range plasmid, PRK 290 by
inserting into its unique Eco RI site a 5.3 kb fragment of
T-DNA of the octopine Ti plasmid (Kemp, J.D. and D.W. Sutton,
in preparation). A 3.1-kb long Bam HI-Hind III fragment of
phaseolus vulgaris DNA coding for phaseolin in French bean
seeds, in two different combinations with a 1.15 kb Bam
HI-Hind III fragment of Tn 5 DNA encoding neomycin phosphotransferase II, has been ligated into the vector PKS III,
linearized by removal of a tiny Hind III fragment. Recombinant cells of Agrobacterium tumefaciens obtained after
transformation with these two types of plasmids separately
were conjugated with Escherichia coli cells harboring a P-1
group incompatible plasmid pPHIJI. This facilitates exchange,
by homologous recombination, of the T-DNA fragment between
the resident Ti plasmid and pKS 111 carrying the phaseolin
gene. Consequently, the phaseolin gene with flanking T-DNA
gets inserted into the native Ti plasmid, and the original
vector, pKS 111 gets eliminated.

One week old sunflower plants were then injected with
50 µl of saturated cultures of the recombinant Agrobacterial
cells to produce crown galls, which were excised out after
three weeks and cultured on solid medium.

Southern blot hybridizations of restriction digests of
phaseolin DNA with in vivo labelled RNA isolated from the cultured tissues as probe revealed the occurrence of phaseolin-
specific RNA in the crown gall cells produced by one particular clone, although we have not been able to detect phaseolin
protein in these cells as yet. The results thus indicate
the possible integration and transcription of the phaseolin
gene with the plant genome.

*Permanent address: Sections on Biochemistry and Psycho-
genetics, Biological Psychiatry Branch, National Institute
of Mental Health, Bethesda, Maryland 20205

FROM GENE TO PROTEIN:
TRANSLATION INTO BIOTECHNOLOGY

DNA STIMULATES ATP-DEPENDENT PROTEOLYSIS AND PROTEIN-
DEPENDENT ATPase ACTIVITY OF PROTEASE La FROM
ESCHERICHIA COLI

Chin Ha Chung, Lloyd Waxman, and Alfred L. Goldberg
Department of Physiology and Biophysics, Harvard Medical
School, 25 Shattuck Street, Boston, Ma. 02115

Mutations in the lon gene in E. coli result in a
variety of phenotypic alterations including the reduction of the cell's capacity to degrade abnormal polypeptides. Recently, it has been shown that the product of
the lon gene is an ATP-dependent serine protease,
protease La, that also has ATPase and DNA-binding
activities. The ATPase activity is stimulated two to
three-fold in the presence of hydrolyzable substrates
such as casein and globin, while proteins which are not
degraded (albumin, hemoglobin) do not stimulate the
ATPase. Addition of double-stranded or single-stranded
DNA to the protease in the presence of ATP was found to
stimulate the hydrolysis of casein or globin two to
seven fold, depending on the DNA concentration. Native
DNA from several sources (plasmid pBR322, phage T_7 or
calf thymus) had similar effects, but after denaturation
the DNA was 20 to 100% more effective than the native
form. Although poly (rA), globin mRNA and various tRNA
did not stimulate proteolysis, poly(rC) and poly(rU)
were effective. Poly(dT) was stimulatory, but $_p(dT)_{10}$
was not. In the presence of DNA as in its absence,
proteolysis required concomitant ATP hydrolysis, and the
addition of DNA also enhanced ATP hydrolysis by protease
La approximately two fold, but only in the presence of
protein substrates. At much higher concentrations, DNA
inhibited proteolysis as well as ATP cleavage. Thus,
the association of this enzyme with DNA may regulate the
degradation of cell proteins in vivo.

FROM GENE TO PROTEIN:
TRANSLATION INTO BIOTECHNOLOGY

METHYLATION IN CELLULAR MEMBRANES OF FRIEND VIRUS TUMORS
OF MICE

W.E. Cornatzer, Dennis R. Hoffman and Judy A. Haning. Guy
and Bertha Ireland Research Laboratory, Department of
Biochemistry, University of North Dakota School of Medicine,
Grand Forks, ND 58202

Phosphatidylcholine (PC) is the major lipid in all
cellular membranes. PC biosynthesis is known to occur by
two major different pathways. One pathway involves the
methylation of phosphatidylethanolamine (PE) to phosphatidylcholine which is catalyzed by endoplasmic reticulum membrane
enzyme, phosphatidylethanolamine methyltransferase (PEMT) and
requires three S-adenosylmethionine (AdoMet) for methyl
groups to produce three S-adenosylhomocysteine (AdoHcy). It
has been established that AdoHcy is a competitive inhibitor
of PEMT. The second pathway of PC biosynthesis is catalyzed
by choline phosphotransferase (CPT). Methylation of PE to
PC was measured in the spleen of BALB/c male mice infected
with Friend virus for 14 days. There is a significant
stimulation of PC biosynthesis during tumor growth, 5, 10,
14 days following inoculation of the virus by the CDP-
choline pathway (pmoles PC formed/min/mg protein). However,
there is a decrease or no stimulation in the methylation of
PE to PC. The specific activities (pmol of PC formed/min/mg
protein) for PEMT, phosphatidylmonomethylethanolamine
methyltransferase (PMMEMT) and phosphatidyldimethyl-
ethanolamine methyltransferase (PDMENT) which measures
progressive methylation of PE to PC show no stimulation over
controls whereas the choline phosphotransferase was stimulated significantly 10 fold. There was a significant 5 fold
increase over controls in the concentration of AdoHcy (nmol/
g tissue) in the viral tumors. The AdoMet/AdoHcy ratio
decreased signifiantly to a level which inhibits PEMT
transferase. During this rapid growth of the tumor the
major phospholipid of cellular membrane, PC is synthesized
by the CDP-choline pathway rather than the methylation of
PE to PC. Phospholipid methylation of PE to PC can alter
structure and function of biomembranes. The phosphatidyl-
cholines produced by the methylation pathway have more
polyunsaturated fatty acids than PC from the CDP-choline
pathway.

Supported by grant from American Cancer Society, N.D.
Affiliate.

IDENTIFICATION OF A NOVEL, H-2-LIKE GENE BY ANALYSIS OF cDNA CLONES

David Cosman, George Khoury and Gilbert Jay
Laboratory of Molecular Virology, National Cancer Institute, National Institutes of Health, Bethesda, Maryland 20205.

The classical H-2 transplantation antigens of the mouse are encoded by a family of genes that has been mapped to chromosome 17. As membrane-associated proteins, they are responsible for rapid allograft rejection and are involved in the associative recognition of foreign antigens by cytotoxic T cells. These H-2 antigens are composed of extracellular, transmembrane, and cytoplasmic domains.

A series of H-2 cDNA clones has been isolated by screening a mouse liver library with a human HLA-B cDNA probe. Restriction enzyme mapping, cross-hybridization, and partial DNA sequence analysis have shown extensive similarities among the various clones, and a high degree of homology with the available protein sequences. However, extensive differences were detected in the 3' non-coding region of different clones. These differences were used to define three distinct classes of cDNA clones and to detect three subclasses of H-2 mRNA in mouse liver.

Sequence analysis of one class of H-2 cDNA clones revealed unusual features within the coding sequence for the transmembrane region of the protein. The nucleotide sequence predicts several charged or polar amino acids in this region that would destroy its hydrophobicity. In addition, there is a small frameshift deletion that results in a termination codon towards the cyptolasmic side of the transmembrane region.

The presence of this unusual sequence in liver mRNA predicts the existence of a novel, shortened form of H-2-like antigen which contains the entire biologically active portion of the molecule but lacks the normal hydrophobic transmembrane region and the hydrophilic cytoplasmic domain. It is unlikely that such a molecule would be inserted into the plasma membrane, and it might instead be secreted.

FROM GENE TO PROTEIN:
TRANSLATION INTO BIOTECHNOLOGY

A GENETIC ANALYSIS OF THE GLUCOCORTICOID RESPONSE

Mark Danielsen and Michael R. Stallcup

We are undertaking a genetic analysis of the glucocorticoid responses of WEHI 7 mouse lymphoma cells by selecting and characterizing phenotypic variants. The glucocorticoid-induced cytolysis of these cells has been used previously to select steroid resistant variants. However only receptor variants were isolated. We have introduced a glucocorticoid inducible, selectable marker, mouse mammary tumor virus (MMTV) into these cells. Cells expressing MMTV proteins can be selected for by binding the cells to a plastic petri dish coated with antibody against MMTV. Using this technique we have been able to obtain populations of cells which are glucocorticoid resistant but which have an inducible MMTV response, indicating the presence of functional receptor.

The MMTV marker was also used to assay for the presence of functional receptor in glucocorticoid resistant clones. Resistant clones were generated by mutagenesis with 250 µg/ml EMS and, after 5 cell generations, plating on top of 0.5% agar noble containing growth medium and 10^{-6} M dexamethasone. These clones were then infected and cells capable of a dex induced MMTV response selected. We have obtained 3 populations of infected clones that induce MMTV RNA and protein in response to glucocorticoids, but are not killed by the hormone. Characterization of clones is in progress.

Supported by USPHS grant GM28298.

NUCLEOTIDE SEQUENCE OF THE REGION SURROUNDING THE 5' END
OF 18S rRNA FROM SOYBEAN INFERRED FROM THE GENE SEQUENCE

V.K. Eckenrode*#, R.T. Nagao@, and R.B. Meagher#@,
Departments of Biochemistry*, Molecular and Population
Genetics#, and Botany@, University of Georgia, Athens, Ga.
30602.

Although the rRNA genes are transcribed at high levels
in all eucaryotes, little is known about their transcription
initiation process in plants. To examine this process in
detail, we are characterizing the intergenic spacers of the
ribosomal repeat in soybean.

The 5' end of 18S rDNA from soybean has been
identified. This DNA was originally cloned in Charon 4A as
part of a one-and-a-half rDNA repeat. Those regions which
code for 18S and 25S rRNA were localized by Southern and
Northern analysis of the cloned DNA. These data were
confirmed by S-1 mapping experiments. The fragments of DNA
thought to code for the 1000 bp surrounding the 5' end of
18S rRNA were sequenced. Those sequences which actually
code for 18S rRNA show an average of greater than 70%
homology with the 5' ends of rDNA from both yeast and
Xenopus. Additional comparisons among these sequences show
regions of high homology interspersed with regions of little
homology. Interspersed homology between the 18S rRNA's of
yeast and Xenopus was first reported by Salim and Maden
(1981, Nature 291: 205-208). This same pattern extends to
the homology between the 18S rRNA's of soybean and both
yeast and Xenopus.

Experiments, similar to those performed on the 5' end
of 18S on rRNA, are in progress to pinpoint the 3' end of
25S rRNA and the 5' end of precursor rRNA. It is hoped that
from this information we will be able to find the site of
initiation of transcription of rRNA in soybean.

Supported by USDA research grant #78-69-2133-1-1-068-1.

FROM GENE TO PROTEIN:
TRANSLATION INTO BIOTECHNOLOGY

AN ELISA ASSAY FOR DETECTING CELL SURFACE ANTIGENS ON ADHE-
RENT CELLS. Feit C, Bartal AH, Tauber G, Hirshaut Y. Labor-
atory for Immunodiagnosis, Memorial Sloan-Kettering Cancer
Center; New York, NY 10021.

The effective use of monoclonal antibodies (mAb) in the
analysis of cell surface antigens is critically dependent
upon the availability of a rapid, reproducible, quantitative
assay for their detection and characterization. To meet
these requirements an ELISA assay has been developed that
allows for the detection of mAb recognizing antigens on the
surface of viable, anchorage-dependent cells. Its utility
in identifying potential markers of human sarcoma cells has
been tested.

Suspensions of adherent human cells are placed in a 96-
well tissue culture plate at a concentration yielding a con-
fluent monolayer upon overnight incubation. Plates are
washed 3 times with phosphate buffered saline (PBS) containing 1% bovine serum albumin (BSA). 50 ul of mAb supernate
is placed in appropriate wells, and plates are incubated for
1 hr at 4°C with periodic shaking. Excess antibody is removed by again washing plates with PBS + 1% BSA, this time
with the addition of 1.5mM $MgCl_2$ and 2mM monothioglycerol
(MTG). Affinity purified sheep $F(ab)_2$ anti-mouse IgG (heavy
& light chain) conjugated to β-galactosidase is placed in
the wells for 2 hr at room temperature. Plates are washed
as above, followed by incubation with substrate (ρ-nitrophe-
nyl-β-D-galactopyranoside + 100 mM MTG) for 1 hr at room
temperature. The reaction is stopped by the addition of
0.5M Na_2CO_3, which also enhances the color development.
Quantitative results are easily obtained on an automatic
plate reader. Using this technique more than 1000 wells can
be easily screened in a single day.

This procedure was compared with an immunofluorescent
assay (IF) that utilizes acetone-fixed sarcoma cells as antigen. Using 17 mAb previously obtained from the fusion of
sarcoma-immunized spleen cells, 9 were positive in both
assays, 4 were negative in both and 4 were positive only
with IF. IgG titers were 6-20X greater using ELISA. The 4
ELISA negative mAb presumably recognize internal antigens.
Subsequently, ELISA and IF were used in parallel to screen
600 wells from a new fusion. This allowed the identification of 24 mAb positive only with ELISA. These mAb are currently being characterized.

Thus, ELISA is a rapid, reproducible, quantitative assay
for screening and characterizing mAb. This assay is useful
for identifying cell surface antigens in their native un-
fixed state.

FROM GENE TO PROTEIN:
TRANSLATION INTO BIOTECHNOLOGY

THE INVOLVEMENT OF DNA POLYMERASE α IN ADENOVIRUS
DNA REPLICATION

David A. Foster and Geoffrey Zubay
Department of Biological Sciences, Columbia University,
New York, New York 10027

Adenovirus DNA replication is sensitive to the DNA
polymerase α specific inhibitor aphidicolin only at
levels much higher than that required for the inhibition
of host cell DNA replication (1,2). It is argued here
that this is due to a limited number of nucleotide linkages requiring the α enzyme. This arguement is supported
by observations on KB cells infected with H5ts125 temperature sensitive mutant virus in which all rounds of
virus replication are completed after a temperature up-
shift. Subsequent viral DNA synthesis after a down-
shift to the permissive temperature is initiated at the
ends of the DNA molecules where the origins of adenoviral DNA replication are located. This reinitiation of
adenoviral DNA replication is sensitive to much lower
levels of aphidicolin. The high-sensitivity of initiation of replication suggests that DNA polymerase α is
required for this process.

It is also shown that adenovirus DNA replication in
the presence of high levels of aphidicolin leads to the
appearence of discrete intermediates of replicating virus,
suggesting a further requirement for the α enzyme at other
as yet undetermined sites on the genome.

Supported by a grant from the American Cancer Society
(CD-13).

1. Foster, D.A., Spangler, R. and Zubay, G. J. Virol.
   37: 493-495 (1981).

2. Krokan, H., Shaffer, P. and DePamphilis, M. Biochemistry 18: 4431-4443 (1979).

DEVELOPMENT OF HUMAN GROWTH HORMONE PRODUCED IN
RECOMBINANT BACTERIA AS A THERAPEUTIC AGENT

L. Fryklund, K. Fhölenhag, H. Flodh, B. Holmström,
A. Skottner-Lundin, B. Strindberg and A. Wichman
KabiVitrum AB, S-112 87 Stockholm, Sweden.

Human growth hormone (somatotropin, hGH) is used
therapeutically for the treatment of children with
pituitary deficiency. The hormone is a protein of
MW 22,000 which is isolated from human pituitary glands.
Although the present product is of excellent quality and
efficacy, the supply of pituitaries is a limiting factor.

The gene for human somatotropin has been inserted
into the plasmid PBr 322 in E.coli K_{12} by recombinant DNA
techniques. The recombinant bacteria have been grown in
batch culture and the hGH extracted from the disinte-
grated cells. Conventional biochemical separation methods
such as ion exchange and gel chromatography have been
used for purification on a scale suitable for industrial
production.

The biosynthetic hGH appears to be identical to the
pituitary hormone with regards to molecular weight and
sequence with the exception of an N-terminal methionine.
No contaminating proteins or isohormones are present.
Biological activity, determined in hypophysectomized
rats by increase in epiphysial width of tibia, weight
gain, and sulfate incorporation into connective tissue
appears to be identical to that of the pituitary hormone.

One month toxicity studies in rats and dogs, in a
dose range up to one hundred times the normal therapeutic
dose have revealed no toxic effects.

A two-week safety study in man has indicated no
adverse effects and we are at present starting a two
month study in adult patients with hypopituitarism who
have previously been treated with growth hormone. If
these patients respond satisfactorily a one year clinical
trial in hypopituitary children will be initiated.

FROM GENE TO PROTEIN:
TRANSLATION INTO BIOTECHNOLOGY

Expression of the Genes for the Pst Restriction-Modifi-
cation Enzymes after Cloning into a Temperature-Sensitive
Replication Plasmid

Roy Fuchs and James F. Kane, Division of Molecular
Biology, Bethesda Research Laboratories, Inc., Gaithersburg
Maryland   20877

Our objective was to develop a model system to study the
use of an overexpression plasmid to increase synthesis of
a restriction enzyme in Escherichia coli. We chose the
restriction (Pst I)-modification system of Providencia
stuartii that has been cloned into the Hind III site of
pBR 322 (Walder, et al., 1981, P.N.A.S. 78:1503). One
hybrid plasmid, designated pPst 102, contained a 5.9 kb
insert with an internal Hind III site, producing 1.9 and
4.0 kb fragments. Although both enzymes were specified
by the 4.0 kb fragment, the genes were poorly expressed
in E. coli. We subcloned the Pst genes into a temperature sensitive runaway plasmid (pBEU I) and transformed
E. coli HBIOI. Four independent transformants were
isolated and characterized: pRF I (1.9 kb insert);
pRF 2 (4.0 kb insert); pRF 3 and pRF 4 (5.9 kb insert).
All four strains were subjected to temperature shifts
to allow for overexpression of the plasmid. Crude cell
extracts were prepared and examined for Pst I. When the
Pst I activity of E. coli/pPst 102 is assigned a value
of 1.0 (700 units/mg protein), the following relative
enzymatic activities were estimated in the four strains:
pRF I<0.5; pRF 2<0.5; pRF 3<0.5; pRF 4-10.0. These
results demonstrate that the overexpression plasmid
produces a 10-fold increase in the activity of Pst I in
E. coli. Furthermore, this system provides a basis for
studying the regulation of the Pst I gene.

FROM GENE TO PROTEIN:
TRANSLATION INTO BIOTECHNOLOGY

THE NUCLEOTIDE SEQUENCE OF pACYC 177

J. Galen, D. Kuebbing, L. Runkel, and K. Rushlow.
Instrumentation Division, Bethesda Research Laboratories,
Inc., Gaithersburg, Maryland, 20877

We have determined the nucleotide sequence of the
cloning vehicle, plasmid pACYC 177. This plasmid is derived from the P15A replicon, the kanamycin resistance gene
from the R6-5 plasmid, and the ampicillin resistance gene
from Tn3 (1). The DNA sequence was determined utilizing
both the chemical modification method of Maxam-Gilbert, and
the dideoxy chain termination procedure of Sanger. Template DNA for the latter method was prepared by "shot-gun"
cloning DNA restriction fragments of pACYC 177 into M13-mp8
RF DNA, and subsequent purification of the viral DNAs containing pACYC inserts. The accumulated sequence data was
subjected to computer analysis to determine sequence
overlaps, restriction endonuclease recognition sites, and
putative promoter and protein coding regions. We have
tentatively identified the sequences in pACYC 177 which
encode the genes for kanamycin and ampicillin resistance.

(1)   Chang, A.C.Y. and Cohen, S.N., J. Bacti. 134
      1141-1156 (1978).

FROM GENE TO PROTEIN:
TRANSLATION INTO BIOTECHNOLOGY

THE ISOLATION OF SEA NETTLE LETHAL FACTOR I VIA IMMOBILIZED
MONOCLONAL ANTIBODIES

Pramod K. Gaur, Carrington S. Cobb, A.J. Russo, Gary J.
Calton and Joseph W. Burnett
Division of Dermatology, University of Maryland School
of Medicine, Baltimore, Maryland.

The Sea Nettle (Chrysaora quinquecirrha) kills its prey
with a potent venom which is contained in nematocysts found
on its tentacles. This venom is a complex mixture of enzymes and toxic agents. The venom is thermolabile ($T_{\frac{1}{2}}$ = 24
hr at $4°C$) and may also be degraded by the proteolytic agents it contains.[1] Purification of the venom's lethal factors has been hindered by low yields of pharmacological
activity upon chromatography as well as the large volumes
required to assess lethality in mice.

Monoclonal antibodies (MAB) were prepared using the
crude venom for immunization. An ELISA for the crude
venom was prepared using hyperimmune guinea pig serum.[2]
Fractions from different affinity chromatography columns
were used to determine which hybridomas were secreting
specific antibodies for lethal fractions. Clone C6 produced MAB which reacted with lethal eluates from SP
Sephadex and hexylamine Sepharose columns. C6 MAB neutralized only 3-4 LD_{50} of crude venom.[3]

C6 MAB was immobilized on Sepharose via CNBr and used
for immunochromatography. ELISA with C6 MAB was used to
monitor column fractions. Sea Nettle Factor I (SNLF I)
was eluted with mild conditions (pH 7, 10 mM phosphate,
750 mM NaCl). A 29-fold increase in specific activity
was obtained with complete recovery of lethal activity.
The ELISA indicated no C6 activity in the non-retained
protein fraction and high reactivity to the specific
eluate.

SNLF I produced cardiac abnormalities in the rat.

1   Burnett, J.W. and Calton, G.J., Toxicon 15, 177-196
        (1977).
2   Gaur, P.K., Calton, G.J. and Burnett, J.W., Experientia
        (in press).
3   Gaur, P.K. Anthony, R.L., Cody, T.S., Calton, G.J.
        and Burnett, J.W., Proc. Soc. Exp. Biol. Med.,
        167, 374-377 (1981).

FROM GENE TO PROTEIN:
TRANSLATION INTO BIOTECHNOLOGY

SYNTHESIS OF A SPECIFIC SEQUENCE OF DNA USING RESTRICTION
ENZYMES

Richard Gayle and George Bennett, Dept. of Biochemistry,
Rice University, Houston, Texas 77001

The restriction enzyme MboII recognizes the DNA sequence GAAGA and cleaves the DNA eight bases downstream from
this site, leaving a single 3' protruding base (1). There
are no apparent sequence requirements in the stretch of DNA
between the recognition sequence and the point of cleavage.
Using this property, a segment of DNA from one fragment can
be transferred to another fragment.

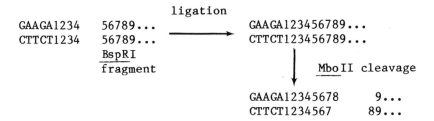

Specially constructed, synthetic DNA duplexes (made in
the laboratory of Dr. P. Gilham, Purdue Univ.) were ligated
onto blunt-ended BspRI fragments, positioning a MboII recognition site so that MboII cleavage would occur in the BspRI
fragment. Subsequent cleavage with MboII resulted in the
transfer of several base pairs from the BspRI fragment to
the synthetic duplex. Utilization of different blunt-ended
fragments yields a wide variety of MboII cleavage products.
Ligation of these cleavage products through complementary 3'
protruding ends results in the formation of a new, defined
sequence of DNA. Incorporation of another restriction enzyme cleavage site in positions 1-4 allows the release of
the MboII recognition sequence from the transferred segment
of DNA. The scheme can be demonstrated by the formation of
a restriction enzyme recognition site from two MboII cleavage products. Extension of the basic scheme would allow the
production of more generalized fragments useful in constructing gene sequences.

Supported by NIH grant GM-26437 and in part by Training
Grant GM-07833 from the National Institutes of Medical
Sciences.

(1) Brown, N.L., Hutchison, C.A., III, and Smith, M., J.
Mol. Biol., 140, 143-148 (1980).

FROM GENE TO PROTEIN:
TRANSLATION INTO BIOTECHNOLOGY

INDUCTION OF STRESS PROTEINS BY HEAT SHOCK AND METAL IONS

L. Gedamu, N. Shworak, J. J. Heikkila, and G. A. Schultz

University Biochemistry Group, Faculties of Medicine and
Science, The University of Calgary, Calgary, Alberta, Canada

Chinook salmon embryonic cells were incubated at either
18°C, 20°C, 24°C or 26°C for 2 hours. ^{35}S-methionine was
added to the culture media for the last hour to label nas-
cent proteins. Electrophoretic examination of the spectrum
of newly synthesized proteins of cells kept at either 24°C
or 26°C revealed an increase in the relative labelling of
a number of species. The most prominent ones have molecular
weights of approximately 84K, 70K, 68K, and 41K. The fact
that the induction of these proteins could be blocked by
pretreatment of the cells with actinomycin D suggests that
the expression of these "stress" proteins are regulated at
the transcriptional level. Furthermore, these stress proteins are detectable in the products of an in vitro translation system programmed with heat shock RNA. Treatment of
these cells with 5 µM cadmium or 100-500 µM Zinc also
induced the major 70K "stress" protein. The protein, metal-
lothionen, was also induced by these metal ions but after a
much longer treatment period. The molecular weights of
these fish cell "stress" proteins are similar to heat shock
proteins found in Drosophila. While the function of these
proteins is unknown, they are believed to play a role in
ceullular homeostasis.

FROM GENE TO PROTEIN:
TRANSLATION INTO BIOTECHNOLOGY

Analysis of the Rous Sarcoma Virus Promoter Region by Site-Directed Mutagenesis

Gregory M. Gilmartin and J.T. Parsons
Department of Microbiology, University of Virginia
School of Medicine, Charlottesville, Virginia 22908

Transcription of the Rous sarcoma virus genome is presumed to be modulated, at least in part, by DNA sequences that lie within the U_3 region of the long terminal repeat. This putative promoter is comprised of a sequence very similar to the Goldberg–Hogness box that resides 24 to 30 bp upstream from the viral RNA cap site. We have attempted to delineate the viral DNA sequences that control viral RNA transcription, employing site-directed mutagenesis techniques. Using a recombinant plasmid (pS1103) containing a biologically active, genome length viral DNA insert (Highfield et al., 1980, J. Virol. 36:271), we have constructed a series of mutants containing deletions or small insertions at a site 53 bp upstream from the RNA cap site. A comparison of the biological activity and DNA sequence analysis of individual mutants reveals two classes of mutants. Mutants of the first class possess deletions extending into the Goldberg–Hogness box and are replication defective. Mutants within the second class contain either deletions or small insertions upstream from the Goldberg–Hogness box. These mutants can be further subdivided into replication competent mutants and partially defective replication mutants. Replication competent mutants are phenotypically indistinguishable from wild type upon transfection of chick embryo cells, while the partially defective replication mutants have a significantly slower rate of replication compared to wild-type virus. Further analysis of the transcriptional activity of these mutants should allow us to define regions of the viral genome involved in the control of RSV transcription and replication.

Supported by NIH grant CA27578

CLONING AND SEQUENCING OF AN INSULIN-LIKE RNA FROM HUMAN BRAIN

Antonio Gonzalez and Lydia Villa-Komaroff
Department of Molecular Genetics and Microbiology
University of Massachusetts Medical School,
Worcester, Massachusetts 01605

We have used a cloned sequence encoding human
insulin to identify RNAs homologous to insulin in
human fetal brain. These sequences do not appear
to encode insulin, but rather a family of sequences related to insulin. To further characterize
these sequences we synthesized double-stranded DNA
using total fetal brain RNA as template and inserted it into the Pst site of pBR322 using the oligo
d(G)-d(C) joining procedure. We used a cDNA clone
encoding the B peptide of insulin to screen the
colonies we obtained. From 300 transformants, we
identified 14 colonies which hybridized to the
human insulin probe under nonstringent hybridization conditions. One, pAG-B6, with an insert
of 1300 base pairs, was chosen for further study.
The region homologous to insulin was localized by
restriction enzyme digestion and blot hybridization. The inserted DNA was sequenced by the
chemical method. The clone appears to represent
the 3' portion of an RNA of 4000 bases. The putative untranslated region contains the simple
sequences $(AC)_{22}$ and $(GACA)_{10}$. The repeat of
(TG) has been found in the intron of the
$^G\gamma$-globin and the $^A\gamma$-globin (1). To
characterize the RNA transcripts corresponding to
the sequences in this clone, we have used
restriction fragments of pAG-B6 to probe RNA from
fetal brain. Our data indicate that this clone
may correspond to a 4000 bases RNA transcript
which hybridizes to our insulin probe and that
this sequence is present in brain RNA and not in
liver RNA.

(1) Slightom, J.L., Blechl, A.E. and Smithies, O.
Cell 21 627 - 638 (1980).

Supported by the NIH, the American Diabetes
Association and the Scientific Council of UMMC.

FROM GENE TO PROTEIN:
TRANSLATION INTO BIOTECHNOLOGY

VASOACTIVE INTESTINAL POLYPEPTIDE—FROM GENE TO PROTEIN TO PEPTIDE

I.Gozes, R.J.Milner, F.-T.Liu*, R.Benoit, E.Johnson, E.L.F.Battenberg, R.A.Lerner*, D.H.Katz* and F.E.Bloom
The Salk Institute and Scripps Clinic and Research Foundation*, La Jolla, California 92037

Vasoactive intestinal polypeptide (VIP) was originally isolated from the intestine but has subsequently been shown to be widely distributed throughout the body, with a high concentration in the hypothalamus and cerebral cortex. In addition, this 28 amino acid polypeptide has been localized to and released from nerve terminals in the brain, suggesting that VIP may function as a neurotransmitter. Biochemical experiments suggest that VIP may regulate energy metabolism in the brain. In view of the potential importance of VIP, we have begun to examine its synthesis and phenotypic expression.

Methodologically we have taken two independent complementary approaches. First, we have produced monoclonal antibodies against VIP that will allow the detection of newly synthesized precursors. These antibodies could also be used for localization and structure-function relationship studies. We have triggered spleen cells in vitro to produce antibodies against VIP covalently coupled to keyhole limpet hemocyanin (KLH). We have further fused these spleen cells with SP2/Ø tumor cells in the presence of polyethylene glycol. Hybrid cells were screened for antibody production and VIP immunoreactivity, and subcloned at limiting dilution. These antibodies are currently used for immunohistochemistry. They have been shown to specifically stain pancreatic islet cells and, possibly, nerve endings to the islet cells. We are also using these antibodies to identify presumptive VIP precursors by in vitro translation of brain mRNA. Indeed, anti-VIP antibodies show binding activity against high molecular weight proteins of 50,000 M_r and less. Our second approach is molecular cloning of the VIP gene using a synthetic undecaoligodeoxynucleotide and poly (A) mRNA from brain to generate VIP specific cDNA.

These experiments were supported by grants from the Sun Oil Company, Mobil Foundation, Inc., McNeil Pharmaceuticals, and USPHS grant AI-13874.

DETECTION OF SPECIFIC MOUSE HISTONE mRNAs BY S1 NUCLEASE
MAPPING

Reed A. Graves, Donald B. Sittman, Ing-Ming Chiu, and William F. Marzluff
Dept. of Chemistry, Florida State Univ., Tallahassee,Fl. 32304

The histone genes are repeated and dispersed in the mouse genome.
Each histone (H2a, H2b, H3, H4) gene is present in about ten copies. We
have used genomic clones (1) of two different H3 genes(H3-1 &H3-2)
and one H2b gene to determine the structure of the respective RNAs
and have developed a quantitative assay for specific histone RNA
transcripts.

The DNA sequence of the two H3 genes is nearly identical in the
protein-coding region (only one amino acid change) while the 3'-
untranslated regions are highly divergent. This suggests that the
coding regions of the histone region are highly conserved while the
flanking regions may vary considerably.

Both H3 genes have a SAL I restriction site at amino acid 58.
This site can be 5'-end labeled and used in the S1 nuclease method
of Weaver & Weissman (2) to analyze and quantitate the H3 mRNAs.
Since we are labeling the 5'-end of the coding strand,hybridization
to histone mRNA followed by S1 nuclease digestion will result in
a protected fragment composed of two parts:
> a.) A 174 nucleotide piece representing the coding
> portion of the RNA (58 amino acids X 3=174)
> b.) A 30-50 nucleotide piece representing the 5'-
> untranslated region from the intiation site to
> the AUG codon.

Both H3 genes result in protection of a major fragment (representing
about 85% of the total protected radioactivity) of approximately
180+ 15 nucleotides. The minor protected fragments differ between
the two probes. Using H3-2 DNA as probe gives two minor fragments:
one of 220 + 5 nucleotides,which presumably represents the RNA
transcript from this gene, accounts for about 10% of the total protected. The other minor fragment is about 190+ 10 nucleotides and
accounts for less then 5% of the total protected radioactivity. In
contrast the H3-1 gene protects a minor fragment of 200+ 10 nucleotides
representing about 10% of the total protected. We believe that about
10% of the H3 mRNA in 3T3 and myeloma cells is derived from each of
these H3 genes we have selected.The intiation site of the mRNA from
each H3 gene (H3-1 &H3-2) has been determined by DNA sequencing coupled
with S1 mapping.

A similar result has been obtained with the H2b gene. Here
we used a AVA I site at amino acid 99,which should yeild a fragment
of 297 nucleotides (coding region). We observed a major fragment
of 290+15 nucleotides and a minor fragment of 320+15 nucleotides.

FROM GENE TO PROTEIN:
TRANSLATION INTO BIOTECHNOLOGY

ISOLATION AND CHARACTERIZATION OF HUMAN IMMUNE INTERFERON
(IFN-γ) MESSENGER RNA

Patrick W. Gray, David Leung, Pamela J. Sherwood,
Shelby L. Berger*, Donald M. Wallace*, and David V. Goeddel
Department of Molecular Biology, Genentech, Inc., South
San Francisco, CA 94080 and *NIH, Bethesda, MD 20205.

Messenger RNA (mRNA) was isolated from human peripheral
blood lymphocytes stimulated with the mitogen Staphylo-
coccal Enterotoxin B. Sucrose gradient centrifugation and
acid-urea gel electrophoresis was used to fractionate mRNA.
Following fractionation, mRNA was recovered, injected into
Xenopus oocytes, and the translated fractions were assayed
for antiviral activity in a cytopathic effect inhibition
assay. Two activity peaks were observed in sucrose
gradient fractionated RNA. One peak sedimented with a
calculated size of 12S and contained 100-400 units/ml of
antiviral activity (compared with an IFN-α standard) per
microgram of injected RNA. The other peak of activity
sedimented as 16S in size and contained approximately half
the activity of the slower sedimenting peak. Only one
peak of activity was observed in gel fractionated mRNA.
This activity comigrated with 18S RNA and had an activity
of 600 units per microgram of RNA. Each observed activity
peak appears to be due to IFN-γ, since no activity was
observed when the same fractions were assayed on a bovine
cell line (MDBK), which is not protected by human IFN-γ.
The size discrepancies among activity peaks observed on
sucrose gradients (12S and 16S) and acid-urea agarose gels
(18S) may be the result of fractionation under conditions
which do not totally denature the mRNA.

FROM GENE TO PROTEIN:
TRANSLATION INTO BIOTECHNOLOGY

MONOCLONAL ANTIBODIES THAT INHIBIT THE RAT LIVER RECEPTOR FOR
ASIALOGLYCOPROTEINS

Joe Harford, Mark Lowe and Gilbert Ashwell. NIADDK, NIH,
Bethesda, MD. 20205

A specific receptor exists in the plasma membrane of
mammalian hepatocytes that mediates endocytosis of desialy-
lated glycoproteins. The receptor has been isolated from
Triton X-100 extracts of rat liver by affinity chromatography
on ligand-Sepharose. The binding activity appears as a high
MW (>200,000) entity by gel filtration, but SDS-PAGE reveals
a major band corresponding to 42,000 daltons and two less
prominent bands of slightly higher apparent MW. It has been
suggested that the observed polypeptides are structurally
related (1,2) but the nature of this relationship remains
unclear.

Spleen cells from BALB/C mice immunized with a soluble
receptor preparation were fused with P3X63-Ag8653 myeloma
cells. Media from the resultant hybridomas were screened for
their ability to bind radioiodinated receptor. After subcloning, positive ascites was produced in BALB/C mice by i.p.
injection of 10^6 cells. Anti-receptor could be isolated on
receptor-Sepharose. A clone was identified that produced
antibody directed toward a determinant related to receptor
function, i.e. it blocked binding of ligand by plasma membranes. The SDS-PAGE pattern of the radiolabeled receptor
preparation bound by this antibody was indistinguishable from
the Coomassie blue staining pattern of the preparation used
as antigen. However, due to the high apparent MW of the
Triton X-100 extract, this result was viewed as inconclusive.
No reaction could be detected with the individual polypeptides recovered by a preparative SDS-PAGE using the gel system of Laemmli (3). Accordingly, a modified preparative
SDS-PAGE system was devised which proved to be less destruc-
tive of antigenic reactivity. Using this method, the monoclonal antibody was shown to react most strongly with the two
minor polypeptides of the receptor preparation and only
weakly with the 42,000 dalton polypeptide. These results
indicate that the recognized determinant is shared, at least
in part, by the three polypeptides.

1. Warren, R. and Doyle, D. (1981) J. Biol. Chem. 256: 1346-
    1355.

2. Schwartz, A.L., Marshak-Rothstein, A., Rup, D., Lodish,
    H.F. (1981) Proc. Nat. Acad. Sci. USA 78: 3348-3352.

3. Laemmli, U.K. (1970) Nature 227: 680-685.

FROM GENE TO PROTEIN:
TRANSLATION INTO BIOTECHNOLOGY

SOLID PHASE NUCLEOTIDE AND PEPTIDE SYNTHESIS BY ADAPTATION OF LABORATORY HPLC EQUIPMENT

Derek Hudson and Ronald M. Cook

Biosearch, 1281 F Andersen Drive, San Rafael, CA   94901

HPLC Equipment and supports can be used for the rapid and economical synthesis of oligonucleotides and peptides. Efficient coupling is achieved without need for recycling, and excess reagents from previous steps in the synthetic procedures were removed easily by washing.

For nucleotide synthesis a Varian 5020 HPLC was used charged with dry acetonitrile or other wash solvent in reservoir A and DMT- removal reagent in reservoir B. Activated nucleoside intermediates and other ancillary reagents were added via microprocesser controlled solenoid valves. The procedure is compatible with existing phosphite and phosphate triester chemistries. Activation of the appropriately protected deoxynucleoside residues with methoxyphosphine ditetrazolide was an effective and economical procedure for coupling. The 3' terminus of the desired nucleotide sequence was attached to HPLC silica supports either via hemisuccinate derivatives or by readily prepared novel dimethylsilyl linkages.

The advantages of microprocesser control and flow methodology were further exemplified by peptide sythesis. Excellent yields were obtained using a 30 minute cycle consisting of coupling of Fmoc- amino acids with diisopropyl carbodiimide and deblocking with piperidine in dimethyl formamide.

The system described forms the basis of the Biosearch Synthesis Automation Module (SAM).

FROM GENE TO PROTEIN:
TRANSLATION INTO BIOTECHNOLOGY

CLONING AND SEQUENCING OF A cDNA FOR αA CRYSTALLIN mRNA OF
THE MOUSE LENS: PRESENCE OF AN EXTENSIVE 3' UNTRANSLATED
REGION

Charles R. King, Toshimichi Shinohara and Joram Piatigorsky
Laboratory of Molecular and Developmental Biology, National
Eye Institute, National Institutes of Health, Bethesda, Md.

The ocular lens of vertebrates synthesize and accumulate large amounts of structural protein called crystallins.
There are four immunologically distinct classes of crystallin called α-, β-, γ-, and δ-crystallins. The 14S mRNA
(1300-1500 nucleotides) for the αA chain of α-crystallin of
all lenses examined is nearly three times larger than required to code for the polypeptide (1), which contains 173
amino acids (2). In order to understand this anomaly, a
cDNA clone for the mouse αA crystallin mRNA has been constructed in pBR322 and sequenced. Translation of the coding
region showed that mouse αA crystallin is very similar to
that of other organisms. The mRNA contains 536 nucleotides
located on the 3' side of the coding region. Computer
analysis showed that the 3' sequence does not carry information for any of the crystallins and has multiple termination
codons in the three possible reading frames, indicating that
they are not translated. Knowledge of the sequence of the
mouse αA-crystallin mRNA is a first step towards explaining
the conservation of the unexpectedly large size of this mRNA
relative to the size of the encoded polypeptide.

1. L.H. Cohen, D.P.E.M. Smits, H. Bloemendal. Eur. J.
    Biochem. 67, 563 (1976)

2. F.J. van der Ouderaa, W.W. de Jong, H. Bloemendal.
    Eur. J. Biochem. 39, 207 (1973)

FROM GENE TO PROTEIN:
TRANSLATION INTO BIOTECHNOLOGY

MONOCLONAL ANTIBODIES RECOGNIZING COMMON ANTIGENIC
DETERMINANTS ON LYSOSOMAL ENZYMES

David A. Knecht, Robert C. Mierendorf, Jr. and Randall L.
Dimond.
Department of Bacteriology, University of Wisconsin,
Madison, Wisconsin.

We have previously demonstrated that rabbits immunized
with either of two purified lysosomal enzymes
(N-acetylglucosaminidase or B-glucosidase) from
Dictyostelium produce antisera that cross-react with
several other lysosomal enzymes as well as a wide spectrum
of cellular and secreted proteins. The antigenic
determinant shared by these proteins appears to be a
carbohydrate modification (1).

To investigate the number of determinants shared by
these proteins and to obtain a catalog of antisera specific
to each, we have begun screening mouse monoclonal
hybridomas secreting antibodies which recognize these same
enzymes. In addition to the commonly used indirect Elisa
assay for screening clones we have developed a second
direct screening method. S. aureus Protein A is bound to
microtiter plates followed by consecutive reactions with
monoclonal supernatants and purified or crude lysosomal
enzyme. Fluorescent substrate for one or more enzymes that
might be recognized by the common antisera is then added.

Hybridoma clones from mice injected with either
purified enzyme or a crude lysosomal enzyme mixture were
screened using both direct and indirect assays. The two
assays showed some overlap but each assay also detected a
unique set of clones. Using these methods, we have
isolated hybridomas that secrete antibodies which
cross-react with lysosomal enzymes in a manner similar to
the rabbit antisera. We are currently analyzing the
specificity of each clone to determine how many different
subsets of common antigen containing proteins can be
recognized. These antibodies will be useful in studying
the role of modification in the cellular localization and
secretion of lysosomal enzymes.

Supported by NIH Grant #GM 29156 and NSF Grant #PCM79-02692

(1) Knecht, D.A. and Dimond, R.L., J. Biol. Chem.
256:3564-3575 (1981).

FROM GENE TO PROTEIN:
TRANSLATION INTO BIOTECHNOLOGY

PROTEINS OF A MINIPLASMID DERIVATIVE OF THE BROAD HOST RANGE
PLASMID RK2.

Jon A. Kornacki and William Firshein. Biology Department,
Wesleyan University, Middletown, CT. 06457.

The plasmid RK2 (56.4 kb) belongs to the P-1 incompatability
group and confers resistance to kanamycin, ampicillin, and
tetracycline. There has been much interest in the use of this
plasmid as a cloning vehicle because it is capable of replication in a wide range of gram-negative bacteria. Also, because
RK2 is maintained at a low copy number (stringently controlled
replication), it serves as a useful model system for understanding the regulation of DNA replication in bacteria. Unlike
most other plasmids, the regions of the genome necessary for
replication are not clustered together. Three separate regions
of the plasmid genome, totalling 5.4 kb, are essential for replication. They include the origin and two others (trf A, trf
B) which code for trans-acting factors necessary for replication (for review, see 1).

We have been interested in elucidating the nature of proteins involved in initiation of RK2 replication. A miniplasmid
derivative of RK2 (1) that contains only those regions described above was used in order to determine the types of proteins synthesized. However, since the plasmid is stringently
controlled it was necessary to develop a system in which plasmid-specific proteins could be enriched over the background
of bacterial protein synthesis. For this purpose, the maxicell
system of Sancar et al (2) was chosen. This involves the use
of a mutant E. coli strain (maxicells) extremely sensitive to
ultraviolet light that, in addition, is unable to be photore-
activated (rec A, urv, phr 1). When this mutant is transformed
with the RK2 miniplasmid or other plasmids containing the essential regions, irradiated, and incubated further with a mixture of radioactive amino acids, it is possible to enrich for
plasmid-specific radioactive proteins because extensive degradation of bacterial DNA occurs concurrently with amplification
of plasmid DNA. Using this procedure, we have identified a
number of proteins specifically synthesized by the miniplasmid
derivative of RK2, the first such demonstration from this low
copy number plasmid. Some of these proteins are encoded by
the trf A and trf B controlling regions.

Supported by NSF Grant PCM 8110082.

1. Thomas, C.M. (1981) Plasmid 5:277-291.

2. Sancar, A., Hack, A. and Rupp, W.D. (1979)   J. Bacteriol.,
   137:692-693.

FROM GENE TO PROTEIN:
TRANSLATION INTO BIOTECHNOLOGY

MONOCLONAL IgM ANTIBODIES THAT INDUCE T CELL—MEDIATED
CYTOTOXICITY OF MURINE SARCOMA CELLS

Eddie W. Lamon and John F. Kearney. Departments of Surgery
and Microbiology, Veterans Administration Hospital, Cancer
Research and Training Center, University of Alabama in
Birmingham, Birmingham, AL 35294

Ten days after intramuscular injection with 0.1 ml of
Moloney—murine sarcoma virus (M—MuSV) into BALB/c mice,
spleen cells from a tumor bearing animal were fused to a
nonproducer myeloma cell. Tissue culture supernatants
from the hybridoma colonies were tested for antibody
binding to glutaraldehyde—fixed monolayers of tumor cells
in microplates by an enzyme—linked immunoassay. Colonies
were selected that produced antibodies specific for two
different MuSV transformed tumor cell lines (Ha2 cells of
CBA origin and MSB cells of C57Bl/6 origin) but were negative when tested against polyoma—transformed cells (PyBl/6
of C57Bl/6 origin). Seven colonies of IgM producing
hybridomas were obtained whose antibodies reacted specifically with the MuSV transformed cells. Tissue culture
supernatants from all seven produced significant antibody-
dependent cell—mediated cytotoxicity (ADCC) against the
target cells transformed by murine sarcoma virus but not
against polyoma—transformed cells. Active effector cells
included spleen cells, lymph node cells and thymocytes.
Two of the hybridoma colonies producing IgM antibodies
specific for murine sarcoma cells have subsequently been
cloned and expanded in ascites form. Similar results have
been produced by the monoclonal antibodies including 19S
fractions containing only IgM. Enzyme—linked immunoassays
using isotype-specific antisera have shown that these
monoclonal antibodies are μκ. No other isotype was detected.
Immunofluorescence studies with these monoclonal antibodies
showed that they react with Moloney and Rauscher lymphoma
cells but not with Gross virus induced lymphoma cells.
Thus, it appears likely that the antibodies have specificity
for a Friend—Moloney—Rauscher group specific cell surface
antigen. Thus, monoclonal IgM antibodies can induce
specific tumor cell destruction by normal lymphocytes
including thymocytes.

Supported by NCI grant CA—17273—07, grant AI 14782 from NIH
and project 5132—02 from the Veterans Administration
Hospital, Birmingham, Alabama

FROM GENE TO PROTEIN:
TRANSLATION INTO BIOTECHNOLOGY

DEVELOPMENT OF METHODS FOR THE LARGE-SCALE CULTURE OF
OXYTRICHA NOVA.

Tommie J. Laughlin, J. Michael Henry, Edward F. Phares,
Mary V. Long, and Donald E. Olins. The Univ. of Tenn.-Oak
Ridge Grad. Sch. of Biomed. Sci. and the Biol. Div., Oak
Ridge Natl. Lab., Oak Ridge, TN 37830

Hypotrichous ciliates, such as Oxytricha nova, have
two types of nuclei. The transcriptionally inactive micronuclei have chromosome-size DNA molecules and divide mitot-
ically. The transcriptionally active macronucleus has gene-
size DNA fragments and divides amitotically. This segregation of active and inactive DNA into separate nuclei
provides a unique model system for the study of gene expression. However, such studies have been limited due to the
inability to grow the quantities of cells needed for biochemical analysis of chromatin and nuclear proteins.

Our laboratory has been developing methods for the
large-scale culture of hypotrichous ciliates. These cells
are bottom feeders normally found in the shallow water of
still ponds. In culture, they grow best in unaerated, un-
agitated Petri dishes or fernbach flasks. Cultures of this
type are limited in volume (usually to about one liter)
and have a very nonuniform distribution of cells, with the
majority of the cells at the very top or bottom of the
media. We have found that Oxytricha nova (kindly supplied
by David Prescott, Boulder, CO) can be grown in specially
designed 50 liter fermentation vats with very gentle agita-
tion and aeration (enough to maintain a saturated oxygen
level). First, Chlorogonium are grown to a Klett of 90-100
units in 48 liters of steril Oochromonas medium (1 g/l each
of glucose, liver extract, Bactotyptone, yeast extract, and
sodium acetate). Then, two liters of mature Oxytricha cultures are added. The cultures are grown at 25°C and under
fluorescent lights. Within seven days the Oxytricha have
consumed almost all of the chlorogonium and have reached a
steady-state density of about 8,000 cells/ml. The cell distribution is very uniform. Using this cell culture method,
along with methods we have developed for the rapid harvest
of these cells using a continuous flow centrifuge, we are
able to isolate quantities of macronuclei and micronuclei
sufficient for comparative biochemical and biophysical
analyses of active and inactive chromatin. (Res. supported
by grants from NIH to DEO [GM 19334] and the Office of Hea.
& Environ. Res., U.S. Dept. of Energy, under contract W-
7405-eng-26 with UCC. TJL is a postdoc. invest. supported by
subcontract No. 3322 from the Biol. Div., ORNL to Univ. of TN)

CLONING OF CDNA ENCODING THE SWEET-TASTING PLANT PROTEIN THAUMATIN AND ITS EXPRESSION IN ESCHERICHIA COLI

J. Maat, L. Edens, A.M. Ledeboer, M. Toonen, C. Visser, I. Bom and C.T. Verrips.
Unilever Research Laboratory, Vlaardingen, THE NETHERLANDS

Thaumatin is a sweet-tasting protein originating from the berries of the West-African shrub Thaumatococcus daniellii Benth. Its mRNA was isolated from the arils, converted into double stranded cDNA and cloned in E. coli HB 101 using pBR 322 as vector. Thaumatin encoding transformants were detected by in vitro translation of mRNA, selected by specific hybridization to isolated plasmid DNA preparations. Nucleotide sequence analysis of a thaumatin encoding cDNA clone revealed that thaumatin represents a maturation form of "preprothaumatin". The latter protein contains extensions at both the amino and carboxy terminus. The amino terminal extension of 22 amino acids is very hydrophobic and very much resembles an - excretion related - signal sequence. The six amino acid long carboxy terminal extension is very acidic in character in contrast to the overall very basic thaumatin molecule. We hypothesize its biological function as a compartimentalisation related signal.

The cloned DNA was engineered as to code not only for preprothaumatin but also for its maturation forms. The thus constructed DNAs were brought under transcription control of E. coli lactose and tryptophan promotor/operator systems. Expression of the various proteins was then monitored by specific immunoprecipitation followed by SDS-polyacrylamide gel electrophoresis and by a specially developed ELISA.

FROM GENE TO PROTEIN:
TRANSLATION INTO BIOTECHNOLOGY

AMPLIFICATION OF THE METALLOTHIONEIN-I GENE IN
CADMIUM- AND ZINC-RESISTANT CHINESE HAMSTER OVARY
CELLS

*Kenneth S. McCarty, Sr., +Gregory G. Gick and
**Kenneth S. McCarty, Jr.
Departments of *Biochemistry, **Pathology and
Medicine, Duke University Medical Center, Durham,
NC; +Sloan Kettering Memorial Institute, New
York, NY

A subclcone of Chinese hamster ovary cells
(R40F) was selected for its unusually high resistance to lethal concentrations of cadmium and
zinc. These cells demonstrate 80% tetraploidy, a
generation time of 33 h, and an increased capacity for metallothionein synthesis. Metallothioneins I and II (MT-I, II) were identified using
G-75 Sephadex, DEAE-Sephacel, ultraviolet spectrum, polyacrylamide gel electrophoresis, HPLC,
binding affinity to $^{109}Cd^{++}$, incorporation of
labeled cystine, and amino acid composition.

When compared to wild-type cells, R40F cells
cultured in the presence of 200 µM cadmium for 48
h demonstrated a 200-fold induction in the levels
of metallothioneins I and II (MT-I, II).

A ^{32}P-labeled mouse MT-I cDNA was employed
in solution hybridization studies to measure the
level of MT-I mRNA. R40F cells challenged with
0.5 m or 200 µM cadmium for 48 h accumulated 1598
and 7020 MT-I mRNA molecules per cell. Wild type
cells challenged with 0.5 cadmium for 48 h accumulated only 95 MT-I mRNA molecules per cell. In
the absence of exogenous cadmium, wild-type cells
accumulated only 40 MT-I mRNA molecules per cell.

Gene copy number estimated by Southern blot
analysis of EcoR1, BAM, Hind III and Hinc II-
cleaved DNA indicated that the MT-I gene was amp-
lified approximately 60-90-fold in R40F cells.

Supported in part by CMB-NIH-ST-32-6N07184,
NO1-CB-84223 and NO1-CB-63996.

CONSTITUTIVE THYMIDINE KINASE ACTIVITY IN DIFFERENTIATING
MOUSE MYOBLASTS TRANSFORMED WITH CLONED HERPES VIRUS THY-
MIDINE KINASE GENES

Gary F. Merrill and Stephen D. Hauschka. Department of
Biochemistry, University of Washington, Seattle, WA.

Mouse myoblasts were transformed with cloned herpes virus
thymidine kinase genes (HSV-TK), and TK activity was monitored as cells differentiated in response to mitogen depletion.
The TK⁻ line used (TK⁻A1A) was derived from an established
myoblast line (MM14) by EMS mutagenesis, BUdR/light selection and extensive serial subcloning. TK⁻A1A cells were
transformed with pBR322 containing HSV-TK (gift of Dr. D.
Galloway) using the Ca^{++} $PO_4^=$ precipitation method. Transformation frequency was 1-2/µg DNA/10^6 cells (50-fold less
than achieved using TK⁻L cells). No HAT-resistant colonies
appeared in cells exposed only to carrier DNA. Like MM14
and TK⁻A1A parentals, transformed myoblasts withdraw from the
cell cycle and commit to terminal differentiation within one
cell cycle time of mitogen depletion (12.5 hours in MM14; 17
hours in TK⁻A1A and transformants).

Whereas in non-transformed myoblasts, endogenous TK activity plummets to less than 2% of proliferative levels within
12.5 hours of mitogen depletion; in HSV-TK transformants, TK
activity remain high. Even after 60 hours, when extensive
myotube formation has taken place, TK activity in transformants has declined only to 65% of proliferative levels (this
moderate decline may be due to non-specific detrimental
effects of prolonged incubation in serum-free mitogen-depleted
medium). The molecular basis for constitutive expression of
HSV-TK, and conversely, for repression of endogenous TK, is
unknown. Experiments in progress are aimed at studying this
problem with TK⁻A1A cells transformed with cloned TK genes
from eucaryotic sources.

FROM GENE TO PROTEIN:
TRANSLATION INTO BIOTECHNOLOGY

Cloning of the Gene for alpha-Amylase from Bacillus
stearothermophilus into Escherichia coli

Jonathan R. Mielenz, Biotechnology Dept., Moffett Tech.
Center, CPC International, Inc., Argo, Illinois

The gene for thermostable alpha-amylase from Bacillus
stearothermophilus has been cloned in Escherichia
coli. The alpha-amylase produced by B. stearothermophilus is highly heat resistant compared to the alpha-
amylase from Bacillus subtilis or Bacillus amylo-
liquefaciens. Heat stability is an enzyme characteristic important for industrial application of alpha-
amylase. The alpha-amylase produced by the E. coli
clones retains the thermostability of the B. stearothermophilus enzyme, demonstrating the heat stability resides in the protein structure alone.

E. coli colonies producing amylase contain a common
sized piece of DNA cloned in pBR322. This fragment
has been sub-cloned in both orientations onto four
different plasmids with retention of the amylase
activity. Sequential trimming at both ends of the
cloned DNA with various restriction enzymes has removed 60% of the cloned DNA with retention of the
amylase activity. These results confirm the cloned
amylase gene is self-contained and E. coli is successfully expressing an industrially important gene
from a Gram-positive thermophilic bacterium.

FROM GENE TO PROTEIN: Copyright © 1982 by Academic Press, Inc.
TRANSLATION INTO BIOTECHNOLOGY   All rights of reproduction in any form reserved.
                                                         ISBN 0-12-045560-9

EXPRESSION OF FOREIGN GENES IN RECOMBINANT PLASMIDS UNDER
THE CONTROL OF T7 AND T3 LATE PROMOTERS

Claire E. Morris and William T. McAllister, Dept. of Microbiology, CMDNJ- Rutgers Medical School, Piscataway, New
Jersey 08854.

Bacteriophages T3 and T7 code for structurally simple
RNA polymerases. These enzymes are highly selective and
will initiate transcription efficiently only from homologous late promoters. DNA fragments containing the late
promoters have been inserted into recombinant plasmids.
When cells harboring these plasmids are infected by the
appropriate phage, the promoters are recognized by the
phage RNA polymerase resulting in transcription of plasmid
DNA sequences. The amount of RNA synthesized depends upon
the orientation of the promoters within the plasmid. At
maximal levels, a 6-10 fold increase in the level of plasmid transcription is observed upon infection of the cells.
Since infection by T3 or T7 results in inactivation of the
host RNA polymerase, the only genes transcribed at late
times after infection are those preceded by a late promoter.
The highly selective nature of transcription suggests that
these plasmids will serve as useful cloning vectors.

To examine this possibility, we have constructed a
recombinant plasmid in which the gene coding for β-galac-
tosidase is preceded by a phage promoter. Infection of
cells harboring these plasmids results in the production
of β-galactosidase. The maximal level of β-galactosidase
achieved at 20 minutes after infection is approximately
50% higher than that of a fully induced strain of E. coli
K12 which does not contain the plasmid. When cells harboring the test plasmids are used as a lawn on media containing Xgal, blue plaques are produced; clear plaques are produced when the plasmid does not contain a late promoter.
These plasmids will therefore prove useful in selecting
or screening for polymerase and promoter mutations that
result in altered transcriptional activity.

FROM GENE TO PROTEIN:
TRANSLATION INTO BIOTECHNOLOGY

THE INABILITY OF IN VITRO TRANSFORMING SV40 SUBGENOMES TO
CAUSE TUMORS IN VIVO

M.P. Moyer[1] and R.C. Moyer[2].   [1]Surgical   Oncology   Lab,
Department   of   Surgery,   The   University   of   Texas   Health
Science Center at San Antonio, and [2]The Thorman Cancer
Research Laboratory, Trinity University, San Antonio, Texas
78284.

Simian virus 40 (SV40) DNA isolated from purified
virions was cleaved with the restriction endonucleases
Eco RI, Bam I, Hha I, Hpa I, Bgl I, or Hpa II.   The
resultant linear or subgenomic DNAs were separated and
twice-purified by agarose gel electrophoresis. They were
transfected into cells from a variety of species (including
rodent, dog, muntjak, and monkey).   The DNAs were also
injected subcutaneously into neonate Syrian hamsters for
tumorigenicity testing.

The "early region" restriction endonuclease-generated
subgenomes of SV40 were capable of transforming cells in
vitro. Although the "late region" of the virus was unable
to transform cells, it could alter the growth properties and
persist in some permissive monkey cells.   Complete,
uncleaved, circular genomes or complete linear genomes could
transform nonpermissive and semipermissive cells and were
infectious for permissive cells.

Tumors were not formed in neonate hamsters upon
injection with subgenomic SV40 DNA. Tumors did result upon
injection of complete (circular or linear) genomes or
co-injection of complementary subgenomes. Infectious virus
could be rescued from the resultant tumors.

These results suggest that SV40 tumor formation in vivo
may require a complete genome, and that in vitro
transformation by SV40 DNA may be mechanistically different
from in vivo tumorigenesis.

Supported by the Thorman Trust and the Robert J. Kleberg and
Helen C. Kleberg Foundation.

MODEL RIBOSOMES

Tadayoshi Nakashima and Sidney W. Fox.   Institute for
Molecular and Cellular Evolution, University of Miami

Basic proteinoid (lys-rich thermal polyanhydro
amino acid) has the ability to synthesize peptides from
amino acids and ATP, especially in basic solution (1,2).
Imidazole can be substituted for proteinoid (3).   Selective incorporation of preformed aminoacyl adenylate into
suspensions of microparticles comprising polyribonucleo-
tide and basic proteinoid has been observed (4).

The syntheses of peptides were tested at pH 7.2 in
a suspension of acidic+basic proteinoid particles, and
in a suspension of polyribonucleotide+basic proteinoid
particles, the latter as model ribosomes.   The activity
of the complex is several times as large as that of the
basic proteinoid alone.

The contribution to peptide bond synthesis of basic
proteinoid is consistent with the knowledge that trans-
peptidation by ribosomes depends upon groups having a pK_a
of 7.5-8.0 (imidazole or N-terminal α-amino) and perhaps
one of pK_a 9.4 due to ε-amino of lysine (5).   Basic ribosomal proteins might provide a basic environment for
peptide synthesis by interaction with ribosomal ribonucleic
acid.

When the amino acids in the suspension of particles
are mixtures of glycine and phenylalanine, substantial
proportions of mixed depeptides are formed.   The sequences
in the major fraction of such dipeptides were determined
indirectly by dansylation.   In a suspension of poly U+ or
poly C+ basic proteinoid particles, phegly is formed predominantly, but in poly A+basic proteinoid, glyphe mostly
results.   The affinity difference of purine and pyrimidine
for amino acids, or the hydrophobicity of polynucleotides,
may account for these priorities in peptide sequence.

Supported by NASA grant No. NGR 10-007-008.

(1) Fox, S. W., Jungck, J. R. and Nakashima, T., Origins Life 5
227-237 (1974); (2) Nakashima, T. and Fox, S. W., J. Mol. Evol. 15
161-168 (1980); (3) Weber, A. L., Caroon, J. M., Warden, J. T., Lemmon,
R. M. and Calvin, M. BioSystems 8 277-286 (1977); (4) Nakashima, T.
and Fox, S. W., Proc. Nat. Acad. Sci. U.S.A. 69 106-108 (1972); (5)
Harris, R. J. and Pestka, S., in Molecular Mechanisms of Protein Biosynthesis, H. Weissbach and S. Pestka, eds. (Academic Press, New York)
pp. 418-442 (1977).

SELECTIVE INDUCTION BY PEPTIDE GROWTH FACTORS OF THE SYNTHESIS OF SECRETED PROTEINS IN 3T3 CELLS

Marit Nilsen-Hamilton, Richard T. Hamilton, and W. Ross Allen.
Cell Biology Laboratory, The Salk Institute, P.O. Box 85800, San Diego, CA 92138

The changes in cellular metabolism needed for the initiation of DNA synthesis in quiescent cells almost certainly require the synthesis of new proteins and probably involve the activation of genes. We have found that peptide mitogens such as epidermal growth factor and fibroblast growth factor selectively increase the extracellular appearance of two glycoproteins secreted by 3T3 cells. These proteins are, "mitogen regulated protein" (MRP) (Mr 34,000), and "major excreted protein" (MEP) (Mr 37,000).

The mitogens regulate at transcription or translation, not at glycosylation or secretion. In Balb/c 3T3 cells, maximum extracellular levels of MEP occur about five hours after adding mitogen to quiescent cells. MEP and MRP turn over rapidly inside the cell; a large proportion of each is degraded before being released into the medium. The amount of extracellular MRP and MEP is increased at least 2- to 3-fold by agents, such as ammonium chloride and chloroquine, that inhibit proteolysis in the lysosomes.

We speculate that these mitogen-induced proteins could participate in the regulation of cell growth in one of several ways. First, the properties of these proteins - selective increase in their synthesis in response to mitogens, and rapid turnover - are the properties of the proteins postulated by Rossow et al. (PNAS, 76, 4446-4450, 1979) as regulating the rate of initiation of DNA synthesis in 3T3 cells and other cells. Therefore, MEP or MRP could be internal growth regulating proteins. Second, these proteins, being secreted, could be communicatory molecules for coordinating the growth of a tissue, such as occurs during wound healing.

Supported in part by ACS No. CD-9.

FROM GENE TO PROTEIN:
TRANSLATION INTO BIOTECHNOLOGY

DELETION OF DNA SEQUENCES BETWEEN PRE-SELECTED BASES

Nikos Panayotatos and Kim Truong
Biogen, S.A. 3 Rte de Troinex, 1227
Carouge/Geneva, Switzerland

Blunt-end ligation of a "filled-in" HindIII, Sal I, Ava I or Bcl I restriction site with a DNA fragment having A, G, C, or T as the terminal 3' nucleotide regenerates the corresponding restriction site. A combination of this property with the action of BAL 31 nuclease which progressively removes base-pairs from the ends of linear DNA, can generate deletions extending to desired pre-selected nucleotides and introduce unique restriction sites at those positions. Similarly other restriction sites can be used to select for the deletion of sequences between specific di-, tri-, tetra- and penta-nucleotides. Using this method, 10 base pairs were deleted from the end of a restriction fragment carrying the late promoter for bacteriophage T7 gene 1.1 to create a molecule with a unique restriction site at the initiation codon for translation.

FROM GENE TO PROTEIN:
TRANSLATION INTO BIOTECHNOLOGY

EXPRESSION OF HERPESVIRUS THYMIDINE KINASE GENE UNDER CONTROL OF EARLY PROMOTER OF SV40

Mary M. Pater(1), Alan Pater(1), Giampiero di Mayorca(1) and James R. Smiley(2). Department of Microbiology, College of Medicine and Dentistry of New Jersey, New Jersey Medical School, Newark, New Jersey 07103; and Department of Pathology, McMaster University Medical Center, Hamilton, Ontario, Canada L8N325.

Studies of the regulation of gene expression at the level of transcription in mammalian cells has been difficult due to the lack of regulatory mutants. We have taken advantage of several recent developments to overcome this problem. First is the availability of a 0.31 Kilobase pairs (Kb) DNA fragment of SV40 virus which contains the promoter for the in vivo expression of the early genes of the virus (1). Second is the availability of herpes simplex virus 1 (HSV-1) cloned thymidine kinase (TK) gene, whose promoter can be removed (2) and the availability of efficient selection systems for the expression of the TK gene (3). We have engineered a plasmid in which the promoter region of the TK gene has been replaced by the early promoter of SV40 and have used this plasmid to transform TK-deficient mammalian cells to a TK+ phenotype. This plasmid is integrated into high molecular weight cell DNA in both of the transformed colonies that we have studies. Also the mRNA in these cells is a hybrid molecule containing both TK SV40 sequences. It thus appears that the TK gene without its promoter is an excellent vehicle by which mammalian promoters required for gene expression can be identified and used for further analysis.

Supported by NIH Grant No. R01CA25168 and by MRC of Canada and a CMDNJ grant to MMP.

(1) Mulligan, R.C. and Berg, P. Proc. Natn. Acad. Sci. USA 78:2072-2076 (1981).

(2) Culbere-Garapin, F., Chousternan, S., Horodniceanu, F., Kourilsky, P. and Garapin, A.-C. Proc. Natn. Acad. Sci. USA 78:3755-3759 (1979).

(3) Dubbs, D.R. and Kit, S. Virology 22:493-502 (1964).

FROM GENE TO PROTEIN:
TRANSLATION INTO BIOTECHNOLOGY

REGULATION OF HEAVY CHAIN GENE EXPRESSION IN MOUSE
MYELOMA MOPC 315 CELLS

Phyllis Ponte, Michael Dean, Vincent H. Pepe and Gail E.
Sonenshein
Department of Biochemistry, Boston University School of
Medicine, Boston, Massachusetts

To study the regulation of immunoglobulin gene expression, we are characterizing wild-type and mutant cell
lines of the mouse plasmacytoma MOPC 315. MOPC 315 cells
produce an IgA with a λ2 light (L) chain and an α heavy (H)
chain. One variant line, DMC, synthesizes and secretes
approximately 5 fold greater than w.t. levels of α H chain.
Production of L chain is normal. In a rabbit reticulocyte
cell free system, DMC cytoplasmic RNA will direct the
translation of similarly higher amounts of H chain polypeptide. Measurement of H chain mRNA levels by RNA blot
hybridization to a cDNA α chain constant (C) region
probe demonstrated that DMC contains approximately 10
times more messenge for α chain. As expected, hybridization with a light chain probe detected normal levels of
λ 2mRNA.
To examine the structure of the gene, Bam H1 digested DNA from w.t. cells was hybridized to the α H chain
cDNA probe in a Southern blot. Two bands of hybridization at 12.5 and 9.3 kb were detected. Both of these
fragments are rearranged from the germ-line configuration
of 16 kb. By hybridization to a J and switch region
probe, the 12.5 kb fragment has been determined to be
the productive allele. The 9.3 fragment, which does not
hybridize to the J region probe, is therefore the "excluded" allele. DMC DNA, digested with Bam H 1, and
hybridized to α C region probe, displays the 9.3 and
12.5 kb alleles and an extra band at 21 kb. The DMC line
was recloned. Six individual clones were tested by
Southern blot analyses. They all gave the identical profile as the parent DMC line. Hybridization to a J region
probe detected only the 12.5 kb fragment. Therefore, the
extra band in DMC has not been productively rearranged, and
it is presumably not responsible for the overproduction of
H chain polypeptide. Genomic DNA libraries have been
constructed from both w.t. and DMC cells for comparison of
DNA structure and for in vitro transcription analyses of
the H chain gene.

Supported by N.I.H. grant AI-16051 and American Cancer
Society IM214.

AMPLIFICATION AND RAPID PURIFICATION OF EXONUCLEASE I
FROM ESCHERICHIA COLI K-12

Douglas Prasher and Sidney R. Kushner. Department of
Molecular and Population Genetics, University of Georgia
Athens, Georgia 30602

Exonuclease I from Escherichia coli K-12 specifically
hydrolyzes single-stranded DNA in the 3'-5' direction.
Although its specific role in genetic recombination and
DNA reapir is not completely understood, its usefulness as
a biochemical reagent has been demonstrated. Wider use of
this enzyme has been limited by the difficulty in purifying
large amounts of the protein. The structural gene for
exonuclease I was originally cloned as a 17kb HindIII
fragment into the vector pMB9. This plasmid (pVK10 [1]) was
too unstable, however, to be useful. Accordingly, a 3.7kb
ClaI- BamHI fragment has been subcloned into a runaway-
replication plasmid pMOB45. At 30° this recombinant plasmid
(pDP16) is very stable while after shift to 40°, crude extracts
contain 20-60-fold more exonuclease I activity. The monomeric
molecular weight of exonuclease I is 56,500 based on SDS
polyacrylamide gel electrophoresis. The enzyme can be rapidly
purified by a combination of Blue Agarose and phosphocellulose
chromatography.

The structural gene for exonuclease I contains EcoRI,
PvuII and SmaI cleavage sites. Two unidentified proteins,
having molecular weights of 15,200 and 13,100 are encoded
in close proximity to the exonuclease I gene. Preliminary
evidence suggests that neither are required for exonuclease
I activity.

Supported in part by USPHS Grants GM21454 and GM26389.

(1) Vapnek D., Alton, N.K., Bassett, C.L., and Kushner, S.R.,
Proc. Natl. Acad. Sci. USA 73 3492-3496 (1976).

ANALYSIS OF β-INTERFERON mRNA IN HUMAN CELLS

N. B. K. Raj, Paula M. Pitha and Kevin Kelley
Oncology Center, School of Medicine, The Johns Hopkins
University, Baltimore, Maryland

Inducibility of human interferon is known to depend
on the nature of the inducer and also the cell type. Thus
poly rI.rC induction of human fibroblast cells produces
mainly or exclusively β type, while those produced by
lymphoblastoid cells (Namalva) consist of about 90% α
and 10% β type. The kinetics of induction of β-interferon mRNA in these two systems was determined by RNA
hybridization with a cloned βcDNA and by translation in
Xenopus oocytes. The time course of accumulation of β-
interferon mRNA in poly rI.rC induced fibroblast cells
and Sendai stimulated Namalva cells corresponded to the
kinetics of β interferon synthesis in intact cells. In
poly rI.rC induced fibroblast cells βmRNA disappears several hours after its appearance in cytoplasm(half life 30
min) reaching undetectable levels at 7hrs after induction.
The degradation of βmRNA requires ongoing protein synthesis; therefore, βmRNA accumulates up to 11hrs in the continuous presence of cycloheximide. The size of the βmRNA
is determined by the stage of the induction; for example,
βmRNA detected at early hours of induction (3hrs) is about
1100 nucleotides long and its size progressively decreases with time after induction (by about 100 nucleotides
between 3 and 11hrs). A decrease in the size of βmRNA was
observed whether the induction was done in the absence or
presence of cycloheximide. βmRNA detected in Namalva
cells induced with Sendai virus has the same size as that
detected in fibroblast cells; however, the kinetics of
its synthesis and degradation does not show the trans-
iency found in poly rI.rC induced fibroblasts. The accumulation of β interferon mRNA in Sendai induced Namalva cells
is maximal at 6hrs after induction and reaches an undetectable level by 24-48hrs after induction. This suggests
that the mechanism of regulation of β interferon gene expression in poly rI.rC induced fibroblast cells and sendai-
induced Namalva cells is different.

FROM GENE TO PROTEIN: Copyright © 1982 by Academic Press, Inc.
TRANSLATION INTO BIOTECHNOLOGY   All rights of reproduction in any form reserved.
                                                        ISBN 0-12-045560-9

THE PURIFICATION OF MICROSOMAL GLUCOSE-6-PHOSPHATASE FROM
HUMAN LIVER: AIDS TO THE STUDY OF GLYCOGEN STORAGE DISEASE

Peter R. Reczek and Claude A. Villee. Dept. of Biological
Chemistry and Lab. of Human Reproduction and Reproductive
Biology, Harvard Medical School, Boston, MA. 02115.

Glucose-6-phosphate phosphohydrolase (EC 3.1.3.9)
(G6Pase), an enzyme of the microsomal fraction, is a
major intermediate in the mobilization of glucose from
calf liver cells. A defect in the hydrolytic function of
this enzyme is thought to be responsible for Glycogen
storage disease Type I (von Gierke's disease).

Lack of a purified preparation of G6Pase has hampered
efforts to understand the molecular details of this autosomal recessive genetic disorder. The present study was
undertaken to purify this membrane-bound enzyme using the
techniques of affinity chromatography.

Microsomes were prepared from human liver samples
obtained at autopsy by the method of Garland and Cori (1).
Wash buffers were supplemented with 0.2 per cent Na deoxycholate which disrupted microsomes and increased enzyme
activity ten-fold. These microsomes were transferred to a
DEAE-Cellulose column (2.5 x 10 cm). Bound proteins were
eluted from this column by a NaCl gradient to 0.5M. Two
elution peaks were resolved. Protein eluted at high salt
was found to have high activity representing a purification of approximately twenty-fold, while recovering some
ten per cent of the total enzyme activity.

Preliminary studies using glucose-6-phosphate covalently linked to a Sepharose 6B column gave an additional
measure of purification. Gel electrophoresis suggests a
molecular weight for this enzyme at 30,000 daltons with
little additional protein contamination.

Current experiments in this laboratory use purified
G6Pase as an antigen in the monoclonal antibody technique.
Such an antibody is important for planned studies on the
structure and regulation of G6Pase in vivo in the diseased
state.

This work was supported by an ongoing grant from the
Children's Hospital Medical Center-Harvard Medical School
GSD Fund from the Miller Foundation.

(1) Garland, R.C. and Cori, C.F. Biochemistry 11
4712-4718 (1972).

FROM GENE TO PROTEIN:
TRANSLATION INTO BIOTECHNOLOGY

TRANSFER AND MAPPING OF GENES USING METAPHASE CHROMOSOMES
FROM ICR 2A FROG CELLS

Barry S. Rosenstein. Department of Radiology, The University
of Texas Health Science Center at Dallas, 5323 Harry Hines
Blvd., Dallas, Texas 75235.

ICR 2A frog cells are haploid and possess five large
chromosomes and eight small chromosomes. Cells were
arrested in metaphase of mitosis using vinblastine sulfate
and chromosomes isolated from these cells. The chromosomes
were then sedimented in a sucrose gradient in order to
separate them into two size classes and used in chromosome
mediated gene transfer experiments (1) to transfer the genes
coding for a thermolabile thymidine kinase associated with a
thymidine-specific saturable transport system and a thermostable thymidine kinase (2). The recipient cells were
deficient in the production of these enzymes and selection
took place in medium containing HAT (hypoxanthine,
aminopterin, thymidine) or HAT containing 60X the normal
thymidine concentration. Treatment of recipient cells with
the fraction containing large chromosomes resulted in the
highest level of transfer for the gene coding for the
thermolabile thymidine kinase while the small chromosome
fraction was most effective for transfer of the gene coding
for the thermostable thymidine kinase. In an additional
series of experiments chromosomes were isolated from several
of these "transferents" and the chromosomes fractionated in
sucrose gradients. In this case both fractions (large and
small chromosomes) were capable of transferring the two genes
indicating that the transferred piece of genetic material was
not restricted to the homologous site of the particular gene,
but may have occurred at several locations.

(1) McBride, O.W. and Ozer, H.L. Proc. Natl. Acad. Sci.
(U.S.A.) 70:1258-1262 (1973).

(2) Freed, J.J. and Hanes, I.M. Exptl. Cell Res. 99:126-
134 (1976).

FROM GENE TO PROTEIN:
TRANSLATION INTO BIOTECHNOLOGY

INTRACELLULAR SELECTION FOR PLASMIDS THAT CARRY GENES FOR HUMAN INSULIN

Paul R. Rosteck, Jr. and Charles L. Hershberger.
Department of Antibiotic Development, Eli Lilly and Company, Indianapolis, Indiana.

Plasmids may be lost from bacterial hosts that are cultured without selection for plasmid-retention unless the plasmid encodes an active partition function to insure distribution to each progeny at cell division. This is particularly true for certain chimeric plasmids that incorporate foreign genes into vectors derived from Col EI-like plasmids. For example, plasmids coding for chains of human insulin can be lost from virtually the entire population when E. coli K12 strains harboring the plasmids are propagated under non-selective conditions. Alternative selection schemes are necessary to force retention of these recombinant plasmids without continuous exposure to antibiotics. We have demonstrated the feasibility of constructing an intracellular conditionally lethal system whereby loss of the plasmid from any progeny cell results in the death of that cell.

A 990 base pair PstI-HincII fragment of the genome of bacteriophage λcI_{857} contains the entire structural gene for the temperature-sensitive λ repressor, cI_{857}, and its promoter, P_{RM}. The fragment does not contain the adjacent gene, cro, and its promoter P_R and does not contain most of the adjacent structural gene, rex. Newly constructed plasmids contain the λ-fragment inserted into the human insulin-coding plasmids, pIA7Δ4Δ1 and pIB7Δ4Δ1, resulting in the new recombinants pPR17 and pPR18, respectively. Separate E. coli K12 RV308 host strains containing the new recombinants have been lysogenized with the repressor-defective bacteriophage λcI_{90}. Loss of the plasmid from a lysogenic host strain causes concomitant loss of λ-repressor and cell death because the prophage is induced to enter the lytic growth cycle. Effectiveness of the intact selective system is demonstrated by comparisons of plasmid retention in lysogenic and non-lysogenic hosts when the cultures are propagated without antibiotic selection.

FROM GENE TO PROTEIN:
TRANSLATION INTO BIOTECHNOLOGY

IMMUNOELECTRON MICROSCOPIC ANALYSIS OF THE IMMUNOGLOBULIN
MOLECULE FOR IDIOTYPES AND ALLOTYPES

Kenneth H. Roux and Dennis W. Metzger. Department of
Biological Science, Florida State University, Tallahassee,
FL, and Division of Immunology, St. Jude Childrens
Research Hospital, Memphis, TN.

We have developed a novel approach for the analysis of
antigenic determinants (idiotypes and allotypes) or intact immunoglobulin molecules. Immune complexes composed
of IgG in combination with anti-idiotype or anti-allotype
antibody were "visualized" by transmission electron
microscopy. Samples were prepared and negatively stained
as described by Seegan et al. (1979). Specific antibody
was used either as intact molecules or as Fab fragments.

Unreacted IgG was observed to be in the classical (Y)
configuration. Individual Fab fragments of anti-idiotype
or anti-allotype Ab, when bound to the IgG, altered the
configuration of the molecule in an interpretable manner.
Fab of anti-idiotype antibody bound to the distal
terminus of the V region of the IgG molecule, thus
extending the long axis of the Fab arms (ͳ , ͳ).
Similarly, whole molecules of anti-idiotype bound to
idiotype-bearing IgG at the distal terminus of the V
region (⃪⃕ , ⃪⃕).

Analysis of a rabbit VH framework allotype (a1) revealed
that the determinant(s) is (are) located on the lateral
portion of the V region of IgG. Binding of the anti-a1
Fab fragments was always at approximately right angles
to the long axis of the IgG Fab arms (Ϥ , Ϥ , ʍ , ʌ).
Fab antibody to an allotype located on the constant
region of the rabbit k light chain (b4) bound to the
lateral portion of the proximal half of the IgG Fab arms
(Ϯ , Ϯ , Ϥ).

These observations are consistent with previous
serological and biochemical evidence as to the location
of these determinants. Therefore, this technique
should be of value in localizing less well defined
immunoglobulin determinants. Supported by NIH #A1-16596.

Seegan, G.W., Smith, C.A., and Schumaker, V.N. 1979.
Proc. Natl. Acad. Sci. USA 76: 907.

FROM GENE TO PROTEIN:
TRANSLATION INTO BIOTECHNOLOGY

CROSS REACTIVITY OF LETHAL VENOMS TO A MONOCLONAL ANTIBODY

Russo, A.J., Cobb, Carrington S., Calton, Gary J. and
Burnett, Joseph W.
Division of Dermatology, University of Maryland School
of Medicine, Baltimore, Maryland 21201.

Monoclonal antibodies (MAB) to specific lethal factors
from the venom of the Portuguese Man-O'war have been prepared in our laboratories.[1]

We have examined the cross reactivity of K5 MAB which
neutralizes the lethal activity of Portuguese Man-O'war
venom Factor I (which comprises approximately 50% of the
lethal activity of the crude venom) with that of a number
of other venoms available to us. A strong cross reaction
was seen with the venoms of the South American rattle-
snake, Crotalus durissus terrificus, the Oriental Hornet,
Vespa orientalis, the Sea Wasp, Chironex fleckeri, the Sea
Nettle, Chrysaora quinquecirrha and a purified component
of the venom of the sea anemone, Anemonia sulcata, ATX_{II}.[2]
K5 MAB did not react with APA and APC from the venom of
the sea anemone, Anthopleura xanthogrammica[3] or ATX_{I} from
A. sulcata. No reaction was seen with the crude venom of
Bolocera tuediae.

To confirm this correlation a quantity of C. durissus
terrificus venom was purified by immunochromatography with
K5 MAB immobilized on Sepharose via cyanogen bromide. Elution was carried out with a gradient of phosphate-saline
buffer, pH 7. The "crotoxin complex" was specifically retained and subsequently eluted by 100 mM phosphate, 50 mM
NaCl buffer. The identification of the complex was made
by SDS PAGE where the complex is resolved into two closely
associated bands at 13,000 molecular weight. This fraction
was essentially devoid of contaminating material.

1   Gaur, P.K., Anthony, R.L., Calton, G.J. and Burnett,
        J.W., Toxicon (in press).
2   Beress, L., Beress, R. and Wunderer, G., FEBS Letters
        50, 311 (1975).
3   Norton, T.R., Kashiwagi, M. and Shibata, S., Drugs
        and Food From the Sea, Univ. Oklahoma, Norman p.
        37-50 (1978).

FROM GENE TO PROTEIN:
TRANSLATION INTO BIOTECHNOLOGY

CLONING OF AAV INTO pBR322: RESCUE OF INTACT VIRUS FROM THE
RECOMBINANT PLASMID IN HUMAN CELLS

Richard J. Samulski, Kenneth I. Berns, Ming Tan, and Nicholas
Muzyczka. Department of Immunology and Medical Microbiology,
University of Florida, College of Medicine, Gainesville,
Florida.

AAV is a defective parvovirus that requires helper virus
co-infection for its own replciation (1). In the absence of
helper virus, AAV DNA can integrate into the host genome to
establish a latent infection (2). In the case of latently
infected human cells, AAV may be rescued with high efficiency
(up to 10% of latent cells yield virus) upon challenge with
either adenovirus or herpes simplex virus (3).

In order to determine essential sequences needed for AAV
virion propagation, we have cloned intact duplex AAV DNA into
the bacterial plasmid pBR322. Transfection of the
recombinant plasmid into human cells with Adenovirus 5 as
helper results in rescue and replication of the AAV genome.
Rescued virus DNA and wild type AAV DNA are indistinguishable
when compared by restriction analysis. The efficient rescue
of viable AAV from the recombinant plasmid should facilitate
the genetic analysis of AAV. In addition, the recombinant
plasmid itself may be a model for studying the rescue of a
latent AAV viral infection.

(1)   Berns, K.I. and Hauswirth, W.W. Adv. in Virus Res. 25
      407-449 (1979).

(2)   Cheung, A.K.M., Hoggan, M.D., Hauswirth, W.W., and
      Berns, K.I. J. Virol. 33 739-748 (1980).

(3)   Hoggan, M.D., Thomas, G.F., Thomas, F.B., and Johnson,
      F.B. In: Proceedings of the Fourth Lepetit Colloquim,
      Cocoyac, Mexico, pp. 243-249. North Holland Publishing
      Co., Amsterdam (1972).

FROM GENE TO PROTEIN:
TRANSLATION INTO BIOTECHNOLOGY

GENERATION, CHARACTERIZATION, AND UTILIZATION OF
MONOCLONAL ANTIBODIES REACTIVE WITH HUMAN MAMMARY
CARCINOMA ASSOCIATED ANTIGENS

J. Schlom, D. Colcher, P. Horan Hand, M. Nuti, D.
Stramignoni, Y. Teramoto, and D. Wunderlich,
National Cancer Institute, N.I.H., Bethesda, MD;
D. Kufe and L. Nadler, Sidney Farber Cancer
Inst., Boston, Mass.

Splenic lymphocytes of mice, immunized with
membrane-enriched fractions of human metastatic
mammary carcinoma cells, were fused with murine
non-immunoglobulin secretor myeloma cells.
Following initial screening and double cloning of
hybridoma cultures, eleven monoclonal antibodies
were further characterized. These monoclonals
could be placed into five major groups based on
their differential reactivities with extracts of
breast tumor metastases, the surface of live
mammary tumor cells in culture, and immunoperoxidase staining of tissue sections of primary
and metastatic mammary tumors. None of the
antibodies bound to the surface of fourteen human
cell lines derived from a variety of normal
tissues, including normal mammary cells. Surface
binding to mammary tumor cells by two of the
monoclonal antibodies was shown to decrease
during density dependent arrest; further cell
cycle analysis demonstrated differential antibody
surface binding at S phase. Prolonged exposure
of mammary tumor cells to antibody showed no
evidence of antigen capping or internalization.
The reactive antigens in tumor cells have been
identified and range from 60,000d to a complex of
proteins ranging from 220,000 to approximately
400,000d. This latter complex has been purified
to homogeneity via traditional and antibody
affinity chromography and represents approximately 0.01% of the protein of tumor cells.
Radiolabeled monoclonal antibodies and antibody
fragments have been shown to selectively detect
human mammary tumors transplanted in athymic
mice.

FROM GENE TO PROTEIN:
TRANSLATION INTO BIOTECHNOLOGY

STRUCTURE OF THE ACTIN MULTIGENE FAMILY OF SOYBEAN (Glycine max)

D.M. Shah, T.D. McKnight, R.C. Hightower, B.R. Dhruva, and R.B. Meagher, The Department of Molecular and Population Genetics, University of Georgia, Athens, Ga. 30602
Soybean actin is encoded by a small multigene family composed of seven or eight genes (Nagao et al., 1981, DNA 1, 1-9). Seven members of this multigene family have been isolated from a library of the soybean genome cloned in Charon 4A. We have determined the complete sequence of the coding regions of three of these genes. The complete nucleotide sequence of one of these genes has been reported previously (Shah et al. 1981, Proc. Natl. Acad. Sci., U.S.A., in press). The 5' ends and flanking regions of several additional soybean actin genes have also been sequenced.

The positions of the three introns in each soybean actin gene examined are identical, however, their sequences and lengths vary. The introns are similar to the introns found in animal genes in that they conform to the GT-AG rule, are AT rich and contain termination codons in all three translational reading frames. Despite the fact that the soybean actin introns occur in the same locations within the members of this multigene family, the introns of the actin genes of other organisms (Drosophila, sea urchin, and yeast) do not occur at these locations. The positions of introns in other actin genes vary widely. The actin genes of Dictyostelium which have been examined do not contain introns.

A comparison of the polypeptide coding regions from these actin genes shows that this multigene family is more diverged within soybean than within Dictyostelium or Drosophila.

As in the Dictyostelium actin multigene family, there is little homology between the 5' flanking regions of the soybean actin genes. Short sequences within this flanking region show some homology with the transcriptional control signals (-70 site, TATA box, and capping site) of other eukaryotic genes. S1 nuclease mapping is being used to position the 5' ends of the actin transcripts on the DNA sequence in order to locate the actual transcriptional control signals.

Supported by USDA research grant #78-59-2133-1-1-068-1

FROM GENE TO PROTEIN:
TRANSLATION INTO BIOTECHNOLOGY

IDENTIFICATION OF DEFECTIVE GLOBIN mRNA IN MEL x FIBROBLAST
HYBRID CELLS. J. A. Shapiro, Y. Chiang, T. Ley, A. W.
Nienhuis, and W. F. Anderson. National Heart, Lung, &
Blood Institute, NIH, Bethesda, MD 20205.

M11X cells, formed by the fusion of mouse erythroleukemia
cells (MEL) and human fibroblasts, contain the gene for
human β-globin. Upon induction with HMBA, 10S or longer
mRNA transcribed from the human β-globin gene is detected by
Northern blot hybridization analysis. No human β-globin
translation product is found using acid-urea polyacrylamide
gel electrophoresis, radioimmune assay, or amino acid
sequence analysis of labeled globins (L. S. Haigh, Ph.D.,
thesis, Geo. Washington Univ., 1981). Although transcription occurs, improper initiation or processing of the
transcript is therefore suggested. The mechanism of this
defect is being investigated by S_1 mapping and sequence
analysis of the mRNA. RNA from M11X cells was prepared by
lysis of the cells followed by $CsCl_2$ buoyant density centrifugation. Normal β-globin mRNA was prepared by lysis of
peripheral erythrocytes and phenol extraction. A portion of
each of the RNA preparations was enriched for polyadenylated
mRNA using oligo-dT cellulose chromatography. DNA oligonucleotide probes complementary to portions of the human β-
globin mRNA were synthesized using the modified triester
chemical technique. These oligonucleotides were used to
directly sequence the mRNA using reverse transcriptase and
dideoxyribonucleoside triphosphates to terminate chain
elongation. Oligo-dT cellulose purification of the mRNA
template is not necessary when using this technique. Using
an octadecanucleotide primer, the 3' noncoding region extending into the mRNA encoded by exon 3 was shown to be
identical in M11X mRNA to that of control human β-globin
mRNA. DNA probes were created by cloning various segments
of the human β-globin gene into the bacteriophage, M13.
These cloned DNA fragments served as templates for the synthesis of radiolabeled probes complementary to human β-
globin gene transcripts. Differences between the control
and M11X human β-globin mRNA were mapped to the 5' region as
determined by the length of the portion of the β-globin
probes protected from S_1 nuclease by the mRNA. Studies are
underway using a tridecanucleotide primer complementary to
the middle of the mRNA to determine the primary sequence of
the 5' end of the M11X mRNA. The primed reverse transcriptase sequencing technique is also being used to
determine the specific primary sequence defects in the mRNA
obtained from patients with β-thalassemia.

FROM GENE TO PROTEIN:
TRANSLATION INTO BIOTECHNOLOGY

A PLASMID CLONING VECTOR FOR INDUCIBLE OVERPRODUCTION
OF PROTEINS IN BACTERIAL CELLS.   A.R. Shatzman and M.
Rosenberg, Laboratory of Biochemistry, National Cancer
Institute, NIH, Bethesda, MD 20205

A plasmid cloning vector containing both transcriptional and translational regulatory sequences derived
from the bacteriophage lambda genome was constructed to
achieve high level expression of prokaryotic and eukaryotic genes.

The system utilizes a plasmid vehicle carrying the
strong, regulatable lambda promoter, P_L, and certain
host lysogens into which this vector can be stably
transformed. The lysogen synthesizes sufficient repress-
or (cI) to control P_L expression and thereby stabilize
plasmids which carry such a highly efficient promoter.
Use of a temperature sensitive repressor permits simple,
rapid induction of P_L transcripts at any given time.
Efficient transcription of essentially any coding sequence
is assured by providing the phage lambda anti-termination
factor, N, and a site on the transcription unit for its
utilization (Nut site). We have demonstrated that N
gene expression in single copy provides sufficient N to
remove all transcriptional polarity within the P_L transcription unit. Several prokaryotic genes have been
inserted into the vector and each has yielded a high
level of gene expression (e.g. from 5-20% of total cellular protein). The ability to control induction and obtain
rapid, high level expression has proven particularly·
useful for synthesizing gene products which are turned
over and/or are lethal to the cell.

The system has been extended to express essentially
any gene coding sequence by providing the information
necessary for efficient translation in bacteria. This
was done by appropriately inserting into the P_L transcription unit the ribosome binding site and initiation codon
of the efficiently translated phage lambda cII gene.
Immediately adjacent to the initiator ATG, we engineered
a unique cloning site which allows any coding sequence
to be fused in frame directly to the cII start site.
This system has been used to overproduce one prokaryotic
(E. coli β galactosidase) and one eukaryotic (SV40 small
T antigen) protein. β-galactosidase is synthesized as
30-40% of cell protein and small T as >5% of cell protein
after only a 60-90 minute induction. The use of this
vector system to overproduce other eukaryotic genes will
also be discussed.

FROM GENE TO PROTEIN:
TRANSLATION INTO BIOTECHNOLOGY

CHARACTERIZATION OF HUMAN GLOBIN $\psi\beta_2$ - A PHANTOM
PSEUDOGENE

Shi-Hsiang Shen, Patricia A. Powers and Oliver Smithies
Laboratory of Genetics, University of Wisconsin,
Madison, Wisconsin 53706

Human globin $\psi\beta_2$, for which no globin chain has been
detected, was described by Fritsch et al. (Cell 19:959-
971, 1980) following hybridization of a human γ-globin
cDNA clone to clones derived from the human β-globin
gene cluster. We have sequenced 3.5 kbp of human DNA
covering the entire $\psi\beta_2$ region. Visual inspection and
computer analysis of the data have shown no significant
homology between $\psi\beta_2$ and any coding sequences in the
human β-globin gene cluster. We conclude that $\psi\beta_2$ is
not a β-globin pseudogene.

In analyzing the sequence data, we found that the
human globin $\psi\beta_2$ region is A+T rich (over 60%) and
contains a number of clusters of poly A and poly T tracts.
The longest of these tracts is $(T)_{28}$. We suspect that
these tracts may account for the hybridization of $\psi\beta_2$
to the γ cDNA clone, since this cDNA clone had been
constructed with poly A-T tails. We obtained very weak
hybridization of the cloned $\psi\beta_2$ region to a γ-globin
cDNA clone in the absence of poly rA, but none in its
presence.

Our nucleotide sequence data and additional hybridization tests show three members of the Alu family of
repeated DNA sequences clustered upstream of the $\psi\beta_2$
region. Two of these sequences are oriented in the
same direction (toward the $\psi\beta_2$ region as judged by
comparison with Alu sequences upstream of the $^G\gamma$ and β
globin genes). The proposal has been made that Alu
family repeats may have a role in the regulation of gene
transcription or DNA replication. It will be interesting
to investigate whether this cluster of Alu repeats, which
is relatively distant from any known genes, has any
function in gene expression.

Supported by NIH Grants AM 20120 and GM 20069.

(1) Fritsch, E. F., Lawn, R. M. and Maniatis, T.
    Cell 19:959-972 (1981).

FROM GENE TO PROTEIN:
TRANSLATION INTO BIOTECHNOLOGY

PRODUCTION OF HUMAN GROWTH HORMONE IN ESCHERICHIA COLI

Roger Sherwood, Jeremy Court, Ann Mothershaw, William Keevil,
Derek Ellwood, George Jack, Harry Gilbert, Richard Blazek,
Jim Wade and Tony Atkinson.

Public Health Laboratory Service, Centre for Applied
Microbiology and Research, Porton Down, Salisbury,
Wiltshire, England.

Bjorn Holmstrom.

Kabi AB, S-112 87, Stockholme, Sweden.

Human growth hormone (HGH) was produced in high yield in
E. coli at 400 litre fermentation scale. The E. coli strain
carries a derivative of the plasmid pHGH 107, developed by
Genentech Inc.[1] and is the subject of a scale-up fermentation
programme at CAMR on behalf of Kabi Vitrum AB.

HGH yield was directly related to cell density and yield
remained static or declined at the end of the logarithmic growth
phase. Culture generation times were in the range 60 to 72
minutes and four to five generations gave log phase cells
containing at least 9×10^5 monomers/cell.

Cells were killed in the fermenter by treatment at pH 12,
which also resulted in partial lysis and release of HGH into
the culture supernatant. HGH was purified in a 4-step process.
Ammonium sulphate precipitation was followed by ion-exchange,
gel filtration and hydrophobic chromatography. Recovery was 40%
and material was homogeneous based on SDS-PAGE. HGH dimer and
24K mol. wt. HGH variant were only detected when extracts were
held for prolonged periods early in the purification.

[1] Goeddel D.V. et al., Nature (1979) 281, 544.

FROM GENE TO PROTEIN:
TRANSLATION INTO BIOTECHNOLOGY

Novel expression of an interferon gene in E. coli in high
yields.

P. Slocombe, A. Easton, P. Boseley and D. C. Burke.

Department of Biological Sciences, University of Warwick,
Coventry, CV4 7AL, England.

An interferon gene has been cloned into the Bam Hl site of

pAT 153 by means of a cDNA library made from a mRNA

fraction from induced Namalwa cells which had been en-

riched for interferon mRNA. The interferon gene sequence

was cut out and treated to produce a labelled fragment with

blunt ends. This product was then partially digested with

Pvu II and the larger fragment, which had a Pvu II site at

the 5'-end and a Msp 1 site at the 3'-end was isolated.

The interferon gene fragment was inserted into an express-

ion vehicle with T4 ligase and transferred into a suitable

E. coli strain.

DNA from 8 isolates was sequenced at the 3' end to deter-

mine orientation and two were shown to contain the gene in

the correct orientation. Sequencing at the 5' end showed

the inserted DNA to be in phase with the expression vehicle

gene.

When E. coli was infected with a recombinant expression

vehicle containing an interferon clone and the culture

induced interferon in high yield was produced by the

bacteria. The lysate was purified by immunochromatography

to yield a modified interferon which showed the expected

antiviral activity on heterologous cells and was recognised

in the immunoradiometric assay as efficiently as Namalwa

interferon.

FROM GENE TO PROTEIN:
TRANSLATION INTO BIOTECHNOLOGY

HMG PROTEINS 1 AND 2 ARE REQUIRED FOR TRANSCRIPTION

Jose A. Stoute and William F. Marzluff
Dept. of Chemistry, Florida State University
Tallahassee, Florida 32306

Chromatin active in transcription was prepared from isolated nuclei. Treatment of the chromatin with 0.35M KCl in the presence of 5 mM Mg^{++} greatly reduced transcription. Transcription could be restored by adding back the extracted material. The major proteins extracted from chromatin by this procedure were HMG 1 and 2. HMG 1 and 2 were purified by differential precipitation with trichloroacetic acid and gel filtration. Addition of pure HMG proteins 1 and 2 to chromatin depleted of HMG 1 and 2 enhanced transcription 3 to 5 fold. The production of specific RNA polymerase III transcripts, 5S rRNA and the 4.5S tRNA precursors, was measured by gel electrophoresis. The HMG proteins enhanced the synthesis of the RNA polymerase III products 5S rRNA and 4.5S tRNA precursors as well as total RNA synthesis. The addition of HMG proteins 1 and 2 stimulated transcription in a concentration-dependent manner with maximum stimulation when the amounts of HMG 1 and 2 added back were equal to the amounts extracted. Addition of HMG 14 and 17 had no effect on transcription. These proteins were not extracted from chromatin under these conditions.

The results suggest that HMG proteins 1 and 2 are associated with regions of chromatin being transcribed and that their presence is required for transcription to occur. This is true even for genes which are one nucleosome in size, for example, 5S rRNA. The RNA polymerases, even RNA polymerase III remain associated with the chromatin under conditions which remove the HMG proteins.

FROM GENE TO PROTEIN:
TRANSLATION INTO BIOTECHNOLOGY

TISSUE-SPECIFIC METHYLATION PATTERNS IN BOVINE
SATELLITE DNA.

K.Sturm,R.A.McGraw,and J.H.Taylor
Institute of Molecular Biophysics,Florida State
University,Tallahassee,Fla.32306

Bovine satellite DNA has been prepared from various
tissues and analyzed with regard to cytosine methylation.
Restriction analysis with methylation-sensitive endonucleases and base analysis by HPLC indicate that tissue-
specific differences in methylation do occur within this
major satellite fraction.Results of this type comparing
DNA from bull sperm and calf thymus have been reported
previously (1).It was observed that the satellite fragment from sperm,in marked contrast to that from thymus,
is essentially free of methylation,suggesting the
action of an initiation-type methylase early in development.

A more detailed study of satellite methylation in
different tissues is being performed by Southern
blotting and hybridization using a cloned satellite 1
fragment from calf thymus.

Also,we have extended our analysis to the DNA sequence
level.Satellite DNA fragments have been prepared from
the genomic DNA of various tissues and sequenced directly
by the method of Maxam and Gilbert.

Supported by grants from the U.S.Department of Energy
(EY-78-S-05-5854) and from the National Institutes on
Ageing (1 RO1 AG01807-01).

(1) Sturm,K. and Taylor,J.H.,Nucleic Acids Research,in
    press (1981).

FOXOMAS: A SECOND GENERATION METHOD FOR OBTAINING STABLE
HEAVY CHAIN IMMUNOGLOBULIN PRODUCING HYBRIDOMAS

R. Thomas Taggart
Department of Medicine, Human Genetics Research,
UCLA Veterans Administration Medical Center, Sepulveda, CA.

An highly efficient method for obtaining stable murine
hybridomas has been developed. Spleen cells from immunized
mice homozygous for a Robertsonian chromosome translocation
[Rb(8.12)] are fused with myeloma cells deficient in adenosine phosphoribosyl transferase (APRT⁻) and exposed to
culture media requiring APRT activity for cell growth.
Since the APRT locus (chromosome 8) and the heavy chain
immunoglobulin loci (chromosome 12) are genetically linked
in the donor Rb(8.12) mouse spleen cells, the retention of
one or both copies of chromosome 12 in hybrid cells is
insured by selective culture media. The heavy chain Ig loci
undergo allelic restriction and therefore the subsequent
growth of hybrid cells producing antibody in selective media
provides for the continuous retention of the active Ig locus
in theoretically 2/3 of all hybridomas.

Several APRT⁻ myeloma cell lines (FOX) of BALB/c origin
have been isolated by negative selection procedures and have
been used in fusion experiments. In one experiment from
176 to 350 or more hybridomas were obtained from a single
spleen. The absolute number of hybrids is not known since
numerous master well cultures contained from 2 to 5 individual hybrid clones. Enzymatic and karyotypic analysis
demonstrated the concurrent retention of APRT isozyme activity, one or both (8.12) chromosomes and antibody production.

The loss of antibody production by hybridoma cell lines
during isolation, cloning and subsequent cell population
expansion is not a trivial problem. The instability of continued antibody production by hybridoma lines has been suggested to result from the chromosomal loss of the relevent
immunoglobulin genes and to overgrowth by subpopulations of
nonproducing cells arising during the propagation of the
hybrids. Initial studies have demonstrated that the capture
ratio (the ability to obtain stable hybridomas to a given
antigen) is significantly higher with the Foxoma method due
to the continuous retention of the 8.12 chromosomes of the
donor spleen cells. The significance of this research is
that it greatly increases the applicability and efficiency
of the hybridoma technology.

FROM GENE TO PROTEIN:
TRANSLATION INTO BIOTECHNOLOGY

IN VITRO AND IN VIVO TRANSCRIPTIONAL ANALYSIS OF MOUSE GLOBIN GENE PROMOTERS

Carol A. Talkington and Philip Leder
Laboratory of Molecular Genetics, NICHD, National Institutes
of Health, Bethesda, Maryland

Using a simplified in vitro transcription system (HeLa cell cytoplasmic S-100 extract), we have previously shown the accurate initiation of transcription of biologically active mouse globin genes (adult α_1, α_2, β^{maj}, β^{min}, and embryonic x and y); in contrast, the pseudogenes α_3 and α_4 are not transcribed. Further analysis has demonstrated that a 44-bp stretch of DNA (spanning nucleotides -55 to -11) is sufficient to direct RNA polymerase II to accurately initiate transcription in vitro. Since the efficiency of transcription is increased if the stretch of α_1 globin DNA from -10 to +7 is included as part of the promoter segment, the sequence at the capping/initiation site appears to modulate the rate of initiation. This result is further supported by the analysis of hybrid promoters constructed from α_1 and α_4. For example, replacement of α_4's initiation site with that of the active gene α_1 (i.e., nucleotides -9 to +7) restores the DNA's function as a promoter. That RNA polymerase II recognizes more than just the TATAAG sequence in vitro is also apparent from the transcriptase's ability to initiate from the β^{maj} promoter three times more frequently than from α_1.

Recently we have also injected the α_1 and β^{maj} globin-gene-containing plasmid DNAs into frog oocyte nuclei and have identified a 112-bp segment (spanning nucleotides -105 to +7) as being sufficient for accurate initiation of transcription in this in vivo-like environment. Analysis of various mutants is being utilized to further delineate the promoter segment recognized in vivo.

FROM GENE TO PROTEIN:
TRANSLATION INTO BIOTECHNOLOGY

PURIFICATION OF PLASMID DNAs AND SYNTHETIC OLIGODEOXYRIBO-
NUCLEOTIDES USING LOW PRESSURE ANION ENCHANGE CHROMATOGRAPHY

John A. Thompson, Paul W. Armstrong, Keith E. Rushlow,
Kathleen Doran, and Robert W. Blakesley. Instrumentation
Division, Bethesda Research Laboratories, Inc.,
Gaithersburg, Maryland, 20877

The purification of plasmid DNAs and synthetic oligodeoxyribonucleotides has been achieved with low pressure,
high performance liquid chromatography utilizing two new
anion exchange matrices. Each has a high capacity for binding nucleic acids (>1mg nucleic acid/gm resin), yielding 98%
recovery and purity of selected species of nucleic acid,
while operating at <50 psi. Elution of specific nucleic
acids is achieved with an increasing linear salt gradient
employing a peristaltic pump.

The biological utility of both plasmid and viral DNAs
(ranging in size from 3K to 49K base pairs) was not impaired
as a result of this purification method as determined by
reactions with numerous restriction enzymes, polynucleotide
kinase, and T4 DNA ligase. Furthermore, the transformation
efficiency of these plasmid DNAs was similar to the
efficiency for those purified by other methods.

The purification of each individual species of a mixture of oligodeoxyribonucleotides (2 to 30 bases in length)
can be achieved with low pressure, high resolution chromatography. For example, a synthetic molecular linker (8
bases long) and a synthetic primer (15 bases long) were
isolated to >98% purity. The specific sequences of these
standard oligodeoxyribonucleotides were determined readily
by chemical sequencing methods.

FROM GENE TO PROTEIN:
TRANSLATION INTO BIOTECHNOLOGY

MONOCLONAL ANTIBODY AGAINST OUABAIN SENSITIVITY RELATED HUMAN CELL SURFACE ANTIGEN

Yean-Kai Tsung, Aubrey Milunsky, and David Reiss
Genetics Division, Eunice Kennedy Shriver Center, Waltham MA

The steroid Ouabain blocks active NA^+/K^+ exchange transport in mammalian cells by inhibiting the plasma membrane Na^+/K^+ activated ATPase (Na^+,K^+ - ATPase). Normal human cells are sensitive to Ouabain at $10^{-7}M$ or lower. We have selected an Ouabain resistant and thymidine kinase deficient double mutant (HSB $^{TK-OR}$) by stepwise selection method from a human T-lymphocyte cell line (HSB). HSB $^{TK-OR}$ has a reduced NA^+,K^+ ATPase activity and resistant to Ouabain at the concentration of $10^{-5}M$.

We also have derived a monoclonal antibody secreting mouse hybridoma cell line (YKT-2) by fusing mouse myeloma cells (NS_1) with spleen cells from a female BALB/C mouse immunized with a human cell line K-562. The monoclonal antibody secreted by YKT-2 does not bind to the cell surface of HSB $^{TK-OR}$ while it binds to the Ouabain sensitive wild type (HSB) and a wide range of human cells including lymphocytes, granulocytes, and skin fibroblasts.

The results of the antibody binding assay indicate that YKT-2 is specific for a cell surface component which is present on the normal human cell surface, but absent or altered on the Ouabain resistant mutant HSB $^{TK-OR}$. Therefore, the YKT-2 failed to recognize the Ouabain sensitivity related cell surface antigen on HSB $^{TK-OR}$. The target molecule of the antibody and its relationship to the Na^+/K^+ - ATPase need to be elucidated.

The plasma membrane of animal cells plays a central role in diverse aspects of cell proliferation, interaction, and differentiation. A logical approach to unraveling the mechanisms of membrane function is to analyze the relationship between structural and functional alteration resulting from specific genetic mutation. Immunological study by the highly specific monoclonal antibody is an extremely powerful tool to monitor the cell surface molecular changes due to such a genetic event.

FROM GENE TO PROTEIN:
TRANSLATION INTO BIOTECHNOLOGY

SELECTION OF MONOCLONAL ANTIBODIES FOR DESIGNATED USES

V. van Heyningen*, L. Barron[†], D. Brock[†] & W.M. Hunter[°]

*MRC Clinical & Population Cytogenetics Unit, [†]Dept. of
Human Genetics, [°]MRC Immunoassay Team, Edinburgh, U.K.

Monoclonal antibodies are becoming important in
immunoassay systems in routine clinical quantitation.
When hybridomas are destined for this use, the required
range and specificity of the assay must be borne in
mind when suitable clones are selected.

We have raised monoclonal antibodies against human
alphafoetoprotein (h-AFP). One clone, AFP 21.2
secretes an IgG_1 immunoglobulin of sufficient avidity
to allow discrimination of serum levels of AFP over the
whole range from low (non-pregnant) to high (open neural
tube defect-associated) levels in a single 4-hour two-
site immunoradiometric assay (IRMA).

Another clone, AFP 21.1, produces immunoglobulin of
low avidity which has proved ideal as a solid phase
reagent for a single step purification of the antigen
under non-denaturing conditions of elution with 2M $MgCl_2$.
Similar gentle immunopurification systems will be
required for a number of products to be recovered from
bacterial culture systems following DNA cloning.

h-AFP is also produced by a number of tumours.
Monoclonal antibodies for immunohistochemistry should
ideally be directed to epitopes not destroyed by
classical fixation methods. Neither AFP 21.1 nor 21.2
are apparently suitable for this.

FROM GENE TO PROTEIN:
TRANSLATION INTO BIOTECHNOLOGY

TOWARD   ANALYSIS OF THE HLA-DR SYSTEM
WITH MONOCLONAL ANTIBODIES

V. van Heyningen, K. Guy, B.B. Cohen, D.L. Deane
and C.M. Steel

MRC Clinical and Population Cytogenetics Unit,
Western General Hospital, Edinburgh EH4 2XU.

Three monoclonal antibodies DA6·231, ·164 and
·147 all precipitate glycoprotein dimers consisting
of 34K and 29K subunits.   231 and 147 recognize DR-
common epitopes.   164 binds to all except DR7
homozygous cell lines.   Protein transfer from SDS
polyacrylamide gels to nitrocellulose paper, followed
by probing with the monoclonal antibodies has
permitted assignment of the 231 and 147 epitopes to
29K and 34K subunits respectively.   Mutual binding
inhibition between 231 and 164 is incomplete in one
direction, suggesting that there is a major proportion
of 231^+164^+ and a minority of 231^+164^- 29K molecules
expressed on all DR positive cells tested (except DR7
homozygous cells which have lost the 164 epitope).
The serological expression of the 147 epitope at the
cell surface is complex.   The 231/147 ratio on
peripheral and activated B lymphocytes is much higher
than on B lymphoblastoid lines or on CLL cells.   No
significant binding of 147 can be detected on the
strongly 231^+164^+ activated cultured T cells.
Preliminary evidence suggests that molecules able to
bind 147 are released on cell lysis even from 147^-
activated T cells.

FROM GENE TO PROTEIN:
TRANSLATION INTO BIOTECHNOLOGY

A HUMAN PROINSULIN cDNA RECOMBINANT CLONE

L. L. Villa and R. R. Brentani
Lab. Experimental Oncology, Faculdade de Medicina da Universidade de São Paulo, São Paulo, Brasil.

A bacterial clone containing DNA complementary (cDNA) to human proinsulin was constructed (1,2).

The 410 base pairs of DNA corresponding to human proinsulin, inserted on the Hind III site of pBR 322, has been manipulated in order to be subcloned in a plasmid containing part of the β-galactosidase operon sequences and transformed in Escherichia coli HB 101.

The isolation and purification of insulin produced by this system will be discussed in detail.

Supported by CNPq grant 40.1821/80 and FAPESP 81/0049/7

(1) Villa, L. L., Bolivar, F., and Brentani, R. R., Eur. J. Cell Biol. 22(1):18 (1980).

(2) Villa, L. L., Bolivar, F., and Brentani, R. R., Arq. Bras. Endocrinol. Metabol. 25(2):61-64 (1981).

FROM GENE TO PROTEIN:
TRANSLATION INTO BIOTECHNOLOGY

PREPARATION OF DNA LABELED WITH HIGH SPECIFIC ACTIVITY [35S] DEOXYADENOSINE 5'-(α-THIO) TRIPHOSPHATE; THE USE OF 35S-LABELED NUCLEIC ACIDS AS MOLECULAR HYBRIDIZATION PROBES.

Mark J. Vincent, William R. Beltz, and Sarah H. Ashton.
Biological Testing Laboratory, Biochemistry Department, New England Nuclear Corporation, Boston, MA 02118

E. coli DNA polymerase I and [35S] deoxyadenosine 5'-(α-thio) triphosphate ([35S] dATPαS) were used in a nick translation reaction (1) to prepare 35S-labeled pBR 322 DNA (>2.2 X 10^8 dpm/µg DNA using 10µM [35S] dATPαS in a three hour reaction); this DNA was then hybridized to immobilized (2) Hinf I restriction fragments of pBR 322. The inclusion of 10mM dithiothreitol in all hybridization buffers was found to be essential in order to minimize nonspecific binding. Comparison of the 35S-labeled DNA with the corresponding 32P-labeled nick translated probe in the hybridization assay showed no qualitative differences. Satisfactory autoradiographs of the 35S-labeled DNA hybridized probe were obtained with a 24-hour exposure without enhancement.

Secondly, avian myoblastosis virus reverse transcriptase and [35S] dATPαS were used to prepare 35S-labeled c-DNA of rabbit globin m-RNA. Gel electrophoresis and subsequent autoradiography of the labeled c-DNA indicated that reverse transcription of globin m-RNA is complete within one hour utilizing dATPαS as a substrate in the reaction protocol described by Friedman and Rosbash (3).

These results demonstrate that high specific activity [35S]-labeled DNA can be prepared using standard procedures and utilized in hybridization analysis. The six-fold longer half-life of 35S (relative to 32P) and the inherent resistance of thiophosphate in DNA to degradation by phosphatase (4) significantly increase the period of time in which such labeled DNA probes are useful.

(1) Rigby, P.W.J., Dieckmann, M., Rhodes, C., and Berg, P. (1977) J. Mol. Biol. 113, 237-251.
(2) Southern, E. (1979) Meth. Enzymol. 68, 152-176.
(3) Friedman, E.Y., and Rosbash, M. (1977) Nucleic Acids Res. 4, 3455-3471.
(4) Beltz, W.R., and O'Brien, K.J. (1981) Fed. Proc. 40, 1849.

FROM GENE TO PROTEIN:
TRANSLATION INTO BIOTECHNOLOGY

DNA TRANSFER OF BACTERIAL ASPARAGINE SYNTHETASE INTO
MAMMALIAN CELLS.

Mary M.Y. Waye and Clifford P. Stanners
Department of Medical Biophysics, University of Toronto
and the Ontario Cancer Institute, Toronto, Canada M4X 1K9.

We have used DNA from a clone M13oriC81 (1) containing
bacterial asparagine synthetase to transform directly a rat
cell line, Jensen sarcoma, which is deficient in asparagine
synthetase activity, to grow in asparagine-free medium
containing β-aspartylhydroxamate. The efficiency of DNA
transfer is about 2 to 20 colonies per 10^6 cells exposed
per μg of DNA, which is about 1000 to 5000x less than we
routinely obtain for transformation of TK⁻ cells with
cloned thymidine kinase (TK) gene from Herpes virus (2).

Southern blot analysis of the transformants showed
that the M13oriC81 DNA was integrated into high molecular
weight DNA. Unstable transformants which had lost their
ability to grow in asparagine-free medium containing
β-aspartylhydroxamate also lost this M13oriC81 DNA.
Transformants that were grown in increased amounts of
β-asparatylhydroxamate had increased amounts of M13oriC81
DNA. The resistance phenotype could be transferred using
DNA from transformants and these second-step transformants
also contained M13oriC81 DNA.

Northern blot analysis showed that the M13oriC81 DNA
was transcribed in that RNA which hybridized with M13oriC81
DNA was detected in a polysome fraction obtained from
transformants.

Western blot analysis of 2 dimensional gels of proteins
of transformants using anti-asparagine synthetase antibody
showed new spots with some of the properties of the
bacterial enzyme.

These results indicate that a bacterial gene can be
introduced and expressed directly, if inefficiently, in
mammalian cells.

Supported by NCIC and MRC. Mary Waye is a research student
of the National Cancer Institute, Canada.

(1) Kaguni, J., LaVerne, L.S. and Ray, D.S. Proc. Natl.
    Sci. USA 76 6250-6254 (1979).
(2) Graham, F.L., Bachetti, S., McKinnon, R., Stanners,
    C.P., Cordell, B. and Goodman, H.M. In: Progress in
    Clinical and Biological Research: Introduction of
    Macromolecules into Viable Mammalian Cells. R. Baserga,
    C. Croce and G. Rovera, eds. A. Liss, New York (1980)
    p. 3-25.

FROM GENE TO PROTEIN:
TRANSLATION INTO BIOTECHNOLOGY

Production and Characterization of Monoclonal
Antibodies to a Calcium-activated Protease.

M. Wheelock, J. Schollmeyer & W. Dayton. Departments of
Genetics & Cell Biology and Animal Science, University of
Minnesota, Saint Paul, Minnesota

We have produced monoclonal antibodies to a Ca^{++}-activ-
ated neutral protease (CAF) from muscle, which has previously been shown to be located in the Z disc of skeletal
muscle (1). This enzyme has been shown to be present in 2
forms, one requiring higher levels of Ca^{++} than the other
(2). The 2 forms migrate together in SDS gels but have
very different mobilities in nondenaturing gels; they are
also immunologically related as evidenced by cross-reactivity with rabbit antibodies. The rabbit antibodies were
produced using SDS-gel purified enzyme whereas the monoclonal antibodies could be made using a less pure enzyme
permitting the use of nondenatured protein. Balb/c mice
were injected with 1 mg protein (80% CAF), the spleen cells
were fused with NS1 cells and selection was done with HAT
media. The hybridomas were assayed for antibody production
with the Elisa test using crude enzyme as the immobilized
antigen. Cells that were positive were 2X cloned by limiting dilution, and again tested. In order to be certain
that positive clones were producing antibody specific for
CAF, these clones were tested by the following method:
crude CAF was run in an SDS gel; the protein from the gel
was transfered to activated paper; strips of this paper
were incubated in monoclonal supernate, washed and then
treated with ^{125}I-sheep anti-mouse Ig's; the strips were
autoradiographed and clones which showed specific binding
to CAF were retained. Thus we have a probe to study the
immunological relatedness of the 2 forms of the enzyme and
to determine the location of the enzyme in myofibrils and
cultured myoblasts using antibody to nondenatured CAF.

Supported by Minn. Ag. Exp. Sta. & grants from the Muscular
Dystrophy Assoc. & the Leukemia Task Force.

(1) Dayton, W.R. & Schollmeyer, J.V. Exp. Cell Research
    (in press).

(2) Dayton,W.R. et al, Biochimica et Biophysica Acta
    659 48-61 (1981).

PHENOTYPIC EXPRESSION OF MAMMALIAN AND FOREIGN PROCARYOTIC GENES CLONED IN BACILLUS SUBTILIS

D.M. Williams,   R.G. Schoner,   Y-W. Chen   and   P.S. Lovett.
Department of Biological Sciences, University of Maryland
Baltimore County, Catonsville, Maryland.

The mouse dihydrofolate reductase (DHFR) gene, a segment of the E. coli trp operon and the Tn9 chloramphenicol acetyltransferase (CAT) gene are expressed in B. subtilis when cloned into the expression plasmid pPL608. pPL608 was constructed in our laboratory to facilitate the cloning of foreign genes (1). pPL608 contains a total of 5.3 kb and consists of DNA from three sources: 1) the replicon and neomycin-resistance gene from pUB110, 2) a chromosomal CAT gene from B. pumilus which is inducible by chloramphenicol and 3) a SPO2 phage promoter. Phenotypic expression of the genes we have cloned into pPL608 from mouse, E. coli and Tn9 is dependent on the presence of the SPO2 promoter.

The E. coli trp genes reside on a 2.8 kb HindIII fragment. The fragment complements mutations in the B. subtilis trp D C and F genes (2). The E. coli trp fragment was inserted into the unique HindIII site on pPL608, which apparently is located within the pPL608 CAT gene. As a result, the expression of the CAT gene was inactivated while the expression of the E. coli trp C gene became inducible by chloramphenicol.

The mouse DHFR and the Tn9 CAT genes were cloned into the unique PstI site on pPL608. (Cloning the Tn9 CAT gene required deletion of the resident CAT gene on pPL608.) Expression of the mouse DHFR gene results in cells resistant to trimethoprim (2). The DHFR activity in extracts of these cells is inhibited 93% by 10^{-6}M methotrexate, but is not inhibited by 10^{-6}M trimethoprim. This inhibition pattern is characteristic of the mammalian DHFR. Expression of the DHFR gene is not inducible by chloramphenicol and cells remain resistant to chloramphenicol. A mutation in the SPO2 promoter fragment has been identified which increases the specific activity of mouse DHFR in B. subtilis ten-fold.

(1) Williams, D.M., Duvall, E.J. and Lovett, P.S., J. Bacteriol. 146 1162-1165 (1981).
(2) Williams, D.M., Schoner, R.G., Duvall, E.J. Preis, L.H., and Lovett, P.S., Gene in press.

FROM GENE TO PROTEIN:
TRANSLATION INTO BIOTECHNOLOGY

THE HUMAN α1-ANTITRYPSIN GENE AND PULMONARY EMPHYSEMA

Savio L.C. Woo, T. Chandra, Margaret Leicht, George Long*, Kotoku Kurachi* and Earl W. Davie*, Howard Hughes Medical Institute, Department of Cell Biology, Baylor College of Medicine, Houston, TX 77030, and *Department of Biochemistry, University of Washington, Seattle, WA 98105

Alpha-1-antitrypsin is a protease inhibitor that accounts for 90% of total anti-protease activity in the blood. It is transported by passive diffusion into the alveolar structure of the lung and protects it from destruction by polymorphonuclear leukocyte elastase. Reduced levels of this protein in certain individuals constitute a genetic disorder known as "α1-antitrypsin deficiency" which is often associated with development of chronic obstructive pulmonary emphysema. The deficiency is characterized by the presence of a mutated α1-antitrypsin gene which gives rize to a variant Z type protein instead of the normal M type protein, and is inherited by an autosomal recessive trait. The gene frequency of the phenotype in Caucasians of Northern European ancestry is 0.02-0.03, and it has been estimated that 1/4000 individuals is a ZZ homozygote in the United States. Alpha-1-antitrypsin is a single polypeptide of 45,000 daltons in molecular weight and is synthesized in the liver. We have purified the α1-antitrypsin mRNA by specific immunoprecipitation of total baboon liver polysomes and constructed a full length cDNA clone. The amino acid sequence of the entire protein was deduced from the nucleotide sequence of the gene. This baboon α1-antitrypsin cDNA clone was used to screen a human chromosomal gene library by cross hybridization and four independent isolates containing the corresponding human gene were obtained. A total of 21 Kb of human DNA is represented in these overlapping genomic clones, with 8 Kb of DNA flanking the 5' and 3' termini of the α1-antitrypsin gene. Electronmicroscopic and restriction mapping analyses have shown that the human chromosomal α1-antitrypsin gene contains 3 intervening sequences and is approximately 5 Kb in length. The entire gene resides within a 9.6 Kb Eco RI fragment, which will facilitate the cloning of the chromosomal gene from deficient individuals. The human chromosomal α1-antitrypsin gene has been used as a probe to analyze the deficiency syndrome at the gene level in order to develop gene mapping methods for prenatal diagnosis of the genetic disorder.

FROM GENE TO PROTEIN:
TRANSLATION INTO BIOTECHNOLOGY

GENE EXPRESSION DURING MAMMALIAN TOOTH ORGAN DEVELOPMENT

Margarita Zeichner-David, Mary MacDougall, Conny Bessem,
Pablo Bringas, Jr. and Harold C. Slavkin. Laboratory for
Developmental Biology, Graduate Program in Craniofacial
Biology, University of Southern California, Los Angeles,
CA 90007

The developing embryonic tooth organ is an ideal
system to study differentiation-specific gene expression.
Our studies are designed to determine when and where
enamel gene expression occurs during rabbit molar organ-
ogenesis. Molar organs were dissected from New Zealand
White rabbit embryos 21-days of gestation through 2-days
post-natal development. Epithelial-specific gene products were monitored using (a) analysis of newly synthesized proteins (labeled with ^{35}S-methionine) by SDS-PAGE
and fluorography, (b) isolation and translation of mRNAs
(1) and (c) specific antibodies prepared against enamel
organ proteins. Enamel proteins first appear at 23-days
gestation and continue to be synthesized and secreted
until 28-days gestation. The mRNA results suggest a
close relationship between transcription and translation
rates of enamel synthesis. The identity of specific
enamel proteins was established by immunoprecipitation.
Immunohistochemical staining showed the localization of
these proteins at 23 days gestation in the tooth organ.
These studies define the developmental stage for synthesis and secretion of enamel gene products during rabbit embryogenesis.

Research supported by DE-02848 and training grant DE-
07006 (NIH, USPHS).

(1) M.Zeichner-David, B.G. Weliky and H.C. Slavkin,
    Biochem. J. 185 489-496 (1980).

INDEX OF AUTHORS

SUBJECT INDEX

This index has been prepared from the keywords supplied by the contributors. The page number is that of the first page of the relevant contribution.

A

Agrobacterium tumefaciens,
  transformation of, 105
  plasmid T_1 from, 105
Alpha-2-macroglobulin and cystic
  fibrosis, 143
Amino acids, in cell culture, 75
Antibodies, monoclonal, 3, 129
Antibodies, monoclonal and genetic
  disease, 143

B

Bacterial contaminants, 445

C

Carbohydrates, in cell culture, 75
Cell culture, 45
  amino acid metabolism, 75
  carbohydrate metabolism, 75
  differentiation, 53
  growth factors, 45
  high density, 75

hormone requirements, 45
microcarriers in, 75
Cells
  extracellular matrix and growth, 53
  hormones and growth, 53
  human, 45
  plant, 105
Cell-surface proteins, 165
Cloned human gene product, 429
Cloning, 265
  human gene product, 429
  polio virus cDNA, 459
Crown gall, role in transformation,
  105
Cystic fibrosis and alpha-2-
  macroglobulin, 143

D

Dispersed genes, 27
DNA
  chemical synthesis of, 235
  recombinant, 367, 391
  synthetic for structural genes, 213